Charles O'Neill

A Dictionary of dyeing and Calico printing

Containing a brief Account of dyeing and printing Textile Fabrics

Charles O'Neill

A Dictionary of dyeing and Calico printing
Containing a brief Account of dyeing and printing Textile Fabrics

ISBN/EAN: 9783337249175

Printed in Europe, USA, Canada, Australia, Japan

Cover: Foto ©Andreas Hilbeck / pixelio.de

More available books at **www.hansebooks.com**

A

DICTIONARY

OF

DYEING AND CALICO PRINTING.

A

DICTIONARY

OF

DYEING AND CALICO PRINTING;

CONTAINING

A BRIEF ACCOUNT OF ALL THE SUBSTANCES AND PROCESSES IN USE IN THE ARTS OF DYEING AND PRINTING TEXTILE FABRICS;

WITH

PRACTICAL RECEIPTS AND SCIENTIFIC INFORMATION.

BY

CHARLES O'NEILL,

ANALYTICAL CHEMIST;

FELLOW OF THE CHEMICAL SOCIETY OF LONDON, MEMBER OF THE LITERARY AND
PHILOSOPHICAL SOCIETY OF MANCHESTER; AUTHOR OF "CHEMISTRY OF
CALICO PRINTING AND DYEING," ETC.

TO WHICH IS ADDED

AN ESSAY ON COAL TAR COLORS AND THEIR APPLICATION TO DYEING AND CALICO PRINTING.

BY

A. A. FESQUET,

CHEMIST AND ENGINEER.

WITH AN

APPENDIX

ON

DYEING, AS SHOWN AT THE EXPOSITION OF 1867, FROM THE REPORTS
OF THE INTERNATIONAL JURY, ETC.

PHILADELPHIA:
HENRY CAREY BAIRD,
INDUSTRIAL PUBLISHER,
406 Walnut Street.

1869.

PHILADELPHIA:
COLLINS, PRINTER, 705 JAYNE STREET.

PUBLISHER'S PREFACE

TO

THE AMERICAN EDITION.

IN presenting to the Dyers and Manufacturers of the United States an edition of so well known and justly esteemed a book as Mr. O'Neill's "DICTIONARY OF DYEING AND CALICO PRINTING," the publisher deems nothing necessary except to call attention to the character and extent of the additions which have been made to it.

PROFESSOR FESQUET has contributed a very comprehensive essay on COAL TAR COLORS AND THEIR APPLICATION TO DYEING AND CALICO PRINTING, which will be found to embody the latest novelties of this branch, besides which, in an appendix, he has given the important features of the report of the International Jury of the Exposition of 1867, upon general improvements in other descriptions of Dyeing.

PHILADELPHIA, *January* 1, 1869.

*

PREFACE.

THIS work is intended by the Author to form a practical hand-book of reference upon all the chemical substances and processes in use among dyers and calico printers.

While written from a practical point of view, it takes a middle course between the generalities of high science and the technicalities of pure practice. Avoiding, on the one hand, the applications of chemical principles which are not yet clearly perceived, and, on the other, a detailed description of processes which would be either unintelligible or unnecessary, the Author hopes he has produced a book which may be profitably consulted by all who are either interested or practically engaged in printing and dyeing.

The claims which the Author has to be heard upon these subjects rest upon his familiar acquaintance with calico printing, acquired by nine years' service in a very extensive establishment, and upon a further professional experience of three years, which has brought him into contact with nearly all styles of dyeing.

This work is founded upon the Author's "Chemistry of Calico Printing," which was published only two years ago, but which has been for some time out of print. The substance of that book has been re-cast in a more popular form, all scientific formulæ and laboratory processes omitted, with the addition of a large amount of matter bearing upon practical operations.

Though appearing in a condensed form, this book contains considerably more matter than the Author's previous production.

In the course of the publication in parts, this book has experienced a most favorable reception; and the Author is obliged to many writers for their kind recommendation of it. From many quarters he has been solicited to extend his plan, and make a more complete treatise; but he has felt obliged to decline these suggestions, and to adhere strictly to the course originally laid out, and finish the work within the space at first specified. Nothing would have been easier than to have doubled the size of the work; if there is credit in its compilation, it lies in rejecting what was not necessary to the plan. For each receipt given, at least five have been withheld, as merely occupying space without presenting any instructive differences in their composition; in articles upon rare or little known coloring matters, the researches and descriptions occupying whole pages of original memoirs have been compressed into as many lines. Nevertheless, there is nothing omitted, and nothing unduly abridged, which is of real interest in practice.

The Author informed his friends, some time ago, that he was engaged upon an Encyclopædia of Dyeing and Printing, which he intended for a great comprehensive work upon the subject; but little progress had been made when it became evident that he could not reasonably hope to accomplish this favorite project; it would require more time than he could spare from the practice of his profession, and it has, in consequence, been indefinitely postponed.

No separate work upon calico printing, by any one practically acquainted with it, had appeared in England for a period of seventy years before the Author's book upon the subject; the dyers have been more fortunate in original and translated works; but it is a matter for regret that out of so many calico printers around Manchester and Glasgow, eminent alike for scientific attainments and practical skill, not one has found leisure to write a treatise upon this important branch of British manufactures. While the Author maintains the correctness of what he has written, no one can be more sensible of the superior advantages possessed by a writer who could have brought to the task that breadth of knowledge and maturity of judgment which is only acquired by the accumulated observations of a long life of practice.

The absence of a good work upon this subject has encouraged an idea among color mixers and foremen that they were the depositaries of a secret art, and they have exercised a jealous guard over their processes, which has operated to shut out improvement, and perpetuate a succession of absurd and empirical processes.

The Author will be glad if the circulation of this work among the responsible servants of dyers and printers causes them to trust less for success to supposed secrets, of doubtful value, and to rely more upon an intelligent comprehension of the nature and uses of the materials at their command. Many conspicuous improvements have originated in the color-shop and the dye-house; and, though the laboratory has carried off the honors and the profit of recent discoveries, there is much yet remaining to be found out by practical workers, if they will study the operations they are engaged upon in the light of true science.

With regard to the receipts scattered throughout the body of the work, the Author can only say he knows them to be genuine, either from his own experience or from friends in whom he has confidence. This will not, however, be inconsistent with them proving failures in other hands; each dyer and color mixer has his own peculiar methods of working, and it frequently happens that the drugs and proportions which work well with one man will not answer with another. The chief value of these receipts will consist in their illustrating various actual and possible means of attaining certain ends, and suggesting probable improvements or modifications upon existing processes.

CHARLES O'NEILL.

92 Grosvenor Street, Manchester,
  November 1, 1862.

ACKNOWLEDGMENT.

During the preparation of this book the author has had the undermentioned works upon his table, and has frequently referred to them, to supplement his own practical knowledge :—

Philosophy of Permanent Colors, by Edward Bancroft, M. D.   *London*, 1794.
Eléments de l'art de la teinture, par C. L. et A. B. Berthollet.   *Paris*, An. xiii. (1804).
Experimental Researches, etc., on Permanent Colors, by E. Bancroft, M. D. *London*, 1813.
Manuel du fabricant d'étoffes imprimées, etc., par L. S. le Normand. *Paris*, 1830.
Manuel du fabricant d'indiennes, par L. J. S. Thillaye.   *Paris*, 1834.
Elements of the Art of Dyeing, by Berthollet.   Translated by Dr. Ure. *London*, 1824.
Leçons de chimie appliquée à la teinture, par M. E. Chevreul.   *Paris*, 1829.
Manuel du teinturier, etc., par A. D. Vergnaud.   *Paris*, 1832.
Traité théorique et pratique de l'impression des tissus, par J. Persoz. *Paris*, 1846.
Précis de l'art de la teinture, par M. Dumas.   *Paris*, 1846.
Hulfsbüch für den gewerblichen chemiker vou M. Gerstenhöfer.   *Leipzig*, 1851.
A Manual of the Art of Dyeing, by James Napier.   *Glasgow*, 1853.*
A Manual of Dyeing Receipts for General Use, by James Napier. *Lond. and Glasgow*, 1858.
Abridgments of Specifications of Patents relating to Dyeing and Printing. *London*, 1859.
Practical Treatise on Dyeing and Calico Printing.   Anonymous.   *New York*, 1860.
Leçons de chimie elementaire appliquée aux arts industriels, par M. J. Girardin.   *Paris*, 1860.
Le teinturier au xixe siècle, par Théophile Grison.   *Rouen*, 1860.
A Manual of Botany, by Robert Lindley, F. L. S.   *London*, 1861.
Chemical Gazette.   17 vols., from 1842 to 1859.
Répertoire de chimie pure et appliquée.   *Paris.*   Now publishing.
Le Technologiste.   *Paris.*   Now publishing monthly.
The Chemical News.   *London.*   Now publishing weekly.
Handwörterbuch der reinen und angewandten chemie.   *Braunschweig.*   Now publishing.
Bulletin de la societé industrielle de Mulhouse.   Now publishing.

* Republished in the United States under the more appropriate title of "A System of Chemistry Applied to Dyeing." 2d edition, 1869.

ERRATUM.

On p. 178, second line from bottom, *for* 164, *read* 168.

COAL TAR COLORS (ANILINE, ETC.),

AND

THEIR APPLICATION TO DYEING AND CALICO PRINTING.

·   # INTRODUCTION.

SINCE the publication of the last edition of Mr. O'NEILL'S DICTIONARY ON DYEING AND CALICO PRINTING, great progress has been made in the manufacture and the applications of coal tar colors.

Aniline remains the great source from whence these colors are derived. Recently, however, carbolic acid and naphthaline have added their share in the production of new dyes, and we shall mention some of them, which have been sanctioned by practice. Many naphthaline dyes have been proposed, but on trial they have not presented the same degree of brilliancy and fastness which aniline colors possess. These are difficulties which, in all probability, will be overcome by further study and experience.

Aniline colors themselves are just emerging from a chaotic state of processes of manufacture, and of speculations relative to their nature. Their theory, at the present time, is far from being complete; but the light which has been shed on the subject has considerably simplified the processes of manufacture, improved the products, and lowered their price.

We now possess a complete gamut of colors from this source alone, and the only difficulty is how to choose among the 150 or 170 names given to these dyes. Several names are often given to the same color, and many colors have not stood the test of practice, so that the number of coal tar dyes in use is not so extended as that of their names.

We shall divide this essay on the applications of coal tar colors to dyeing and calico printing into four chapters :—

2

I. Mordants, thickenings, and discharges, especially employed for coal tar colors.

II. Dissolution and purification of coal tar colors, and their
precipitation from old baths.

III. General methods for dyeing and calico printing, leaving
the special process to the special colors to which they
refer, and which will be found in the next chapter.

IV. Coal tar colors subdivided into Reds, Purples, Violets,
Blues, Yellows, Oranges, Greens, Browns, Maroons,
Blacks, and Grays.

The object and the limits of this work being, not to dwell
on the manufacture of coal tar colors, but only to treat of their
application to dyeing, we have briefly indicated their nature,
from what substances they are derived, and those of their properties which may be useful to the dyer.

Those persons who may be desirous to become thoroughly
acquainted with the manufacture and properties of coal tar
colors should consult the following works:—

P. Schützenberger, *Traité des Matières colorantes.*

M. Reimann, *Aniline and its Derivatives.*

Th. Chateau (Collection Roret), *Couleurs d'Aniline, &c.*

F. C. Calvert, *Coal Tar Colors, &c.*

J. Girardin, *Chimie élémentaire.*

Note.—The Imperial gallon, holding 10 pounds of water,
being the one in use in Mr. O'Neill's Dictionary, we have retained it as the standard in our calculations for receipts.

The wine gallon of New York holds only 8 pounds of water.

The weights are avoirdupois.

CHAPTER I.

MORDANTS.—DISCHARGES.—THICKENINGS.

THE animal fibres, silk and wool, for instance, have such an
affinity for coal tar colors that, in most cases, by a simple dipping in a solution of these colors, they become dyed without
the help of any mordant.

On the other hand, the vegetable fibres are entirely devoid
of such an affinity for coal tar colors (aniline black excepted.)
It is therefore necessary to impart to them the property of fixing these colors by the help of some mordant. This operation is sometimes called animalization.

The mordants in use are :—

Albumen from the white of eggs.

" " blood.

Gluten dissolved in caustic soda (W. Crum process).

" " a weak acid (acetic acid).

Scheurer—Rott process.

Caseine or *Lactarine* (curd of milk) dissolved in caustic soda (W. Crum process).

Caseine dissolved in acetic acid.

Gelatine.

Tannate of Gelatine.

Tannin, pure, or from fresh decotions of gall-nuts, sumach, &c.

Certain oils, such as those used for Turkey red.

Certain acids, such as sulpho-margaric, sulpho-oleic, sulpho-glyceric, &c.

Certain resins, such as gum-lac dissolved in alkalies or borax.

Arsenite of Alumina.

Stannate of Soda.

Note—MM. Depouilly recommend the preparation of the stannate of soda by precipitating the oxymuriate of tin by ammonia, and dissolving the washed precipitate with as little caustic soda as possible.

Lead Salts, &c. &c.

Of all these mordants, the *albumen from the white of eggs* is the best; with it the work is easy, and the brightness of the colors is not impaired. The only drawback is its cost.

The *albumen from the blood,* being always more or less yellow, is inferior to the former.

Before using albumen it is always prudent to pass it through a fine muslin sieve, in order to strain off certain impurities which, otherwise, would appear as dark-colored specks on the dyed cloth.

Gluten dissolved in caustic soda.—Wet gluten is allowed to rest for from 5 to 10 days, according to temperature, until it becomes viscous. Then, to 10 lbs. of this gluten, add 17 or 18 ozs. of a solution of carbonate of soda, having a specific gravity of 1.15, which dissolves certain impurities. The gluten remaining over the filtering cloth is washed in cold water, and mixed with about 14 ozs. of a solution of caustic soda of 1.08 sp. gravity. The gluten becomes dissolved into a mucilage, which, afterwards, is diluted to the proper consistency for printing, with about 8 quarts of water. This gluten is printed first, dried, steamed, and rinsed in pure water before printing with the color, which operation is succeeded by a second steaming.

Lactarine.—Two pounds of dried and pulverized caseine

(curd of milk) are dissolved in ½ gallon of water and ⅓ gallon of a solution of caustic soda (1.08 sp. gr.).

4 pounds of fresh caseine may also be dissolved in the above solutions.

Soap Mordant.—For 20 lbs. of cotton yarn dissolve 1 lb. of tallow soap in sufficient water. The cotton is worked for some time in this hot bath, allowed to dry (without washing), and washed just before it is dyed in the coloring bath. By replenishing the soap bath with small additions of new soap, it may last a long time.

Arsenite of alumina (Schultz process).

Arsenite of soda	12 drachms.
Acetate of alumina (10° B⁶) . .	1 pint.
Magenta	65 grains.

and the whole is thickened with starch.

Instead of acetate of alumina, Mr. Schultz claims that solutions of acetate of zinc or magnesia at 10° B⁶ may be employed.

Phosphite of soda, antimoniate or stannate of soda may also be substituted for arsenite of soda, in the proportion of 3 to 5 ozs. to 1 quart of acetate of alumina.

Arsenite of alumina (A. Paraf process).—The difficulty in the employment of arsenious acid is its feeble solubility in water or acids; but glycerine dissolving its own weight of arsenious acid, Mr. Paraf uses it for preparing a mordant, as follows:—

1 part of arsenious acid is dissolved in 1 part of glycerine, and a solution of acetate of alumina is made in the usual way, except that sulphate of alumina is to be preferred to alum.

Then to a solution of aniline color, already thickened with starch, is added 10 to 12 per cent. each of the above solutions. The next operations are printing, steaming for ½ hour, and washing in tepid soap water.

Discharge by zinc powder (Durand process).—The very fine powder of zinc produced during the distillation of this metal, will reduce the colored salts of rosaniline into white salts of leucaniline, when printed on places which must be white. But if, by repeated washings, these salts of leucaniline are not entirely removed, they may color again under the oxidizing influence of air and light. This difficulty will not be found in the next process, where the color is destroyed, not transformed.

Discharge by permanganic acid (Dangvillé and Gauthier process).—The permanganate of potassa, which may be replaced by permanganate of lime, is mixed with a slight excess of sulphuric acid, and then, by the addition of water, is reduced to a solution holding from 1 to 6 per cent. of permanganate of potassa.

For printing, the thickening materials should be kaolin, silica, or alumina, because an organic thickening would decompose the permanganic acid.

After the operation, the destroyed aniline color is replaced by an oxide of manganese, which will be removed by washing in a sulphurous acid solution. But, if there are colors, such as coralline, which are destroyed by sulphurous acid, the washing is effected by a mixture of muriatic acid and protochloride of tin.

Discharge by tin powder (Hesse process).—Finely-pulverized tin is mixed with carbonate of soda, or other alkaline salt, and thickened with gum. After printing with this mixture, steaming and washing, the printed and previously colored places become white.

Gum thickenings.—It has been observed by Mr. Ach. Bulard that certain gums have the property of changing the shade of aniline reds to a dull reddish violet, when the printing mixture has stood only a few hours; and that a small quantity of albumen added to the mixture prevented this change of color. By further experiments, Mr. Bulard found that the best gums for printing aniline reds were the white qualities of gum senegal, and the worst, the gum Arabic, coming by way of Alexandria, Egypt; excepting, however, those which are entirely white.*

CHAPTER II.

DISSOLUTION OF COAL TAR COLORS.

NEARLY all the coal tar colors are remarkable for their slight solubility in water. On the other hand, they are soluble in alcohol, wood spirit (wood naphtha, methylic alcohol), acetic acid, tartaric acid, aniline, some in glycerine, &c.

The least soluble in water are certain kinds of blue and violet. The reds, such as magenta, azaleine, and roseine, are sufficiently soluble in water.

When we say sufficiently soluble, we do not mean that these colors are very soluble, but that they may be directly dissolved in hot water to form a bath of sufficient strength for dyeing. Indeed, when we consider the affinity of the animal fibres for

* The thickening and mordant called *gum water* and *albumen water* is generally 1 pound of the dry substance dissolved in 1 quart of water (wine gallon weighing 8 pounds of water).

these colors, too much solubility would be a disadvantage; the colors would instantly precipitate upon the fibre, and the fabric would not be dyed evenly. It is to avoid this irregularity that it is often necessary to begin with very weak baths, and to add the coloring substances at intervals, by small quantities, while the bath is well stirred, and the fabric constantly worked.

When these aniline dyes, which have been previously dissolved in alcohol, are poured into the water of the bath, they are generally not dissolved but precipitated, as may be ascertained by filtering through paper. But the precipitate is so fine and light, and, if we may say so, in such a hydrated state, that the bath appears limpid, without deposit at the bottom, and in a condition to dye the yarns or fabrics which are put into it.

Therefore, the previous dissolution of certain coal tar colors in alcohol has for its object to present those dyes to the bath in a minute state of division, which could not have been obtained by mechanical means, and which allows them to be thoroughly incorporated with the water, if not entirely dissolved in it.

Of all the solvents we have mentioned, alcohol is the most in use; acetic acid and tartaric acid are sometimes mixed with it. The use of wood spirit, unless perfectly pure, should be avoided for the red colors of rosaniline, which it turns to violet, and the more so when acetic acid is mixed with the former.

To sum up: when such colors as magenta, &c., are sufficiently soluble in water, they are boiled in it, and filtered. About 200 parts of water for 1 of color. Some soluble blues require less water.

When blues and violets, insoluble in water, are to be dissolved in concentrated alcohol, the color is gradually added to it, stirring all the time, and afterwards allowing it to rest. After a few hours, the whole is heated in a water bath to the boiling point, and the volatilized alcohol is condensed. After boiling it is well to allow the solution to rest one night, and lastly to filter.

The proportions of alcohol are variable:—

30 to 50 parts of alcohol for 1 of blue;
"   "   "   "   "   violet;
10 to 12   "   "   "   Hofmann's violet.

Of these latter violets there are some which are soluble in water, and do not require the use of alcohol.

Instead of pouring the alcoholic solution directly into the bath, it is customary to dilute it with 8 to 10 times its volume of water, acidulated with tartaric or sulphuric acid.

Iron, tin, and zinc vessels are to be avoided in making these ·

solutions, especially the latter, which reduce the colors. (See · *Discharges.*)

In order to avoid the rather expensive use of alcohol, various decoctions of vegetable substances have been tried for dissolving coal tar colors. Soapwort (Radix saponica), Panama bark (Quillaya saponica), lucern root, &c., have been proposed, but we believe that they have not been extensively used.

Concentrated sulphuric acid has the property of rendering soluble in water certain blues and violets which were insoluble.

We believe that Mr. Nicholson was the first to make use of this property, and his process is as follows: Add gradually 1 part of insoluble blue to 6 parts of concentrated sulphuric acid; the mass being thoroughly mixed by stirring, heat the whole up from 284° to 300° F., and when cold pour the whole into cold water (8 to 10 parts of water for 1 of oil of vitriol employed). The color becomes precipitated at the bottom of the vase, is collected and washed over a cloth until the water runs blue. The blue is then soluble in about 50 parts of boiling water, its shade is bluer, but it has lost part of its fastness, which is caused by a chemical transformation due to the powerful acid and great heat to which it has been submitted.

Messrs. Rangod-Péchiney and Ach. Bulard employ also concentrated sulphuric acid, but act in a different way. 1 part of the color is dissolved by small quantities, gradually added to 6 parts of oil of vitriol at 66° Bé. At each addition the mass is thoroughly mixed, allowed to stand, and stirred until all the color is dissolved. During the whole operation care is taken to prevent the acid from becoming heated. The liquid is then poured into cold water (20 times the weight of acid employed), not all at once, but slowly, at different places of the water surface, while a constant stirring is going on, in order to avoid too great an elevation of temperature.

The color becomes precipitated, is collected, and washed on a filter until the water runs blue. The resulting paste, if not immediately employed, is mixed with glycerine which prevents · it from drying.

The acid solution may also be poured into the dye bath, but the excess of acid must be neutralized by some alkali.

By this process the color has not been transformed, as is the case by the Nicholson treatment; it remains as fast as before, and although it requires more water to become dissolved, it is in a state of minute division and of hydration, which is similar to that of the same color dissolved in alcohol, and poured into the boiling bath.

PURIFICATION.

Although the coal tar colors, at the present time, are to be found in the market much purer than formerly, and some of them entirely pure, it may not be out of place to indicate the means of removing their impurities. These are generally re-sinous substances soluble in benzine, and light oils from coal tar or petroleum, while the colors are insoluble in them. The operation consists therefore in washing the impure colors with these hydrocarbons.

Another method, applicable to the aniline colors soluble in water, consists in adding to them 5 times their weight of fine white sand, and dissolving the color in boiling water, while the resinous substances stick to the sand.

For those colors which are insoluble in water, an alcohol can be made weak enough for dissolving the color without acting on the resinous matter; whereas if the alcohol was too concen-trated it would dissolve both the color and the impurities.

Happily for the dyer, these impure colors are now rarely sold.

PRECIPITATION.

Under this head we mean the processes for recovering the color from baths which have not been exhausted.

The first process, used by the manufacturers of aniline dyes, consists in saturating the dye-bath with chloride of sodium (common salt), or any alkaline salt; the color becomes precipi-tated in its primitive state, and can be used again in the same manner.

By the second process, a combination of the color with tan-nic acid is formed, which is nearly insoluble in water, and may be employed for calico printing. In order to precipitate the color entirely, and to obtain a fine product, it is necessary that the bath should not be too acid, and that the liquors containing tannin (pure tannic acid, decoction of gall-nuts, &c.) should be freshly prepared. The excess of free acid is first saturated by carbonate of soda, and the color is afterwards precipitated by the above liquors, taking care not to add an excess of them, which would re-dissolve part of the precipitate.*

The third process produces aniline lakes by adding alum to the bath, neutralizing the alum by carbonate of soda as long as no precipitate occurs, and lastly, by precipitating with tannin.

---

* The precipitate is then washed over a cloth, and is better kept for use in a pasty state, although it can be dried at a temperature not exceeding 170° F.

CHAPTER III.

GENERAL METHODS FOR DYEING AND CALICO PRINTING.

IN laying down in this chapter certain general principles for dyeing and calico printing, our object has been to obviate the unnecessary lengthening of this essay, by repeating for each color what can be said at once for all.

We shall illustrate these general rules by several examples. For the special methods, we refer the reader to those colors which require them, and which will be found in Chapter IV.

Wool and silk are worked nearly alike, the differences being that the temperature of the bath is generally hotter for wool than for silk, and that an acidulated bath is more necessary for silk than for wool.

The temperature of the bath and its degree of acidity have also some influence on the shade produced. The hotter and more acid the bath is, the bluer are the shades. Red shades are obtained by a lower temperature and less acid.*

In dyeing, the fabric ought to be drawn several times in the bath, and the color be added successively in several portions. For certain colors, more soluble than others, this is indispensable, in order to dye evenly.

Although the animal fibres do not require any mordant to fix the aniline dyes, the shades have often been found much faster when these fibres had been previously mordanted with alum, alumina, bichloride of tin, &c.

For printing, the color dissolved in alcohol, or acetic acid, is thickened with starch, or gum senegal, gum tragacanth, albumen, &c.

The steaming is generally begun with a small pressure, which is afterwards increased.

Brighter shades will be obtained if the printed colors do not dry too quickly. An addition of glycerine to the printing mixture will keep it moist.

The aniline reds are not so fast as the blues and violets, especially on calico.

---

* It is sometimes preferable to raise the shade by drawing the fabric in another acidulated bath, directly after the dye-bath, or after a washing. The acid generally employed is sulphuric acid.

Chinoline colors afford beautiful shades, but without stability. In this case there should be no acid in the bath.

ANIMAL FIBRES—(WOOL AND SILK.)

Dyeing with salts of rosaniline, pure fuchsine, magenta, &c.—Dissolve the dye in 200 parts of boiling water, which, after filtration, you pour gradually by portions into the bath. The bath is lukewarm for silk, and acidulated with tartaric or sulphuric acid; but for wool it is gradually brought up to the boil, and no acid is necessary.

Dyeing with violets insoluble in water.—The dye dissolved in alcohol and diluted with water, as we have explained in Chapter II., is added by degrees to the bath which is at a temperature of from 104° to 140° F., and acidulated with some sulphuric acid, remembering that the more acid there is, the bluer is the shade.*

Sometimes, for a blue shade, sulphate of indigo is added to the bath. Other persons use only aniline colors (blue and violet), for shading a ground of indigo or Prussian blue. Weak alcohol will dissolve that part of the dye which imparts a red tinge to certain blues.

Aniline blues insoluble in water are dyed as the corresponding violets.

Dyeing green.—The paste, which is a compound of tannin and the green color, is soluble in water acidulated with sulphuric acid. A higher temperature is necessary for wool than for silk. It is well to leave the fabric in the bath until it cools off. Wool is sometimes mordanted with alum.

Certain greens become soluble in water after having been thoroughly mixed with some sal-ammoniac.

Printing red aniline colors.—The dye is dissolved in acetic acid or alcohol, thickened with gum senegal, printed and steamed. A tin mordant is to be avoided.

For 1 quart of alcohol, 1½ ozs. of magenta crystals are used, or more, if deeper shades are wanted.

This is a general process, which may be applied to many aniline dyes for printing on silk and wool.

VEGETABLE FIBRES—(COTTON, &c.)

The yarn or fabric is to be mordanted with some of the mordants already spoken of in Chapter I., and dyed in a hot and

---

* The shades will be much faster if the wool and the silk have been mordanted in a hot bath (167° F.) containing a mixture of 1 part purified cream tartar and 10 parts alum.

acidulated bath. We shall begin the examples by one where albumen is employed, and where a very pure and bright shade is required.

Dyeing with magenta crystals.—For mordanting, dissolve ½ lb. of albumen in 1 gallon of cold water, work the fabric or the yarn in it. Steam, in order to coagulate the albumen, and then dye in a moderately hot, and slightly acid bath.

Cotton prepared with oil, the same as for Turkey red, takes very well the aniline reds and blues. The oil, olive oil for instance, which has been treated with sulphuric acid, becomes better adapted to act as a mordant. 1 lb. of olive oil is well beaten with 4 ozs. of oil of vitriol, and becomes brown. It is then mixed with 1 quart of alcohol, and when all appears dissolved, it is poured into boiling water. These proportions are sufficient for from 70 to 75 lbs. of cotton, which is mordanted in a tepid bath.

Dyeing with tannate of tin as a mordant (Perkin and Puller's process).—The cloth is soaked for one hour or two in a decoction of sumach or any other tanning substance, and then put into a weak solution of stannate of soda, where it is drawn and worked for one hour. It is then wrung out, dipped into diluted sulphuric acid, and well rinsed before it is dyed in a slightly acidulated bath of aniline color.*

Dyeing with a lead mordant.—Cotton may be mordanted with a basic salt of lead, and dyed afterwards in a hot bath where soap and the color have been dissolved together. Lead mordants are somewhat difficult to be evenly absorbed by the stuff.

Dyeing with aluminate of soda for mordant.—The cotton is allowed to rest for from 10 to 12 hours in a solution of soda marking 4 to 5° B⁴, and without rinsing is put into a solution of aluminate of soda, where it rests the same length of time. Alumina becomes fixed in a hot solution of sal ammoniac. The cotton is then dyed in a coloring bath, at the temperature of 122° F.

Other methods of dyeing.—Mordant with oxymuriate of tin, then with tannin, and dye.

Or, dissolve caseine (curd of milk) in as little ammonia as possible; soak the cotton in this solution diluted with water, and dry it. Then work your cotton in another bath of tannin with some muriatic acid, and dye it, after it has been carefully wrung.

Mr. R. Böttger says that a solution of tannin in alcohol is sufficient for mordanting flax and cotton stuffs before dyeing.

---

* The mordanted fabric is of a light yellow color. It is said that alum may be employed instead of stannate of soda.

MM. Franc and Tabourin have proposed biphosphate of lime for mordanting cotton.

For certain kinds of aniline blues, which have some red in them, the cotton is dyed first with Prussian blue. A violet is obtained by shading a ground of Prussian blue with aniline violet, with or without magenta.

Printing with magenta crystals.—Take 4 ozs. of color, and mix it thoroughly with 1 pint of warm water and 1 pint of glycerin, then boil the whole for 15 minutes. After filtration, thicken the color with 1 lb. of pulverized gum senegal, and pass it through a sieve. On the other hand, $3\frac{1}{2}$ pounds of dry albumen have been dissolved in 3 quarts of water and passed through a muslin sieve; this is added to the former substances. Print with the mixture, steam and wash.

When the cotton fabric has been previously mordanted with alumina (red liquor), holding a trace of iron, a pure red cannot be obtained.

Another method consists in mordanting with oxymuriate of tin, printing with magenta and tannic acid, and steaming, &c.

Printing violet.—50 grains of rosolane in paste are dissolved in $\frac{1}{2}$ oz. of alcohol, and thickened with 4 ozs. of gum water, and $5\frac{1}{2}$ ozs. of albumen water. After printing, the fabric is steamed and washed. The pressure of steam is low at the beginning, and is gradually increased.

The preparations of gluten and caseine (lactarine) are cheaper, but are inferior to albumen, as regards the facility in printing, and the solidity and brightness of the shades.

Printing with soluble blue.—1 part of the color is dissolved in 20 parts of water, which is mixed cold with 18 parts of acetate of alumina marking 15° Baumé. Then thicken with gum water, print, steam, and wash.

If the blue is very soluble, it will be well to add to the mixture some carbonate of soda, about $\frac{1}{8}$ of the amount of the color.

Printing with arsenite of alumina as a mordant (Wischine process).—A mixture of arsenite of soda, acetate of alumina, and red or blue colors, is thickened, and then printed. After steaming, the stuff is washed in a soap bath.

The shades stand washing very well.

Printing violet.—A paste is made by mixing and boiling for 15 minutes 1 oz. of aniline violet, $\frac{1}{2}$ pint of water, and $2\frac{1}{2}$ or 3 ozs. of glycerine, to which we add from 3 to $3\frac{1}{2}$ ozs. of gum senegal. When the whole is cold, it is passed through a sieve and mixed with 7 ozs. of dry albumen dissolved in $\frac{1}{2}$ pint of water. After printing, we steam and wash.

Printing by Mr. E. Kopp's tannin process.—We have already seen (Chapter II., precipitation) how to produce the combinations of tannic acid with the salts of rosaniline, mauveine, and their derivatives. These combinations are insoluble in water, but soluble in alcohol, acetic acid, wood spirit, and diluted sulphuric acid ; remembering that impure wood spirit should not be used with the reds, which are turned violet or blue.

The solution of the dye is thickened with gum senegal, or gum tragacanth, or a mixture of both, or starch-dissolved in acetic acid. After printing, the fabric is steamed, and washed in cold water.

The shades obtained in this way stand the action of soap well, but not so well that of light.* ·

The following is another receipt:—

> 1 pound of violet paste.
> 1 "   " acetic acid No. 8.
> 1 "   " ˙ tannic acid.
> 3 quarts of boiling water.
> 1 gallon of gum water.

In this example the tannate of the color is made directly in the mixture. If the tannate was already at hand, no new tannin, or better, only a small quantity, would be needed.

The calico to be printed may also be prepared with stannate of soda, or alumina, gluten, caseine, gelatine, basic acetate of lead, corrosive sublimate, tartrate of antimony (only for aniline violets), and double chloride of potassium and antimony, which precipitate by tannin, and produce faster shades.

When corrosive sublimate (bichloride of mercury) is employed, it has a tendency to turn the reds to violet.

Sometimes the printing is only made with a thickened solution of tannin, more or less concentrated according to the depth of shade desired. It is thus easy to obtain several shades of various intensity on the printed figures.

The calico printed with tannin is steamed, drawn through a weak solution of gelatine or of a metallic salt, and thoroughly washed.

The next operation will be the dyeing, when the color becomes fixed to the parts printed with tannin, while the ground of the fabric is but very little colored. When aniline reds are employed, a washing with soap will remove all color from the ground not printed.

Instead of pure tannin, gall-nuts, sumach, or tannin mixed with some fatty or resinous substance may be employed; but pure tannin gives the finer shades.

Printing by the tannin process of MM. Javal and Gratrix. 1st

---

* The reds, pinks, and some violets do not succeed well by this process. It is preferable to previously mordant the fabrics with stannate of soda.

method.—Dissolve the color in alcohol or acetic acid, thicken and print it on tanned cloth, steam, and wash in pure water, or with soap if aniline reds have been employed.

While steaming, the pressure begins very low, and is increased up to seven pounds per square inch.

2d method.—Print with tannin alone, thickened, and steam as above, then draw through a bath containing a solution of an alkaline arseniate, phosphate, or silicate, and wash in pure water.

Then dye in a bath at 140° F. acidulated with acetic acid, into which the color dissolved in acetic acid is added by degrees, and which is gradually brought to the boil during the working of the cloth, which lasts about half an hour.

When the white portions have absorbed some of the color, they are bleached by passing the cloth through a hot bath containing some mineral acid, which dissolves the coloring matter not combined with the tannin. Soapsuds, or a weak solution of printing clearing liquor, such as is used for garancine, may be employed. Lastly, rinse in pure water.

Printing by the tannin process of MM. Littlewood and Wilson, or MM. Lloyd and Dale.—To one gallon of gum water mix from 8 to 10 ozs. of pure and dry tannin, and enough of aniline color for the shade required. Print and steam at low pressure, and then pass the fabric through a hot solution (170° to 212° F.) of 2 ozs. of emetic or tartrate of antimony per gallon of water. Then wash and dry. If bleaching is necessary, use very weak solutions of bleaching powder, wash with soap, and rinse in pure water.

Printing by Brook's process.—By this process, madder printed colors may have their brightness increased by aniline colors. For this purpose, the madder mordants receive an addition of tannin and acetate of tin, and become fixed by ageing and steaming. The fabric is then dunged in cow's dung, or in phosphates, silicates, and arseniates of soda, dyed first with madder, and afterwards with aniline colors.

Printing by the process of MM. J. and T. P. Miller. 1st method.—Digest 1 lb. of gall-nuts in 1 gallon of acetic acid No. 8 (Twaddle), which is mixed with a compound of tartaric acid, stannate of soda, a small excess of acetic acid, and a suitable quantity of aniline color, then thicken with gum or starch, print, steam, &c. on a fabric which has been mordanted with a tannin solution.

2d method.—Draw the fabric through a solution of 8 ozs. of soap per gallon of water, and afterwards through a bath containing some sulphuric acid ; then dry it.

The printing mixture is made with about 12 ozs. of acetate

of lead per gallon, to which is added the color dissolved in acetic acid. Then thicken, print, and steam.

The proportion of acetate of lead is variable with the quantity of color.

MIXED FABRICS.

Dyeing.—Mordant with sumach and stannate of soda, dye in a boiling bath and wash.

Printing by the R. Böttger process.—Liquid Violet. 1 part or volume gum tragacanth, water 4 parts or volumes. Before adding the violet, dissolve a little less than one ounce of oxalic acid for every gallon of paste, thick or diluted. The mixture is passed through a muslin sieve, and printed without albumen, which is said not to be necessary. After printing, the fabric is steamed at a low pressure for half an hour; washed and dried.

These numerous examples would be incomplete without the *starching process,* which produces the cheapest prints in every sense of the word. No mordants whatever are required for preparing the cloth or printing it. A little aniline color dyes a great amount of starch. By printing with such starch drying, and selling the fabric *without washing* it, the consumers will be sure to become disgusted with aniline colors.

Such prints will bear no washing, and very little rubbing.

A certain quantity of starch, with the proper mordants, may be useful for thickening ; but too large a proportion has the result of interposing an inactive substance between the fibre and the color.

CHAPTER IV.

COAL TAR COLORS.

It would be difficult to arrange these colors alphabetically, on account of the many and different names which have been given to the same dye. We shall then separate them into reds ; purples, violets, blues, &c. &c.

This variety of names, which, in most cases, does not indicate the composition of the color, necessitates a brief consideration of their nature ; and we shall dwell longer on those which require peculiar methods for their use.

The number of coal tar dyes which we present to the reader

is more considerable than is generally found in dye-houses. We have, however, chosen only those colors which, on a large or small scale, have given sufficiently satisfactory results in the large dye houses of Alsace and England.

REDS.

Arsenite of rosaniline; Crude fuchsine.—The old processes for making aniline reds have all given way to the actual treatment of a mixture of aniline and toluidine by arsenic acid. This color is sufficiently soluble in hot water to produce directly a dye bath. It dyes a dirty red when the solution is only slightly heated, and red-brown when heated to the boiling point.

On account of its arsenic and arsenious acids, it is a highly poisonous substance which ought to be handled carefully. The dyed stuffs, however, do not retain any arsenic after a thorough washing.

Hydrochlorate of rosaniline; Magenta; Fuchsine; Solferino; Aniline red.—Obtained from the above arseniate, and purified by several crystallizations. Is found entirely pure in the trade; sufficiently soluble in boiling water, and very soluble in alcohol.

Nitrate of rosaniline; Azaleine; Rubine; sometimes *Magenta.* —Produced by heating for several hours, at about 240° F., 10 parts of aniline with 7 parts of dry nitrate of mercury. Soluble in hot water, and very pure in the trade.

Acetate of rosaniline; Roseine; sometimes *Magenta.*—Obtained by combining concentrated acetic acid with rosaniline precipitated from the arsenite. The crystals are very pure, and are soluble in water.

Coralline; Peonine.—Discovered by Mr. Persoz, Jr., and produced by the action of ammonia upon rosalic acid, under pressure, and at the temperature of 302° F. This color is not very fast, especially on cotton, is soluble in alcohol, acetic acid, and in alkaline solutions which become brown after being a certain time in contact with the air. Sulphurous acid destroys this color.

(Silk and wool dyeing.)—Dissolve coralline in alcohol, add some soda, and pour the solution into a large quantity of water. Then a small addition of tartaric acid will liberate the color, which will produce shades intermediate between magenta and cochineal. Dye in a cold bath.

(Cotton dyeing.)—Dissolve coralline in a solution of caustic soda of 12° B°, or in a saturated solution of carbonate of soda. 1 gallon of this latter liquid will dissolve 2½ lbs. of coralline, which is then diluted with 2 gallons of water, and neutralized by 1 gallon of sulphuric acid at 10° B⁴.

The fabric, which has been mordanted with tin, sumach or tannin, is then dyed in this solution, for 1½ hour, and at a temperature of about 85° F. raised up to 122°.

The shade obtained resists steaming and washing; but soap, alkalies and light alter it rapidly.

(Calico printing.)—The coralline precipitated by water from its solution in acetic acid is dryed at a low temperature, intimately mixed with chalk or oxide of zinc, and printed with albumen.

Erythro-benzine.—Obtained by MM. Laurent and Casthelaz by mixing 12 parts of nitro-benzine with 12 parts of iron filings, and 6 parts of concentrated hydrochloric acid. The whole is allowed to rest for 24 hours at the natural temperature. The resulting resinous mass is treated by boiling water, and the color precipitated by chloride of sodium. Soluble in alcohol, and fast on silk and wool.

Isopurpurate of potassa; Soluble ruby.—Obtained by Mr. Hlasiwetz by the reaction of cyanide of potassium upon picric acid. It is sufficiently soluble in hot water, and alcohol. It is sold in paste, which should contain some glycerine to keep it always wet, otherwise it is a very dangerous product, because it detonates by the least shock.

It dyes silk, wool, and cotton mordanted with albumen, a garnet and puce color, with the addition to the bath of alum and acetic acid. These shades turn to orange by steaming.

Its solution, heated with acetic acid in a copper vessel, becomes of an orange color.

Wool and silk, when mordanted with corrosive sublimate, are dyed a magnificent purple. If the mordant is zinc, the color obtained is a brilliant yellow. We see, therefore, that different mordants will produce different shades.

The colors are very fast, except against sulphurous acid.

Rubis imperial; Imperial ruby.—Mixture of 300 parts of coralline with 200 parts of magenta. Dissolve 1 pound of the mixture in 12 gallons of alcohol, and dye without mordant. When coralline is in excess of the above proportion, the shade is a yellowish cherry red. If magenta is in excess, the shade is somewhat violet.

Grenat; Garnet-red. (Schultz process.)—It is a precipitate obtained by passing nitrous oxide gas through a solution of magenta or fuchsine in alcohol mixed with ammonia.

Phœnicine; Ponceau d'aniline.—Obtained by Mr. F. Duprey by boiling bioxide of barium with a solution of acetate of mauveine (Perkin's violet). Adding then some carbonate of

3

soda, just enough to separate the carbonate of baryta, the solution is filtered, and phœnicine is precipitated by chloride of sodium, washed, &c.

This dye is soluble in ammonia, carbonate of soda, benzine, &c.

Cerise; Cherry red.—Manufactured by Mr. J. R. Geigy, of Bāle. Dissolve 1 part of the color in 6 parts of acetic acid; let it stand one night in a vessel which is kept hot by being 'put in a hot bath. Then pour the whole solution into 15 to 20 buckets of hot water, mix well, skim, filter, and dye in the hot bath.

The shades are raised on silk, by a washing and a drawing through very weak and cold sulphuric acid.

Wool may be mordanted with cream of tartar or alum, or both.

The shades may be varied by the addition of magenta, turmeric, sulphate of indigo, picric acid, archil, &c.

Chloroxynaphthalic acid.—Obtained by Mr. Casthelaz. It is soluble in alcohol, benzine, sulphuric acid, alkaline acetates, and in boiling water sufficiently for dyeing.

It dyes wool a deep red, without mordants. Various shades are produced by the admixture of other coloring substances. It does not succeed with cotton, even animalized, on account of its great acidity.

Rosolic acid.—It is now scarcely used, the shades being without fastness, either on wool, silk, or on cotton.

PURPLES, VIOLETS, AND BLUES.

Sulphate of mauveine; Perkin's violet; indisine; aniline purple; mauve.—This color, discovered by Mr. Perkin, was the first aniline dye introduced to the trade. It is produced by heating a mixture of sulphates of aniline and toluidine with bichromate of potassa, and sulphuric acid, and separating the resinous impurities by benzine. Little soluble in boiling water; but soluble in alcohol, wood spirit, acetic acid, acetone, glycerine, tartaric acid, sulphuric and muriatic acids.

Aniline violet.—Obtained by MM. Depouilly and Lauth, by gradually adding bleaching powder to a solution of hydrochlorate of aniline with acetic acid until the desired shade of violet appears.

This violet has the same basis as Perkin's violet, that is to say, mauveine, which can be precipitated and dissolved in muriatic acid or acetic acid.

Similar violets are produced by using chlorine, bi-oxides of lead and of manganese, permanganate of potassa, &c., instead of bleaching powder or bichromate of potassa.

Bleu and violet de Mulhouse.—Obtained by MM. Gros-Renaud and Schoeffer, by boiling a solution of white shellac and soda crystals with some azaleine (nitrate of rosaniline) previously dissolved in alcohol. For the violet a larger proportion of soda crystals and azaleine is required than for the blue. Thicken the colored liquid with gum tragacanth, and print.

Violet impérial.—Obtained by MM. Girard and de Laire by heating for 5 or 6 hours, and at a temperature of about 330° F., a mixture of equal parts of aniline and hydrochlorate of rosaniline; the excess of aniline and magenta is removed by diluted muriatic acid, and the remaining violet is soluble in alcohol, acetic acid, wood spirit, and boiling water with some acetic acid.

(Silk and wool dyeing.)—Dissolve 1 part of this violet in 2 parts of alcohol, and 1 of acetic acid; let it rest and filter. Dye in a bath acidulated by sulphuric acid, cold at the beginning, and brought up to the boil. The shade will be bluish, if the fibre is taken immediately from the hot bath, and reddish, if allowed to remain in it until it cools off.

Aniline or rosaniline blues.—There are a great many processes for manufacturing these blues. A blue will be produced by a mixture of aniline red and aniline, heated with an organic acid or an organic salt (acetic acid, or acetate of soda, tartaric acid, benzoic acid). By the Nicholson process, rosaniline, aniline, and acetic acid are employed.

All these blues are nearly insoluble in boiling water, but are soluble in alcohol, &c., or in water after a transformation by oil of vitriol.

Soluble blues.—We have already seen in Chapter II. how to render soluble the above blues and many violets. By the process of cold sulphuric acid, the color becomes sufficiently soluble in an acid bath; by the process of hot sulphuric acid, the dye is very soluble in water; but this great solubility requires certain precautions in dyeing, otherwise the shades would be uneven. Some persons add carbonate of soda as a corrective.

MM. Lachmann and Breuninger use two baths; the first contains 1 part of soluble blue in 500 parts of water, and no acid at all; the second bath contains water acidulated with sulphuric acid. The cloth dyed in the first bath is of a light grayish-blue color, and the pure blue shade appears only in the second bath.

Night blue; bleu de nuit; bleu lumière.—It is so called on account of being free from violet, and of keeping its true blue color in artificial light. Made by heating 4 parts of magenta or roseine, 8 parts of aniline, and 2 parts of acetate of soda, during 2 hours, and at a temperature of 392°, raised up to 482° at the end of the operation. The last trace of violet is removed by several washings with diluted sulphuric, or muriatic acid.

It is sufficiently soluble in boiling water with some acetic acid.

Bleu de Paris.—Obtained by MM. Persoz, de Luynes, and Salvetat, by the reaction of 9 parts of anhydrous bichloride of tin on 16 parts of aniline, under pressure, and at the temperature of 356° F. This blue is more expensive than other kinds, but is very fast, soluble in water, alcohol, &c., and keeps a pure blue shade under artificial light.

Bleu de Lyon.—Prepared in a similar way as night blue, only acetate of potassa takes the place of acetate of soda. MM. Girard and de Laire use the process for making arsenite of rosaniline, only the quantity of arsenic acid is considerably increased.

(Silk dyeing.)—1 lb. of the blue is dissolved in 2 gallons of alcohol, 2 or 3 ozs. of sulphuric acid are added, and after standing some time the whole is filtered. The silk is drawn 5 or 6 times through the bath, which is brought up to the boil, and receives the color by portions. After dyeing, the silk is worked in a hot soap bath, rinsed, and the shade raised in acidulated cold water.

(Wool dyeing.)—The same as for silks, but no washing in soap.

(Printing.)—1 part of blue is dissolved in 33 of alcohol, and 1 part of this solution is thickened with 5 parts of gum water.

Dahlia or blue violet.—Many violets are obtained, the same as blues, by the reaction of aniline red with aniline and acetate of soda, and by varying the proportions, the temperature, or the length of the operation. For this dahlia color, the operation lasts longer, and benzoic acid or benzoates are gradually added, until the desired shade is obtained. The mass is then cooled off rapidly, and purified in the usual way. All these violets are scarcely soluble in water alone.

Ethyl or *methylrosaniline violets ; Hofmann's violets ; primula ; iodine violets.*—These colors, remarkable by their beauty and fastness have been obtained by Mr. A. W. Hofmann by treating under pressure, for 3 to 4 hours, and at a temperature a little above 212° F., a mixture of a salt of rosaniline, iodide of ethyl or methyl, and strong alcohol. With equal parts of rosaniline and iodide of methyl or ethyl, the product is a *red violet ;* by doubling the quantity of iodide, the color is a *blue violet.*

When these violets are not separated from their iodine, they are to be dissolved in alcohol.

When free from iodine, and combined with acetic or muriatic acid, they are soluble in about 50 parts of water.

(Silk dyeing.)—Use an acidulated and lukewarm bath, gradually raised to the boil. The color is added by degrees.

(Wool dyeing.)—Hot bath, without mordant or acid.

Printing as by the usual methods.

Aniline violet obtained by Mr. Perkin by heating together equal parts of mauveine and iodide of ethyl, for several hours. The process bears some analogy to that of Mr. Hofmann.

Ethylmauvaniline blue and *violet*. Obtained by MM. Girard, de Laire, and Chapoteaut, by the Hofmann process, and by substituting mauvaniline for rosaniline.

Methylaniline violet; Paris violet.—The originators of this color are MM. Greville-Williams, Poirier, Chappat, and Ch. Lauth. Soluble in water.

Violaniline and *Mauvaniline.*—These dyes with chrysotoluidine have been extracted by MM. Girard, de Laire, and Chapoteaut, from the residues of the treatment of hydrochlorate of rosaniline.

The salts of mauvaniline are soluble in water, and their color is a beautiful violet mauve.

The salts of violaniline are soluble in alcohol and are blue black, with violet reactions.

Regina purple.—Obtained by Mr. Nicholson by carefully heating pure magenta at a temperature of from 390° to 420° F., until the substance appears a dark and thick mass. Ammonia is evolved. Soluble in acetic acid, alcohol, &c.

Azuline.—Obtained by MM. R. Richoud and J. Persoz by gradually oxidizing aniline by coralline. Soluble in alcohol, acetic acid, &c.

For dyeing silk and wool, azuline is dissolved in alcohol, and added to a hot bath acidulated with sulphuric acid, or better, tartaric acid. Heat to the boil.

Azurine; Dark indigo blue.—This color, by the process of MM. F. C. Calvert, Ch. Lowe and S. Clift, is directly produced upon the cloth by dyeing or printing cotton goods with a mixture of tartrate or hydrochlorate of aniline, and acetic acid. After an exposure to the air, of 2 or 3 hours, a green color appears (see greens—emeraldine); the fabric is then drawn through a bath holding a weak solution of soap and caustic soda, or better still, one ounce of bichromate of potassa per gallon of water. The color turns indigo-blue, and even black, if too much of the aniline salts has been employed.

No mordants of alumina or others are required.

Aniline purple.—Produced by MM. J. Dale and H. Caro by boiling in water a mixture of a salt of rosaniline with a soluble copper salt and chloride of sodium. The precipitated color is purified by weak and boiling alkaline solution, and is soluble in alcohol.

For dyeing cotton, mordant with tannin, dye in the color, and fix in a bath of tartarized antimony.

Chinoline blue and violet; Cyanine.—Extracted from cincho-nine. The shades are beautiful, but without any solidity, being acted upon by light, acids, and alkalies.

When dyeing, the bath should not contain any acid.

Harmaline (violet).

Toluidine blue.

Rosolane (violet).

Parme—Bluish purple.

Phenylamine blue.

Rosotoluidine blue.

YELLOWS, ORANGES.

Picric acid and *Picrates.*—The picrates are highly explosive, and their dyeing power is much less than that of picric acid. This color is produced by the action of nitric acid upon car-bolic or phenic acid. It dyes wool and silk a yellow color, with a green tinge. The shade is faster when these fibres have been mordanted with a mixture of alum and cream of tartar.

Cotton mordanted with albumen and lactarine may be dyed with picric acid, although not very fast. That color does not bear steaming.

Chrysaniline yellow; Phosphine; Victoria Orange; Yellow fuch-sine.—This color was separated first by Mr. E. C. Nicholson from hydrochlorate of rosaniline in the manufacture of the lat-ter. Soluble in water acidulated by acetic acid. Dyes silk and wool a beautiful golden yellow. The sulphate of chrysaniline is very soluble in water.

Chrysotoluidine—Extracted by MM. Girard, de Laire, and Chapoteaut are from the residues of the manufacture of aniline red, where it is found associated with mauvaniline and violani-line.

The salts of chrysotoluidine are soluble in water.

Binitronaphthalic acid; Naphthylamine yellow; Jaune d'or; Manchester yellow.—This color is due to Dr. C. A. Martius, and dyes a magnificent golden yellow, without the greenish shade of picric acid. It will support steaming. Used extensively for dyeing wool and leather.

Yellow Coralline. (Printing on Wool).—Dissolve 5 lbs. of coralline in 2 gallons of caustic soda of $10°$ Bé., and at the tem-perature of $140°$ F. Dilute with 20 gallons of water, heat again, and add about 1 quart of bichloride of tin of $55°$, diluted with 1 gallon of water. After filtration, there remain 4 gallons of lake. Then, taking 2 gallons of this unwashed and semifluid lake, mix it with 4 lbs. of pulverized gum, and about 12 ozs. of oxalic acid; heat until all is dissolved, pass through the sieve, and print. After 12 hours' standing, the print is steamed, and is of a bright orange color.

Aniline orange and *yellow*. *Azotileine; Zinaline.*—By passing more or less of nitrous oxide gas through aniline kept cold, these colors are produced. Hydrochloric, or acetic acid is added, and the solution is ready for the dye-bath.

Aniline yellow.—This appears to be a salt of leucaniline, and has been obtained by Mr. Durand by the slow action of nascent hydrogen, during 12 or 24 hours, upon the residues of the manufacture of aniline red. After purification, the color is sufficiently soluble in boiling water to dye wool, silk, and leather without mordant. The color is a nankeen yellow, which will turn ponceau if the fabric is drawn through another bath containing a solution of bichromate of potassa.

Aniline yellow, obtained by Messrs. Simpson, Maule & Nicholson, by the action of nitric acid on aniline. It dyes wool and silk a bright lemon yellow. If picric acid is added, the shade on wool will approach a cochineal color. These colors being volatile, do not bear steaming, and are not fast.

Aniline orange.—Obtained by Mr. E. Jacobsen from the residues of the preparation of azaleine (nitrate of rosaniline). It is soluble in alcohol, sufficiently so in boiling water, and dyes wool and silk a fine gold yellow. Alkalies, ammonia, for instance, change the shade to a brimstone yellow, but acids restore the primitive color.

Chloroxynaphthalate of ammonia.—Very soluble in water, according to Mr. Perkin, and dyes silk a gold color, which light does not affect.

Safranine is a new coal-tar dye, obtained by Mr. Carvès, of St. Étienne, which is said to have twice the coloring power of picric acid, and to afford yellow or red shades, according to treatment.

GREENS.

Mixed green.—Obtained by a mixture of picric acid and aniline blue. It appears grayish under artificial light.

Vert printemps; Spring green.—Picric acid and carmine of indigo produce a beautiful green, which appears violet under artificial light.

Night green; Vert lumière.—Picric acid and Prussian blue. Keeps its true green color in artificial light.

Aniline green; aldehyd green; night green; vert lumière; Usèbe green; viridine.—To 4 parts of aniline-red in crystals, add 6 parts of oil of vitriol and 2 parts of water. When this solution is cold, gradually add to it pure aldehyd (about 6 parts), and heat the whole in a water bath to the boiling point, until a few drops of the solution, put into weak sulphuric acid, or acetic acid, produce a blue coloration. When this point is reached, the liquid is poured into a bath of water (200 times the weight of aniline red employed), containing some hyposulphite of soda, a weight about equal to that of the sulphuric acid entering into the composition of the liquor. Two different colors are produced : a green in solution, and a grayish substance called *argentine*, which remains in suspension in the liquid, and may be separated by filtration.

Mr. Lauth obtains better results by using alkaline polysulphides (liver of sulphur), instead of hyposulphite of soda.

This green bath should be used at once, as it does not keep well.

(Silk dyeing.)—Put the silk into the lukewarm bath, which is gradually brought up to the boil, and let it cool off with the silk in it.

(Wool dyeing.)—Less acidity, and less heat than for silk. It is well to mordant the wool with alum, but this green succeeds better on silk than on other fibres, and requires some practice in its use.

Aniline green in paste.—The above bath is employed for precipitating the color by acetate of soda, or more generally by tannin. The precipitate is collected, washed, and the resulting paste is soluble in water acidulated with sulphuric acid. An addition of sal ammoniac, about $\frac{1}{10}$ of the weight of the paste is said to greatly facilitate the solution.

For dyeing, follow the preceding rules for heat and acidity of the bath.

(Calico printing.)—Print on cloth mordanted with tannin. The color does not stand steaming well.

(Wool and silk printing.)—Mr. Sevez says that the green will bear steaming, if bisulphite of soda is added to the paste, which is made as follows :—

Gum water	1 quart.
Green in paste . . .	10 ozs.
Crystallized bisulphite of soda .	5½ "

Heat in a water bath until the salt is dissolved, let stand for 3 or 4 days, print and steam. If the mixture is not allowed to stand, the color is too light.

This process does not succeed well on cotton.

Iodide of ethyl green.—By boiling Hofmann's violets with water and carbonate of soda, a precipitate is produced, which is re-moved by filtration. The filtered liquor is then treated by picric acid, and another precipitate is obtained of a green color, which is washed and sold in powder. Soluble in alcohol, and possibly in sal-ammoniac.

Emeraldine.—This is one of the few aniline colors which are directly produced on a cotton fabric, and has been discovered by Messrs. F. C. Calvert, Lowe & Cloft.

(Cotton dyeing.)—Draw the fabric through a bath containing 4 oz. of chlorate of potassa per gallon of water, and dry. Then draw it through another solution containing 1 per cent. of hy-drochlorate or tartrate of aniline, and acidulated with hydro-chloric or tartaric acid.

(Cotton printing.)—Print with a mixture of tartrate or hydro-chlorate of aniline   .   .   .   .   .   3 lbs.
    Starch paste   .   .   .   .   .   6 "
    Chlorate of potassa   .   .   .   .   1 lb.

The salt of aniline is added only when the mixture is cold. After printing, and a little steaming, the color becomes developed in a few hours. Then wash the fabric.

If these green prints were passed through a solution of bi-chromate of potassa, the color would be transformed into a dark indigo blue. (See Blues—azurine.)

This green color is fast under the action of light. Alkalies and soap turn it blue; but acids restore the green color.

Rosaniline green.—Obtained by Messrs. J. A. Wanklin and Paraf, by repeatedly treating Hofmann's violets with equal parts of wood spirit and iodides of ethyl or methyl, under pressure, for 3 to 4 hours, and at a temperature of from 230° to 240° F. The coloring matter of the product is dissolved by weak solutions of carbonate of soda.

The shade of green is very fine, is not changed by artificial light, but is not very fast.

Toluidine green.—Produced by a process similar to that of aldehyd green.

Olive green or *aniline olive.*—Mr. Sacc gives the following mixture for printing cotton or silk (?):—

Water . . .	300 parts.	
Farina . . .	36 "	
Chlorate of potassa .	15 "	
Acetate of copper .	15 "	
Nitric acid . .	10 "	} Previously mixed
Aniline . . .	20 "	} together.

This mixture seems very thin.

The olive shades are somewhat brown, and this formula is also used for the latter color.

BROWNS, MAROONS.

Havana brown.—The impure arsenite of rosaniline (crude fuchsine), in a solution heated to the boiling point, dyes silk and wool a reddish-brown. The higher the temperature, the browner is the shade.

Brown maroon.—Obtained by Messrs. Girard and de Laire by melting 4 parts of anhydrous hydrochlorate of aniline with 1 part of arsenite of aniline, or 1 part of aniline blue or violet. The temperature is gradually increased to 465° F., and maintained for about 2 hours, until the mixture evolves yellow fumes, and turns brown or maroon. The color is soluble in water, alcohol, acetic acid, &c., and produces beautiful shades on silk and leather.

Leucaniline brown; puce fuchsine.—Mr. H. Koechlin recommends the following mixture for wool printing:—

Dissolve ¼ oz. of magenta crystals (hydrochlorate of rosaniline) in 2 gills of alcohol; thicken with 1½ pint of gum water, and add 2 ozs. of oxalic acid, and 5 grains of chlorate of potassa.

By decreasing the quantity of chlorate of potassa, a red brown is obtained.

For a yellow shade, add some yellow lake free from protoxide of tin.

All these shades are fast, and resist acids, alkalies, and soap.

Dark brown.—A paste containing tartrate of leucaniline with sulphide of copper, will print shades of a pure dark brown by a process similar to that of aniline black.

Brown.—A brown precipitate obtained by treating a solution of magenta by hydrochloric acid and chlorate of potassa, is soluble in alcohol and sulphuric acid, and may be fixed upon cotton mordanted with albumen.

Olive brown.—See olive green.

Rothine or *Phenicienne.*—Obtained by Mr. J. Roth by the action of a mixture of nitric and sulphuric acid upon carbolic acid. But slightly soluble in water, soluble in alcohol, acetic acid, tartaric acid, and alkalies.

The alkaline solutions are violet blue, and become brown by a slight excess of acid.

The shades are fast on silk and wool and even resist bleaching powder.

(Wool and silk dyeing.)—No mordants are required. The shade will turn from a brown yellow to a garnet color by the addition of bichromate of potassa and some sulphuric acid.

(Cotton dyeing.)—Cotton mordanted with stannate of soda or tannin, will be dyed a dark wood color by the addition of bichromate of potassa to the hot bath. This color on cotton is turned blue by the alkalies, and is dissolved by soap.

(Printing.)—Phénicienne prints do not succeed on either silk, wool, or cotton, the color being changed and altered by steaming.

The results are more satisfactory on mixed fabrics, especially when wood shades are desired. Phénicienne is then dissolved in acetic acid, thickened, and some chlorate of potassa and tartaric acid are added to the mixture.

Mixed brown; cerise brown.—Dissolve 1 part of Geigy's cerise (see reds) in 6 parts of acetic acid, and add 2½ parts of sulphuric solution of indigo. Add also some tartaric acid or cream tartar at the end of the operation, that is to say, at the last boil of the dye bath.

Aniline brown.—Obtained by Mr. E. Jacobsen by gradually heating up to 300° F., and as long as ammonia is evolved, 1 part of picric acid and 2 parts of aniline; or by boiling a concentrated solution of chromate of ammonia with aniline, and adding formic acid.

This color is soluble in alcohol with sulphuric acid or glycerine, dyes wool a brown color, and silk a peculiar shade of brown, called corinth.

BLACKS, GRAYS.

Chloroxynaphthalic black.—Wool is dyed a fine black, by mixing chloroxynaphthalic acid (see reds) with sulphate of indigo.

Aniline black.—(Cotton and silk dyeing.)—According to Mr. Cam. Kœchlin, these fibres may be dyed in a solution made of:—

Water	20 to 30 parts.	
Chlorate of potassa . .	1 part.	
Sal-ammoniac . . .	1 "	
Chloride of copper . .	1 "	
Aniline	1 }	previously
Hydrochloric acid . .	1 }	mixed together.

The fabric or yarn is dried in ageing rooms at a low temperature for 24 hours, and washed afterwards.

(Wool dyeing or printing.)—Mr. J. Lightfoot prepares the wool by a kind of oxidation made as follows: 1 part of bleaching powder is dissolved in 10 parts of water. Then for 1 pound of wool, take about a pint of the above solution, dilute it with

6 gallons of water, and add 3 ozs. of muriatic acid. In this bath, which is at the temperature of 100° F., work the wool during 20 or 30 minutes, and until it has acquired a yellowish tint. Then wash it thoroughly, and let it dry.

Wool and mixed fabrics thus prepared may be dyed and printed in the usual way.

(Silk printing.)—In this case, silk is to be vegetablized (we have already the word animalized) by an immersion in a bath of cellulose dissolved in ammoniacal copper oxide. We think this process quite delicate, on account of the action of ammonia on the silk.

(Calico printing.)—The first application of aniline black to calico printing was made by Mr. John Lightfoot. One of the early printing mixtures was made of:—

Water	5½ quarts.
White starch . . .	1 lb. 14 ozs.
Chlorate of potassa . .	6 ozs.
Hydrochlorate of aniline . .	1 lb.
Sulphate or chloride of copper .	5 ozs.

The aniline black obtained was very fine and fast; but the great quantity of copper salt employed was found to be injurious, both to the fabric and to the metallic printing rollers.

Subsequent experiments made by Messrs. C. Kœchlin, Cordillot, and Lauth, have led to the substitution of sulphide of copper for the sulphate and chloride of this metal, whose presence seems indispensable to the production of aniline black. A good printing paste, which does not weaken the fabrics, and does not corrode the scrapers and the rollers of the printing apparatus is made as follows:—

Heat and digest—

Water	1 lb.
Starch	2 lbs.
Sulphide of copper . . .	8 ozs.

On the other hand, mix and heat—

Torrefied starch . . .	2 lbs. 6 ozs.
Water	4 "
Gum tragacanth water . .	1 quart.
Hydrochlorate of aniline .	1 lb. 9½ ozs.
Sal ammoniac . . .	3½ ozs.
Chlorate of potassa . .	9½ ozs.

Then mix the two compositions, print, and expose the fabric in the ageing-room for 24 hours, and at a temperature from 77° to 104° F.

Here is another paste by Mr Käppelin :—

Starch paste	2½ gallons.
Chlorate of potassa	7 oz.
Gum tragacanth water	5½ lbs.
Sulphide of copper	14 ozs.
Sal ammoniac	9 ozs.
A salt of aniline (tartrate)	2½ lbs.

which is added last.*

After 24 hours' standing in the ageing room, the prints are drawn through a bath containing 2 per ct. of carbonate of soda, steamed and washed.

Acids will turn the color to green, but alkalies will restore the black. A solution of bichromate of potassa intensifies the color; but an excess of this salt is apt to impart a reddish hue.

The best aniline for black is the one which contains a mixture of aniline and toluidine, and which is sought for in the manufacture of reds.

The sulphide of copper is made by dissolving at the ordinary temperature 2 parts of sublimed sulphur in 2 parts of caustic soda, at 38° Baumé. After 24 hours' standing and frequent stirrings, the solution is complete, and is thrown into a warm solution of 10 parts of sulphate of copper in 250 parts of water. The precipitate is washed and drained until about 10 pints are obtained, each pint therefore corresponds to 1 pound of sulphate of copper.

Lucas paste.—It contains acetate of copper and hydrochlorate of aniline, without sal ammoniac, and has been submitted to a peculiar process. When used, this paste is mixed with 6 to 8 times its volume of starch paste. The temperature of the ageing room is about 104° F.

Paraf's paste.—It is a mixture of hydrochlorate of aniline, chlorate of potassa, hydrofluosilicic acid, and a thickening. It produces a very fine black when applied with copper or brass rollers, which furnish the copper necessary to the development of the color. If no copper is present, the shade is only a dirty blue.

All these aniline blacks are remarkable as being very fast, unalterable by acids and alkalies, and even by chlorine to a certain point. If chlorine is not used in great excess, the black color will reappear; if in excess, the color remains fallow.

---

* Tartrate of aniline does not corrode the steel scrapers, and is gradually transformed into hydrochlorate of aniline by the sal ammoniac of the mixture.

Nitrate and hydrochlorate of aniline are the only salts of aniline which can produce the black.

Aniline black may also be printed simultaneously with madder and most steam colors.

All the compositions for producing aniline black must be acid, and the more acid there is, the more rapid is the production of the black. We ought, however, to remain within proper limits, otherwise the fibre may be weakened.

The degree of acidity of the paste will also vary with the thickenings employed. Gum senegal requires more acidity than torrefied starch, and the latter more so than white starch or gum tragacanth.

In printing aniline black care should be taken not to print upon, or too near other places previously mordanted; the mordant would be acted upon, and if it contains acetic acid, this acid once liberated would prevent the formation of the black, which will be only gray.

There is also danger of spontaneous combustion, so rapid is the oxidation going on, when the printed piece is allowed to remain folded and wet. It should be immediately spread out in the ageing room.

Aniline grays.—(Calico printing.)—By diluting the above blacks with an increased proportion of thickening, nothing will be produced, unless by the addition of a mineral acid. The whole paste, in proportion to its volume, should contain as much acidity as the former black paste had.

The true colors of aniline blacks and grays appear only after washing.

Mauveine gray.—Obtained by Mr. J. Castelhaz by dissolving 10 parts of mauveine, in paste, in 11 parts of oil of vitriol (66° B*); 6 parts of aldehyd are then added, and the whole mass is allowed to stand 4 or 5 hours. By washing the product, the gray passes through the filter, and is precipitated by chloride of sodium. Soluble in water, alcohol, &c.

Muréine grays.—Obtained by MM. F. Carvès and Thirault by treating hydrochlorate of aniline by a mixture of bichromate of potassa, an iron salt, water, and sulphuric acid.

By varying the proportions of the reagents, different shades are produced, which are soluble in boiling water, and stand acids and soap well.

DICTIONARY

CALICO DYEING AND PRINTING.

Absorbent.—A term borrowed from the French. It signifies a composition for discharging mordants after padding or printing and before dyeing; but like the English word *discharge*, to which it is nearly equivalent, it often signifies a discharge in the widest meaning of the word. It is also used, but less frequently, to indicate a simple resist. (See DISCHARGE, RESIST.)

Acetate.—All compounds of acetic acid with metals or oxides are called Acetates. They form a very important class of salts, and are extensively used in dyeing and calico printing. The more important acetates are those of Alumina, Copper, Iron, Lead, and Soda, which are treated upon under the head of Acetate.

General Properties of Acetates.—All acetates are soluble in water; the acetates of soda, potash, magnesia, zinc, and lead, though reckoned neutral salts in chemistry, have an alkaline reaction, that is, turn red litmus paper blue, and in other ways act as alkalies; for example, if acetate of potash be mixed with a solution of a per-salt of iron, a salt of the green oxide of chromium, or with bi-chloride of tin, it causes, upon boiling, a precipitate of the oxide of the metal, or a basic salt, much the same as would be produced by weak caustic potash or crystals of soda. When sulphuric acid, hydrochloric acid, and other strong acids, are mixed with an acetate, the acetic acid is set free, and may be expelled by heat, because it is a weaker acid and volatilizes or passes off in vapor. The affinity of acetic acid for metals is weak, and consequently many of the acetates are decomposed spontaneously, the metallic oxide separating from the acetic acid; thus, acetate of peroxide of iron, of chromium, of tin, and of alumina, are decomposable by simply drying in the air, the acid passing away and the oxide of the metal remaining behind; this is what takes place when red liquor or iron liquor, which are respectively acetates of alumina and iron, are printed on cloth, and the cloth dried, the acid flies away and the alumina and iron are left adhering to the cloth. Sometimes a small portion of the acetic acid remains combined with the metallic oxide, forming what is called a sub-salt or a basic acetate, and this form of acetate is generally insoluble in pure water.

Preparation of Acetates.—Acetates are made in two general ways; firstly, the direct method, by taking acetic acid and mixing it with the substance to be acted upon; thus, to make acetate of soda, take any quantity of commercial acetic acid and dissolve in it soda crystals or soda ash until the sourness of the acid is neutralized; a solution of acetate of soda is produced, which, by boiling down or mixing with water, can be brought to any required strength; ground chalk or slaked lime thus mixed and dissolved in acetic acid would give acetate of lime; litharge or oxide of lead thus dissolved would give acetate or sugar of lead. The second, or indirect method, consists in taking an acetate ready formed, so as to produce another by the process of "double decomposition," so called because two salts are decomposed or destroyed at the same time, giving rise to two new ones. To illustrate this, by a practical example, suppose it is required to make some acetate of alumina or red liquor; now the first method is not applicable, simply because the alumina which would be required is not a commercial article; but sulphate of alumina and acetate of lead are readily procurable, and by dissolving them in water and mixing them, the double decomposition spoken of takes place, and there is formation of acetate of alumina and sulphate of lead; this last not being soluble in water settles down as a white sediment. Sulphate of iron and acetate of lead mixed together give rise to acetate of iron and sulphate of lead; sulphate of manganese and acetate of lead give rise to acetate of manganese and sulphate of lead, and so on. When an acetate is mixed with any soluble sulphate, nitrate, or chloride, there will be production of acetate of that sulphate, nitrate, or chloride, although no visible decomposition may take place; thus, if acetate of soda be mixed with bi-chloride of mercury or corrosive sublimate, no visible change takes place, but there is no doubt that acetate of mercury is produced; so also a red liquor may be made by mixing acetate of soda and sulphate of alumina, but it would be less regular in its results, because containing other salts, as sulphate of soda.

Analysis of Acetates.—The value of commercial acetates always depends upon the acetic acid present, and the quantity of this acid can only be ascertained by laboratory methods. I use two processes: first I liberate the acetic acid from a weighed quantity of the acetate to be examined, by acting upon it with sulphuric acid in a deep-bellied retort, heat to drive over all the acetic acid which is condensed in a well cooled receiver; near the end of the process I drive a current of steam through the retort to remove the last traces of acetic acid. The distilled acid is then tested with a standard solution of caustic soda as explained in ACIDIMETRY, and the quantity of acetic acid calculated.

The second method consists in turning the acetate under examination into acetate of soda, and then changing this by a red heat into carbonate of soda, the quantity of which is accurately ascertained by the test acid and process described in ALKALIMETRY, and from that the quantity of acetic acid calculated. Thus, in testing acetate of lime, take 100 grains, dissolve in water, add solution of sulphate of soda until no more precipitate of sulphate of lime takes place, then add one-fourth of the bulk of liquor of methylated spirit, leave for a time, filter and wash the precipitate, evaporate the clear in a platinum vessel to dryness, raise the heat to a good red, stirring the mass until no more vapors are evolved, cool, warm with water, and test with the standard acid. The first process is very good in its results, but on account of a little sulphuric acid finding its way into the receiver the acetate appears better than it is; the second plan I have also found good, and its results are usually on the other side, or against the acetate.

Acetate of Alumina, commonly called *Red Liquor* or *Red Mordant,* sometimes also *Acetite of Alumina* and *Pyrolignite of Alumina.*

Acetate of alumina first began to be used as a mordant towards the end of the last century; long previously, perhaps even by the Hindoos, at the time of Alexander the Great, a more or less impure acetate, mixed with other ingredients, was employed in calico printing, without any suspicion that the acetate was the really useful part of the mixture called red mordant.

Preparation.—The direct production of acetate of alumina from acetic acid and alumina yields the pure salt; the commercial acetate of alumina, which will hereafter be called *Red Liquor,* is always made by the process of double decomposition. I give the proportions for producing red liquor of various qualities:—

4

Red Liquor from Alum and Acetate of Lead.

		1.	2.	3.	4.	5.	6.
Water	gals.	45	45	45	45	45	45
Alum	lbs.	100	100	200	190	190	129
Acetate of Lead .	lbs.	100	129	200	190	129	100
Crystals of Soda .	lbs.	10	10	10	19	—	—

The above red liquors were a short time ago used in Mulhouse, in France, by one of the most successful houses there; they are upon the plan laid down by M. Daniel Koechlin, in his celebrated paper on the Red Mordant, in the *Bulletin of the Industrial Society of Mulhouse*, 1827. The alum, in a crushed state, is dissolved in the water, heated to 140°; the crystals of soda are next dissolved with stirring, and then the acetate of lead, in coarse powder, added, and the whole well stirred for a considerable time, and afterwards at intervals during two or three days. Mr. Koechlin states that if the crystals of soda are added after the sugar of lead, the liquors are neither so strong nor so good. Nos. 1 and 2 are common reds for calico. No. 2 being better adapted for a gum color and for blocking than No. 1; Nos. 3 and 4, strong mordants, suitable for muslin or light goods; Nos. 5 and 6 will do for garancine, and are suitable for mixing with crystals or muriate of tin for forming a resist red.

Red Liquors from Acetate of Lime.

		No. 7.	No. 8.
Acetate of Lime Liquor at 24° . . .	galls.	50	90
Alum	lbs.	200	—
Sulphate of Alumina	lbs.	—	272
Ground Chalk	lbs.	12	34

On account of the cheapness of acetate of lime, it is more used than acetate of lead. The above two red liquors are put together by first heating the acetate of lime liquor in a copper boiler, to a temperature of 140°, then adding the alum of sulphate of alumina, and stirring until all the lumps have disappeared; and, lastly, the chalk is added by small portions at a time, to avoid loss by the effervescence which would be caused if all were put in at once. The whole is then well stirred up until nearly cold, allowed to settle, the clear drawn off, and the bottoms drained upon a woollen filter and washed with water until the washings fall as low as 2° Tw., when they are not worth washing any more. The red liquor, No. 7, used at 20°

Tw., gives the darkest red obtainable on calico for madder and garancine; the No. 8 liquor, at 16° Tw., is used for resist red and for mixing with iron liquor to produce chocolates for garancine work, and also reduced for light reds. These two liquors are quite sufficient for the ordinary run of madder and garancine work. I give receipts for several other red liquors, all of which have been in use to my own knowledge either in England or France.

Red Liquors for Madder Pink.

		No. 9.	No. 10.	No. 11.	No. 12.	No. 13.
Water . . . . . galls.		4½	1	20	3	60
Alum . . . . lbs.		16	3½	75	13	125
Acetate of lead . . . lbs.		12	2½	—	10½	100
Acetate of Lime (dry) . . lbs.		—	—	30	—	—
Ground Chalk . . . lbs.		—	—	5	—	—
Common Salt . . . lbs.		—	—	10	—	—
Nitrate of Zinc 15° Tw. . . galls.		—	—	—	1	—

No. 9 is for dark and No. 10 for light pink, being reduced, when used, with three parts water to one of liquor; No. 11 has done good work, the common salt having an attraction for moisture, has a tendency to make the color age well; the nitrate of zinc, in No. 12, has a still stronger attraction for water, it is usually only added to the color after thickening, as it has a tendency to make starch and flour run thin if boiled with them. No. 13 is only for reducing for light pinks; it will not give a good dark pink.

Miscellaneous Red Liquors.

		No. 14.	No. 15.	No. 16.	No. 17.
Water . . . . . . galls.		3	3½	2½	1
Alum . . . . . . lbs.		30	5	10	5
Acetate of Lead . . . . lbs.		29	5	7½	2½
Vinegar . . . . . galls.		—	1½	2½	—
Crystals of Soda . . . . lbs.		—	—	1	¼
Acetate of Copper . . . . oz.		—	—	12½	—

These are all of French origin. No. 14 is an excessively concentrated liquor, used for printing very small objects required to be well defined; in making it the drugs are finely ground and mixed up without heat. No. 15 is intended for pinks and light reds; the water is partly replaced by vinegar, with doubtful advantage. No. 16 is intended for light reds; neither theory nor practice, so far as experiments in England go, indicate any use for the acetate of copper. No. 17 is a red liquor for garancines; it will stand from 8 to 12 ounces of crystals of tin per gallon, and works well for a resist red.

Remarks on making Red Liquors.—Red liquor is never a pure acetate of alumina; it is found by experience that if the quantity of acetate of lead or lime required by theory to form a pure acetate of alumina be employed, the resulting liquor is no better for giving colors than if about two-thirds of that quantity were used, while it is worse for keeping and more irregular in its results. Theory indicates that for every 10 lbs. of potash alum, 12 lbs. of white acetate of lead are required to change all the alumina into acetate; for ammonia alum 12½ lbs. are required; and for sulphate of alumina, or patent alum, about 17 lbs. would be required. (See EQUIVALENTS.) Practice has shown that if three-fourths only of these amounts be taken, the best results are produced. The use of ground chalk or crystals of soda consists of neutralizing a portion of free acid and strengthening the mordant. I prefer chalk to the crystals, using about an ounce of it for every pound of alum; it has the effect of withdrawing a portion of sulphuric acid from the liquor as insoluble sulphate of lime, and apparently, but not actually, decreasing the strength of the mordant; crystals of soda neutralize the acid but leave the sulphate of soda in the liquor, making it seemingly stronger. I do not think that there is any difference between using potash or ammoniacal alum, if both are of equal purity; sulphate of alumina requires special care, because it is usually more acid than alum, and cannot be so easily made into a first-class red liquor. Brown sugar of lead gives as good results as the white, and acetate of lime, if the right proportions are known, will yield excellent red liquors.

Properties of Red Liquor.—The commercial acetate of alumina or red liquor is of a tawny or brown color, smells of wood tar or pyroligneous acid, has a taste of alum; when heated to about 160° it coagulates, becoming nearly solid; it liquefies again upon cooling; thus reds often go thick in boiling and thin upon cooling. It parts with its alumina readily to cloth, and more especially when the acetic acid is at liberty to escape as it is upon printed goods. The affinity of the acid and alumina is not strong, so that if a quantity of red liquor be boiled to dryness most of the acid would leave the alumina, which would not dissolve again in fresh water. Red liquor sometimes loses alumina by standing. The quality of red liquor for a mordant can only be satisfactorily tested in the practical way, by making colors from it. By chemical analysis, the amounts of acetic acid, alumina, sulphuric acid, and other bodies can be accurately ascertained when necessary; but this information would give no indication of the value of the liquor as a mordant, without actual trial.

Applications of Red Liquor.—Red liquor is used in madder and garancine dyeing as a mordant for red and pink; mixed with iron liquor it is a mordant for shades of chocolate; in logwood dyeing, combined with iron liquor, it gives blacks. It is used in a few steam colors, and also in silk and cotton piece dyeing.

Acetate of Copper, known also as *Verdigris.*—The common verdigris of commerce is a basic acetate of copper which requires an additional quantity of acetic acid to make it soluble in water. Crystallized acetate of copper is all soluble in water. For calico printing purposes acetate of copper is always used in the liquid state, and is prepared by the method of double decomposition from a mixture of sulphate of copper and acetate of lead. The following receipt gives a liquid acetate of copper suitable for catechu browns:—

> 1 gallon of water at 160° F.,
> 4 lbs. white acetate of lead,
> 4 lbs. sulphate of copper.

The whole is stirred until all lumps have dissolved; the clear liquor only is used. Another practical receipt gives only 2 lbs. acetate of lead to 4 lbs. sulphate of copper. Theoretically, 4 lbs. common sulphate of copper or blue vitriol require about 6 lbs. sugar of lead; so it is evident that what is called acetate of copper in a print-works is really a mixture of sulphate and acetate of copper.

Applications.—This salt is chiefly used in catechu colors, in some indigo blue resists, and in a few steam colors where it appears to exercise an oxidizing action. It was used in the black dye for silk, it forms the basis for the Schweinfurt or Scheeles' green, and is sometimes prescribed in iron and red liquors, where its utility seems doubtful.

Acetate of Indigo.—Under this name a purified extract of indigo is spoken of in some dyeing treatises. It is prepared by taking common sulphate of indigo, dissolving in water, filtering, and adding acetate of potash to the filtered liquid; a precipitate takes place, which is called the acetate of indigo; it is in reality a combination of potash with an indigo acid, called sulphindylic acid by Dumas. This precipitate is collected on a filter; and, if required very pure, may be a second time precipitated from a watery solution. Other salts less expensive than acetate of potash give similar results, such as common salt and sulphate of soda. The name of carmine of indigo is also given to these purified extracts. They may be replaced in foreign receipts by English refined neutral extract. (See INDIGO.)

Acetate of Iron, commonly called *Iron Liquor,* also *Black Liquor, Pyrolignite of Iron, Tar Iron Liquor;* French, *Bain noir* and *Bain de fer.*—This liquid, so extensively used in dyeing and printing, is of very ancient origin; but under its present form it has only been in use about eighty years. It is made by steeping old iron of all sorts, such as hoops, worn-out tin plate, etc., in warm wood acid or pyroligneous acid, which is an impure kind of acetic acid. By continually moving the acid, and keeping up a moderate heat, the acid saturates itself with iron in a few days; if not strong enough it is concentrated. Formerly the process was worked cold in very large vats, and lasted forty days or more. Some color mixers consider that cold made iron liquor is best. I have found no reason to think so. The iron liquor is sent out at various strengths, from 18° to 28° Tw.; it is a black fluid by reflected light, but in narrow bottles it has a greenish-olive color; a peculiar smell, chiefly due to tarry matters in it, and an inky taste. Iron liquor can also be made, by the process of double decomposition, from sulphate of iron, or green copperas, and the crude acetate of lime. As the acetate of lime is a drug of uncertain strength, it will require some trials to find out the best proportions to use. Here are two receipts for preparing iron liquor in this manner. The first taken from Muspratt's Chemistry (i. 42) is as follows:—

400 lbs. copperas dissolved in
100 gallons hot water, then add
75 gallons acetate lime liquor at 16° Tw.

The second, used by myself with good results is as follows:—

20 gallons acetate lime liquor at 24° Tw.
65 lbs. green copperas,
2¼ gallons wood acid, at 7° Tw.

The acetate of lime liquor is heated to 140° in a copper, the green copperas in coarse powder added and stirred till dissolved, and then the wood acid; these quantities yield 16 gallons iron liquor at 24°. There is no economy in making iron liquor in this way, but it is frequently advantageous to be able to make various qualities.

The acetate of iron made from copperas and white acetate of lead is not so well adapted as a mordant for dyeing as that made from impure acetates containing tarry matters. The tarry matters appear useful by impeding the action of the air upon the iron, and so enabling it to form a close combination with the cloth before the oxygen of the air changes its nature. (See AGEING and BUFF LIQUOR.)

There is another acetate of iron called the per acetate of iron, but it has not yet received any applications. The essential salt in iron liquor is the proto-acetate of iron.

Applications.—Iron liquor serves as a mordant for madder, garancine, logwood, and other coloring matters; its chief consumption is in madder, and garancine dyeing. Iron liquor at about 6° Tw. gives a black with madder; from 4° to a very diluted state it gives various shades of purple, lilac, or violet; mixed with red liquor it gives chocolates. In piece dyeing iron liquor is not much used; it serves, however, for all purposes in which green copperas is used, and will generally give better and quicker results.

Iron liquor is best tested in the practical way. Chemical analysis can determine exactly the quantity of acetic acid, iron, water, and other matters in it, but cannot tell whether it will give good shades or not.

Iron Liquor Improvers.—There are generally some substances in the market purporting to enable iron liquor to give better results when mixed with them. I have made a very great number of experiments upon this point, and have tried all the substances recommended, but I never found that really good iron liquor was improved by any additions. Arsenic, under various forms, and copper salts are strongly recommended by French authorities; but upon good Lancashire iron liquor they do more harm than good. M. Henri Schlumberger, in an elaborate paper in the *Bulletin of Mulhouse* (xiii. p. 399), gives the results of his experiments upon the addition of various chemical substances to iron liquor. The only decisively advantageous results were with the addition of copper salts, and these only when gum senegal was used as the thickening; their effect being apparently to prevent a coagulation to which gum senegal of certain quality is liable.

Acetate of Lead, commonly called *Sugar of Lead,* also *Salt of Saturn.*—There are two kinds of sugar of lead in trade, called white sugar of lead and brown sugar of lead. The white is in soft crystalline lumps, easily crushed and very soluble in water. The brown is usually in fused lumps, much more compact than the white, of a deep mahogany color, does not dissolve so readily in water, and generally leaves a residue not dissolved. The difference between the two acetates consists in a portion of tarry matter from the wood acid being left in the brown; beyond this there is no essential difference. The brown is, however, poorer in acetic acid than the white, and somewhat richer in lead; so that as a matter of choice the white should be used for making acetates, and the brown for a lead mordant. An analysis of two average samples gave me the following results :—

	White Acetate.	Brown Acetate.
Acetic acid	27.6	21.8
Oxide of lead	58.4	59.9
Water	14.0	15.5
Insoluble matter . . .	0.0	2 8
	100.0	100.0

For the general use of acetates their value is in ratio of the quantity of acetic acid present; the white acetate analyzed would consequently be worth much more than the brown. Sugar of lead may be contaminated with copper and iron, the latter metal being a dangerous impurity when the acetate is used for red liquor mordants. Impure white sugar of lead will have a reddish hue if it contains iron, and a bluish if it contains copper; the best method of testing consists in adding sulphuric acid to throw down all the lead, and then applying the characteristic tests for the other metals, which are given in their proper places. Sugar of lead in the English market is generally free from these metals.

Applications.—The chief use of acetate of lead is in making acetates of alumina, iron, copper, and manganese, by the way of double decomposition. On account of the cheapness of acetate of lime, which acts quite as well, it is not so much used as formerly. It is used as a mordant for chrome orange and in several indigo resists.

Basic Acetate of Lead.—When acetate of lead is shaken up with powdered litharge (oxide of lead) it dissolves a portion of it, forming what is called basic or subacetate of lead; if boiled together a still greater portion of lead is dissolved. This compound has been employed both as a mordant and as a resist; but it has so powerful an action in coagulating all kinds of thickenings that it cannot be generally used. Only the darkest kind of calcined farina or sugar can thicken it without curdling.

Acetate of Lime, known also as the *Pyrolignite of Lime.*— This compound is sold either as a solid or liquid. In the solid state there are three varieties, called respectively white, gray, and black acetate. I analyzed samples of each, and found in one hundred parts of the solid 82, 71, and 69 parts of pure acetate of lime respectively, which was about the ratio of the prices. The liquid acetate of lime seems to be of no particular quality, I have found it of all degrees of purity, and containing muriate of lime or common salt, evidently to make it stand higher on Twaddle. Both in the solid and liquid state acetate of lime can only be accurately valued by chemical analysis. The only use of acetate of lime is in making mordants, for which purpose it answers quite as well as the more expensive acetate of lead. If any sulphate, such as sulphate of iron, be mixed with acetate of lime, the lime takes the sulphuric acid, while the acetic acid goes to the iron or other metal previously combined with the sulphuric acid. The sulphate of lime being insoluble in water settles down as a pasty mass, while the acetate remains clear above.

Acetate of Manganese.—This compound has been used to a small extent for bronze colors and for producing some shades in combination with catechu. It can be made from sulphate of manganese or from bronze liquor (the chloride of manganese) by mixing with acetate of lead.

Acetate of Soda.—This compound can be prepared by neutralizing acetic acid with crystals of soda or caustic soda, it is a commercial article, being sold in small crystals. It is very little used; for some applications of it, see MUREXIDE and SHADED STYLES.

Acetate of Tin.—This compound has been slightly used in calico printing. It may be formed by first making a pulp of tin with a mixture of muriate of tin and carbonate of soda, draining the pulp and leaving acetic acid upon it for twenty-four hours. Or it may be made by dissolving 2 lbs. crystals of tin in a gallon of cold water, and mixing 2 lbs. acetate of soda and stirring, using all the mixture; another way is to use acetate of lead instead of acetate of soda and strain off from the bottoms.

Application.—Only used in producing an orange color in garancine work. (See ORANGE.)

Acetic Acid, known also as *Vinegar, Wood acid, Pyroligneous acid, Tar acid, Acetous acid,* etc.—Commercial acetic acid is made by distilling wood in close retorts; in its first stage it is a crude black tarry looking liquid, which is called wood acid or pyroligneous acid; by several complicated processes the tarry matters or other impurities are removed and the acid left tolerably pure. Acetic acid should be as clear and colorless as water, should leave no residue when a portion is boiled away, should not blacken a piece of calico dipped in it when the calico is dried and made pretty warm by holding before a fire, neither should the calico be tendered; if either blackened or tendered, mineral acids, as vitriol or spirits of salts, are present. It should have an agreeable smell; a particular mawkish odor shows some fault in rectification; but notwithstanding this, the acid may be still good for manufacturing purposes. It stands at from 6° to 9° Twaddle; but owing to a strange peculiarity about acetic acid, its value cannot be ascertained by the hydrometer, even if no adulteration has been practised. The only reliable method of valuing acetic acid consists in ascertaining how much caustic soda a given weight will neutralize (see ACIDIMETRY), and testing for mineral acids. Nitrate of silver gives a precipitate, not dissolved by nitric acid, if any muriate acid is present; and chloride of barium gives a similar precipitate if sulphuric acid is present. Average qualities of acetic acid contain from 18 to 22 per cent. of dry acetic acid, sometimes going as high as 24. Crude pyroligneous acid or wood acid, sometimes used in printing, contains only a small percentage of acid and a large quantity of organic matter of a tarry nature.

Applications.—Acetic acid is not largely used in the dyeing arts, and its uses seem all to depend upon its power of keeping bodies in solution; and, by volatilizing, leaving them to their own affinities. Thus, in many steam colors acetic acid is evidently used to restrain the coloring matter and metallic oxide present from forming an insoluble compound before the color gets on the cloth; if the color was not in solution it would be merely deposited *on* the fibre, and not *in* it, as it should be. In many colors acetic acid prevents coagulation and enables a color to work smooth in the machine which would otherwise go rough and curdy. It serves to form acetates by direct combination with metals and oxides.

Acer Rubrum, or *Scarlet Flowering Maple of North America.*—According to Bancroft, the bark of this maple produces with an aluminous mordant a lasting cinnamon color both on wool and cotton; with iron mordants, he says, it gives a more intense pure and perfect black than even galls or any other vegetable matter within his knowledge, and does not stain whites. It is not mentioned in more recent works upon dyeing, and has probably never been put to use.

Acetometer.—An instrument constructed like a hydrometer, but graduated for acetic acid only. On account of peculiarities alluded to as attending the relation between the density and percentage of acid in acetic acid, the indications of such an instrument are not trustworthy.

Acid.—An acid in chemistry originally signified anything of a sour acid taste; it has a wider meaning now, not easy to give an exact definition of; but an acid may be characterized as a body capable of forming combinations with metals and bases, which combinations are called salts. Many of the acids in chemistry are insoluble in water and have no taste, but all the common acids are sour to the taste. A compound is said to be acid or to have an acid reaction when blue litmus paper dipped into it is turned red. Acidity is neutralized or destroyed by the alkalies as potash, soda, or ammonia, or by lime or chalk. *An acid* in printing is some composition for resisting or discharging mordants or colors, most of which are strongly acid, but some are compounded of salts and acids, and some are not acid at all; the latter are, however, generally distinguished as "neutral pastes," "mild paste," etc. For an account of the different acids in practical use, see their distinctive names as CITRIC ACID, SULPHURIC ACID, etc.

Acidimetry, or *the testing and valuation of acids.*—The hydrometer of Twaddle, though a most valuable instrument, sometimes leads to wrong conclusions upon the strength of liquids. The instrument will only show the density; and though there is in general a direct relation between the density of a liquid and the quantity of solid matter it contains, it is evident that no instrument can be expected to show what kind of matter it is that gives the density. Thus take acetic acid at 8° Tw. and mix it with an equal bulk of water, it will only mark 4°; but by adding common salt to this weak acid, it will be brought to mark 8° again, and as far as the hydrometer shows, it is as strong an acid as before. The practical method of testing acids is to ascertain how much carbonate of soda or caustic soda a given weight of acid can neutralize, and the process may be conducted as follows: Take good crystals of soda neither damp nor white from dryness—the points of the large crystals are very likely to be quite pure and should be selected in preference—crush them into coarse powder and keep in a closely corked or stoppered bottle; this forms an alkaline test powder of a constant and definite strength. Suppose the object is to test the strength of a sample of nitric acid, a certain quantity, say 100 grains, is accurately weighed out and transferred into a porcelain capsule and mixed with a couple of ounces of water, a few drops of solution of blue litmus are added to give the liquor a red tinge; next a quantity of the crushed crystals of soda is weighed, such a quantity is taken as is assumed to be more than the acid experimented upon will require, say 300 grains, and without removing the bulk from the dish or watch-glass in which it was weighed, small portions are taken off with a knife and put into the acid until the red color begins to change towards blue; for greater exactness it is desirable to have the liquor hot, and towards the end boiling, for then the change of color is more satisfactorily seen; by weighing what is left of the soda crystals and deducting it from the original quantity, the amount used is ascertained. The stronger the acid the more soda crystals are required, and the weaker the less; so that without any further reference it is easy to tell which of two acids is the strongest, and how much one is stronger than the other. I have compiled a table which includes the chief acids in use, and by referring to which, it will be easy to calculate the percentage of any of the acids tested. The figures show the decimal parts of a grain of pure acid which is neutralized by a single grain of crystals of soda, so many grains of crystals that 100 grains of the sample under examination has taken so many times this decimal quantity is its percentage of real acid.

One grain of crystals of soda neutralize—

0.36 gr. of Acetic acid dry and pure.
0.38   "   Citric acid   "   "
0.26   "   Muriatic acid   "   "
0.38   "   Nitric acid   "   "
0.25   "   Oxalic acid   "   "
0.28   "   Sulphuric acid   "   "
0.46   "   Tartaric acid   "   "

If, for example, 100 grains of a sample of spirits of salts or muriatic acid had required 120 grains of crystals to neutralize them, the percentage of pure muriatic acid would be found by multiplying 0.26 by 120, the result being 31.2; and so on with the other acids. The percentage is for the pure dry, or as it is called in scientific books, anhydrous acid.

For most purposes of acidimetry a solution of caustic soda, as a test alkali, is preferable to the powdered crystals of soda, and more especially with acetic acid; but the preparation of an accurate test liquor of caustic soda requires great care and many precautions, for which I refer to works on analytical chemistry. The method detailed above is perfectly practicable, and gives results close enough for most practical purposes.

Adjective Colors.—A term used by Bancroft, and after him by other writers upon dyeing. It signifies colors which can only be fixed by means of a mordant, in contradistinction to other coloring matters which are fixed without mordants, and which he called *substantive colors.* Madder is an adjective color, and indigo and safflower are substantive colors.

Adrianople Red.—The same as TURKEY RED, which see.

Aerugo.—An old name for VERDIGRIS, found in some old receipts. Thomson says it signifies carbonate of copper.

Agaric.—A kind of fungus or mushroom growing on putrefying rank vegetation; gives a black dye with copperas.

Ageing; known also as *Stoving* or *Hanging.* The operation of exposing printed or mordanted goods to the action of the air. Formerly the ageing or hanging rooms were kept hot by flues or steam pipes, whence called stoves, a name which they still retain in some places, though heat may not be used. Stoves proper are for simply *drying* heavy piece goods which retain too much water to be well dried over the ordinary steam drying tins. Ageing is mainly intended for *moistening* printed or padded goods which have been dried over the steam chests of the printing machine. The necessity for ageing can be proved by a simple experiment; take a fent printed in dark red, black, and a light shade of purple, straight from the drying tins or steam chests of the printing machine, and having divided it into two equal parts, hang one up in a cool airy place, and the other one dung and dye in the usual manner; after three days, dung and dye the first portion in a similar way; the difference of appearance will be considerably in favor of the aged or exposed part, the unexposed fent will have light uneven reds, the blacks will be rusty and dull, while the light purple, though inferior, will show the least difference in the two fents. The exposure to air has the effect of fixing more mordant upon the cloth, and fixing it more regularly. If we inquire what the nature of this action of the air is, we shall find that it is for the most part attributable to the vapor or steam of water which naturally exists in air, and that the effect of this vapor is to soften the dry color or mordant, to make it moist, and thus to come into closer contact with the fibre of the cloth and enter into combination with it. That it is the moisture of the air more than anything else which acts in ageing, is proved by the fact that in dry air ageing never takes place perfectly; in a long frost the air gets very dry, all the water is frozen out of it, and then there is a complete stop to ageing; on the other hand, steam carefully admitted to the hanging rooms hastens the ageing very much. The quick system of ageing introduced within these three years, simply consists in passing the pieces through a machine full of warm and very moist air, so that the mordant receives all the moisture it possibly can in two or three minutes. If the pieces are folded up in this soft state the ageing goes on rapidly without exposing to air, proving that all that is actually requisite is a thorough moistening of the color and a soaking of it into the cloth. During the penetration of the color some chemical changes take place: the fibre does not combine with all the mordant as it is printed on, but only with a portion of it; thus, acetate of iron is printed on for blacks and purples, the cloth only combines with the iron, not with the acetic acid; and as acetic acid, when set free from the iron, takes the vaporous form, it escapes from the cloth and is carried away by the currents of air. If the printed cloth be packed so close that the air cannot circulate freely between the pieces, then the acetic acid cannot escape, and bad uneven work is produced. If the acetic acid escapes at one part it is retained at another, and the vapor of that which does escape will sometimes condense on other parts, removing some mordant and producing uneven colors. In light colors, as madder lilacs and pale reds, the greatest portion of acetic acid escapes on the steam chests just after the piece has left the printing machine, because there is but little to escape; but in blacks, dark reds, and chocolates, especially in heavy blotch colors, there is a great deal of acetic acid still in the color on the piece which must escape in order to yield a good mordant. Hence these colors require a longer ageing than lighter colors; they require more room, a freer circulation of air, and, if passed through the ageing machine, should not be afterwards laid in folds but hung up freely to the air. Another chemical action accompanies ageing, and this is oxidation. It only affects mordants of iron, and those in a very insignificant degree, so that experiments made by ageing iron mordanted cloth in gases containing no oxygen show as good results as those aged in air or pure oxygen; that iron mordants do absorb oxygen there is no doubt, but this appears the least important result of ageing. Some colors require the absorption of oxygen to make them yield their best shades; catechu colors, as printed for dyeing in garancine and madder, imperatively demand oxygen, and their ageing cannot be forced with safety; steam blues have a very light shade when just steamed, and take twenty-four hours hanging to give them their best color; this is a result of oxidation, but oxidation in this case is generally forced by passing the pieces through bichromate of potash or other oxidizing solutions. Indigo blue dipping is an example of the action of the oxygen of air upon colors; as the piece rises from the vat it is yellowish, the moment it touches the air it becomes green, and in a short time blue; the intermediate green shade is due to the admixture of the original yellow and the newly-formed blue.

Ageing Liquor.—Under the name of ageing liquor several compounds have been sold; the best that I have seen was composed of chlorate of potash and arsenite of soda. It may be prepared in the following quantities:—

> 20 lbs. caustic soda, at 60° Tw.,
> 20 lbs. white arsenic, in powder.

Boil until all the arsenic has dissolved; this forms the arsenite of soda liquor. Make a solution of chlorate of potash, by dissolving 3 lbs. of it in 4 galls. water, and add arsenite of soda liquor to it until it stands at 28° Tw. This takes about three pints. One gallon of this liquor added to 16 gallons of garancine chocolate will enable the iron to fix with a few hours' age, instead of three or four days; but experience shows that it is not regular in its results, and not to be depended upon. It is no assistance to blacks or reds in ageing.

Air.—The common air is mainly composed of two gases, which have very different properties. If a piece of phosphorus be fixed with a wire at the bottom of a bottle and the bottle be turned upside down, with its neck standing in water, it will be found in 24 hours that a portion of the air has been absorbed by the phosphorus, and water has been drawn up in corresponding quantity. Out of every 100 parts of air, 21 parts will have disappeared, neither more nor less; now, upon examining the air left in the bottle, it is found to be quite different to the original air; if a lighted candle be lowered in the bottle it will be extinguished, and if a mouse or bird were put in it they would die almost immediately. The 79 per cent. of noxious air left behind is called nitrogen, and the 21 per cent. of vital air absorbed by the phosphorus is called oxygen. Whenever the air acts chemically upon matters it is the oxygen which acts, and a body so acted upon is said to be oxidized, it having absorbed or combined with oxygen. Nitrogen seems to have no chemically active properties. There is also in the air a small quantity of carbonic acid gas, the actions of which in dyeing or printing are too small to be reckoned of any value. In towns, other gases are found in the air resulting from the combustion of fuel, the putrefaction of animal remains, &c.; and although these things spoil the air so greatly for respiration, they never form so much as one part in a hundred of it. Water, in a state of vapor, is constantly present in the air, but in variable proportion; generally more in summer than in winter, and more with westerly winds than with winds from the east; the vapor of water does not in the least interfere with the clearness of the atmosphere, and there is frequently more in the clear air of a summer day than in a winter fog. Its action upon mordants has been explained in AGEING, and further information will be found under HYGROMETER. For bleaching power possessed by air, reference must be made to BLEACHING, OXYGEN, and OZONE.

Albumen, or *White of Egg;* also *Fish* and *Blood Albumen.*— The glairy white of eggs has long been known as albumen, and from time immemorial has been applied as a vehicle of colors and a varnish in the fine arts, but only applied to calico printing within the last twenty-five years. Besides eggs, a kind of albumen is obtainable from blood, and also from the roe or eggs of fishes. The character which distinguishes albumen from all other animal matters is its property of coagulating by heat. If fluid white of egg be heated it begins to set at about 140° F., and at the boiling point of water it becomes solid. This coagulum does not become fluid upon cooling, nor is it capable of being dissolved by water; only strong acids and alkalies can again reduce it to the fluid condition, and that only by altering or destroying its principal properties. Commercial egg albumen is simply the white of eggs dried by a slow heat; in dissolving it for use the water should be cold or not warmer than 90° or 100° F., if hotter the albumen will be coagulated and injured. Albumen is used in calico printing for two purposes: first, as a vehicle for printing and fixing pigment colors, such as ultramarine blue; and secondly, as a mordant for some few colors like the mauve or aniline purple. For the pigment colors it is the coagulable power alone of albumen which is valuable; when steamed the albumen is coagulated, becomes solid and insoluble in water, grasping the fibre with a closeness and tenacity which fastens all colors it is mixed with, and closely resembles an actual combination. As a mordant, albumen has but few applications: when coagulated it shows an affinity for all coloring matters, but with most gives only dull and worthless shades. Alkalies injure or prevent the coagulation of albumen; acids and metallic salts cause it to coagulate in the cold; acetic acid and phosphoric acid are exceptions. In using albumen it is frequently mixed with gum water; up to a certain extent it will stand this, but there are bounds which if passed cause pigment colors so applied to wash out. Ammonia, oil, and turpentine may be used in moderation to enable the albumen to work smooth and keep longer. The salt called sulphite of soda added to dissolved albumen will keep it sweet for a much longer time than without this addition. About four pounds of egg albumen to a gallon of water, brought to a suitable consistency with gum, will give good results for pigment colors. Blood albumen of good quality works even better than egg albumen; but it is more liable to irregularity in quality, often containing a considerable portion of insoluble matter. (See further ANIMALIZATION, PIGMENT COLORS, and LACTARINE.)

Alcohol, commonly called *Spirits of Wine, Methylated Spirits.*—The low price of methylated spirits, which is alcohol mixed with 10 per cent. of wood naphtha, renders it probable that several uses will be found for it in dyeing and printing. It is already much used for dissolving the coloring matters from aniline. Generally speaking coloring matters are more soluble in alcohol than in water, and several dissolve easily in it which cannot be touched by water. It dissolves resinous bodies and partially greasy and fatty bodies; solution of shellac in methylated spirits is used in finishing velvets and velveteens, in some colors for printing the spirit is also used. It is very inflammable, both itself and its vapor.

Alder Bark (*Betuna Alnus*).—This bark is used in several parts of the world as one of the materials for dyeing black along with copperas or iron liquor; it serves to economize galls, and seems to yield satisfactory results. With tin and aluminous mordants it gives brownish-yellow or orange shades of no particular value. It is used in combination with sumac, logwood, and fustic, in some receipts for brown fixed with copperas.

Alloxan and **Alloxantine** are chemical compounds produced during the manufacture of murexide or Roman purple. They are both colorless, but on exposure to air and ammoniacal vapors they assume a fine red color. Woollen cloth dipped in solution of either of these bodies becomes colored of a deep and beautiful purplish-red by hanging in air containing ammonia, or by passing over a heated iron; it is not a very durable color. (See MUREXIDE.)

Algaroba.—A coloring matter yielding brown and other dark colors, apparently of an astringent nature, is described under this name. It is obtained from Buenos Ayres, and is called after the name of the tree from which it is obtained. According to the description of the imported product, it resembles catechu in appearance.

Alizarine.—Alizarine is the name of the pure coloring matter of madder. It can be obtained in beautiful needle-shaped crystals of an orange red color. These crystals when properly dissolved are capable of dyeing up all the colors which madder root itself dyes, and it is consequently considered that alizarine is the real coloring principle of madder. Pure alizarine is not yet an article of commerce.

Commercial Alizarine is a concentrated preparation of madder, first prepared by Messrs. Pincoffs and Schunck. Their patented process consists in washing madder so as to free it from soluble and non-coloring principles, and then exposing it to the action of high-pressure steam for a certain period. The product dyes up first class purples, does not dye up blacks very well without assistance from garancine or logwood, is not well adapted for pinks or reds, it hardly stains the whites, and pieces can be very well cleared without soap. The name of alizarine is now frequently given in the market to qualities of garancine fitted to dye purples; sometimes the so-called alizarine is simply garancine mixed with ground chalk, but generally it is garancine finished off in a peculiar manner. After washing from the acid the garancine is boiled in very dilute caustic soda for some time,

5

and then a quantity of muriate of lime added; this has the
effect of precipitating a quantity of lime in a very minute state
of division all through the garancine. In France, I believe,
garancines intended for purples are neutralized before taking
from the washer by means of milk of lime. The presence of
lime in some form or other appears beneficial to purple dyeing.

Alkali.—An alkali is the opposite to an acid, which it can
neutralize or kill; alkalies turn red litmus blue, and yellow
turmeric brown. Potash, soda, ammonia, and lime are alkalies.

Alkalimetry.—This term signifies the measuring or testing
of alkalies, such as potash, soda ash, soda crystals, etc. Very
complete and accurate methods are to be found in good chemical treatises; but as a method suitable for practical men, the following will be found to answer: Procure a quantity of pure
oxalic acid, a fair quality of commercial acid is usually pure
enough, if not moist; powder it and keep it in a well-corked
bottle. When going to test a sample of soda ash or other
alkali, weigh out first 100 grains of the alkaline substance and
dissolve it in water, then weigh out 100 grains of the powdered
oxalic acid in a watch glass or little dish, and with a knife blade
or thin strip of metal keep putting portions of the acid to the
alkali until it is neutralized, which can be ascertained by test
paper or by solution of litmus. Now weigh the oxalic acid
that is left, and note how much it has taken to neutralize the
sample of alkali. The more acid it takes the stronger the ash,
and the less acid it takes the weaker it is. The comparative
value of any two samples of potash, soda ash, ammonia, etc.
can be pretty closely ascertained by this method. By consulting the following table the actual percentage of soda, potash,
ammonia in a sample of alkali may be found from the quantity
of oxalic acid it has taken to neutralize the alkali; because the
oxalic acid is of constant composition, and will always neutralize just the same amount of an alkali.

One grain of oxalic acid neutralizes.

0.75 grain pure caustic potash,
0.50   "   pure caustic soda,
0.27   "   pure ammonia,
0.84   "   pure carbonate of soda,
1.10   "   pure carbonate of potash.

If 100 grains of a sample of soda ash had taken 96 grains of
oxalic acid, the percentage of caustic soda would be obtained
by multiplying this figure by 0.5, showing 48 per cent.; by
multiplying the same figure by 0.84, we obtain 80.6 as the percentage of carbonate of soda, and so on, using the other figures
in case of testing potash or ammonia.

Alkaline.—Having the properties of an alkali, of caustic soda for example, though these properties may be very weak. A substance is said to have an alkaline reaction when it turns red litmus paper blue. If caustic soda be poured into lime juice, the first portions are neutralized by the acid, and the liquor tastes acid and reddens blue litmus paper. A stage is reached when all the acid properties of the juice are hidden and also the alkaline properties of the soda; the liquor is then neutral, neither acid nor alkaline; the addition of a further quantity of soda makes the liquid alkaline, it now tastes like weak soda and turns red litmus blue. Borax, phosphate of soda, silicate of soda, and numerous other salts are said to be salts of an alkaline reaction, because their solutions turn red litmus blue.

Alkaline Pink Mordant, or *Aluminate of Potash.*—This mordant is a solution of alumina in caustic potash. If a strong clear solution of alum be put in a glass and strong caustic potash added by degrees, the first result will be the precipitation of the aluminous basis in the form of a pulp; by addition of a further quantity of potash this pulp dissolves up to a clear fluid, which may be called aluminate of potash, being actually a solution of the alumina in the excess of potash used. On the large scale this mordant is prepared by taking strong caustic potash, making it hot in a copper or iron boiler, and adding to it crushed alum or sulphate of alumina, stirring well. The following proportions will be found to yield good results:—

Alkaline Mordant for Dark Pink.

40 gallons caustic potash at 54° Tw.,
140 lbs. sulphate of alumina.

The sulphate of alumina added by portions, and finally the whole boiled for twenty minutes. It should yield about 45 gallons of liquor, at from 32° to 36° Tw., which, thickened with dark British gum or calcined farina, will yield full dark pinks when properly fixed and dyed.

Alkaline Mordant for Light Pinks.

50 gallons caustic potash at 41°,
180 lbs. potash alum.

Dissolved in the same manner; liquor should stand at about 30° Tw., to be reduced according to shade. The chief bulk of common alum is ammoniacal, and will not do for making this mordant. This mordant does not fasten upon the cloth without some fixing agent; the fixing matter is usually mixed with the dung in dunging; sal ammoniac is the most certain material to use, muriate of zinc has also been used. (See PINK.)

Alkanet, *Alkanea; orcantte.*—This is a root growing in warm climates; it contains a considerable quantity of coloring matter of a resinous nature which is not dissolved by water, but is readily extracted by oil, turpentine, bisulphide of carbon, alcohol, and similar solvents. It was used by the ancients for dyeing wool; its principal consumption now is in tinting oils of a pinkish-lilac color. Dissolved in alcohol and mixed with water it dyes cotton mordanted with alumina of an agreeable bluish-lilac; with iron mordants it gives darker shades. It is not a fast color, and is only a little used for dyeing sewing thread and cotton. Its pure coloring matter is called ANCHUSINE.

Aloetic Acid.—An acid derived from aloes, and which seems capable of yielding some good colors, but not yet applied.

Aloes.—The aloes which are used in medicine when treated with nitric acid undergo some change, and communicate a purple color to silk and woollen cloth, which appears to have a fair amount of fastness. A process for obtaining this color was patented January 26, 1847, but I am not aware that it has even been practically applied. (See CHRYSAMMIC ACID.)

Alloy.—The mixture of two metals is called an alloy, except when mercury or quicksilver is one, when the compound is called an *amalgam.* The alloy used for block work is usually made by melting together equal weights of bismuth, tin, and lead; it melts at a low temperature, and when cold resists pressure tolerably well.

Alterant.—Term invented by Bancroft, to designate any substance employed to modify or change the hue of a dyed color; as for example, cotton mordanted in tin and dyed in logwood, acquires a very dull color, but if passed through weak chloride of tin it assumes its proper violet color; the chloride of tin last used would be called an *alterant.* Alum, acids, soda, ammonia, and other bodies may thus at times become alterants by altering or changing shades already produced. The term "raising," very frequently used in dye-houses, sometimes expresses the use of alterants, but has more frequently a wider signification.

Alum.—There are two kinds of alum besides the patent alum, which is more correctly called sulphate of alumina. The old kind of alum, called potash alum, is a double salt, compounded of sulphate of alumina and sulphate of potash; it is frequently called rock or roach alum and Roman alum. The other kind of alum is called ammonia alum, and is compounded of sulphate of alumina and sulphate of ammonia. There is no distinguishing between these two kinds of alum by their external appearances; they have the same shape of crystal, the same taste and solubility; but they can be easily tested by means of caustic soda or potash, for when the ammonia alum is mixed with caustic it gives off a strong smell of ammonia, while the potash alum gives no smell, except sometimes when a little ammonia is accidentally contained in it, and then it gives a faint smell. These two alums are as nearly as possible of the same strength, and, for nine cases out of ten, it does not signify which is used. For making alkaline mordant, ammonia alum is very unsuitable; in two or three other cases preference is to be given to the potash alum, very little of which, however, is to be found in trade. The only dangerous impurity in alum is iron, and this will show of itself, if the alum be old, by a reddish or yellowish tinge. The taste of the alum containing iron is quite different from good alum, and it may be tested by prussiate of potash; if it gives a blue instantly it is bad. In fresh alum crystals iron can exist without showing itself; it must then be tested for by decoction of galls, which will cause it to turn black or bluish-black; and by logwood liquor, which will show a distinctly different hue with good and bad alum. A mixture of red and yellow prussiate is also a good test; but even good alum will show blue after an hour or two; but if a blue be produced instantly upon mixing there can be no doubt of iron being present. Alum sold in the state of *flour* or small crystals may contain too much water by five or ten per cent. Alum by itself is only a weak mordant; it has a strong acid reaction, and parts with very little of its base, unless something be added to neutralize the acid in part. Soda in the state of crystals is mostly used for this purpose; but it is found in practice that the acetate of alumina is by far the best mordant where deep shades are required, so that now alum is only used for light shades or in combination with copperas. The preference which was formerly given to particular species of alum, as the Roman alum, is proved to have arisen from the processes of manufacture favoring the production of a more neutral or basic compound.

Alumina.—This is the earthy base of alum and of all the salts of alumina. It can be made by dissolving alum in hot water and adding soda crystals; so long as they give any pulpy precipitate, the alumina pulp can be drained on a filtering blanket and washed with water. It may be used to make oxalate, tartrate, and nitrate of alumina from by dissolving as much of it in these acids as they can take up; it dissolves in caustic potash forming the alkaline pink mordant. It has been used

as the basis of colored lakes for calico printing; if a certain
quantity of this pulp be diffused through water, and logwood
liquor mixed with it and heated, the pulp will abstract all the
color from the liquor and form a colored pulp or lake, which
mixed with acids, etc., can be printed as a steam color.

Alumina Nitrate.—This salt is prepared by mixing nitrate
of lead and alum, sulphate of alumina may be used instead of
alum. The following proportions yield a nitrate of alumina
well adapted for indigo chromed styles, that is for converting
the chrome orange into yellow wherever printed on :—

> 7 lbs. alum,
> 4½ gallons water at 140°, dissolve and add
> 8 lbs. nitrate of lead ;

take the clear only. Nitrate of alumina is but little used in
general printing; in some few cases of delaine and woollen
colors it is employed, when it appears to have an oxidizing
action owing to the nitric acid it contains.

Alumina Oxalate.—This salt may be prepared by dissolving the moist gelatinous alumina in warm solution of oxalic
acid until saturated. It has been used in some steam reds from
peachwood, along with chlorate of potash as an oxidizing
agent.

Alumina Sulphate, or *Patent Alum.*—This salt is of comparatively recent introduction in the manufacturing arts; it
contains all the essential principles for which alum is valued,
differing from it only by the absence of the sulphate of potash
or sulphate of ammonia, which is an invariable constituent of
alum. It is not possible, however, to use sulphate of alumina
in every case where alum has been employed ; probably because
the commercial article has not yet been produced of a corresponding degree of purity and saturation ; probably also, because the
neutral sulphate in alum exercises some modifying action in its
application. But in a great many cases a good quality of sulphate
of alumina can be advantageously used in place of alum ; it is
more liable to contain impurities than alum ; it is more irregular
in its composition, not crystallizing like alum in clear well-
defined crystals, but being boiled down until it solidifies into a
white opaque cake. The ordinary good qualities contain more
alumina by one-third than alum crystals, and are consequently
stronger as mordants; but the amount of water and acid it contains are subject to fluctuations, which have frequently produced
great losses and irregularities in printing and dyeing. Chemical analysis is necessary to show the amounts of acid, water, and
alumina a sample may contain.

Alumen Ustum, or *Burnt Alum.*—This substance is mentioned in some old receipts; it appears to be alum which has been heated in earthen vessels until it has become dry and white. Modern chemistry does not show that it could possess any special properties.

Amber Colors.—Certain shades of yellow having some resemblance to the hue of amber are so called. On dyed goods they are all derived from a lead basis raised or dyed in chrome. The amber shades may be looked upon as yellows slightly tinged with red. On woollen the shades are obtained by modifying a fustic yellow with cochineal (see ORANGE); on calico the following processes may be followed for 100 lbs. cloth: 10 lbs. acetate and 10 lbs. nitrate of lead dissolved in a sufficient quantity of cold water, work the goods in for thirty minutes, and then in warm water containing 8 lbs. of chrome, for twenty minutes, pass finally through the lead wash, and dry. By another process the goods are mordanted in a plombate of soda bath formed by adding caustic potash or soda to solution of acetate of lead until the white pulp at first formed is dissolved up clear, having a care not to add more caustic than is just necessary; after the goods have been worked in this they are worked in warm chrome liquor. Napier states that sulphate of zinc added to the chrome improves the effect (see further CHROME COLORS). Amber on silk may be obtained from annotta modified by other coloring matters.

The following receipt is a specimen of what may be used in printing to obtain amber shades:—

Steam Amber or Gold.

7 quarts berry liquor at 6°,
1 quart cochineal liquor at 4°,
3 lbs. starch; boil, and when nearly cold add
6 oz. crystals of tin,
2 oz. oxalic acid.

Ameline.—The name given to a dyeing matter of the aniline species very lately introduced. It is a pansy color, or a blue mauve, applied in the usual manner upon delaines. Its method of manufacture is kept secret.

Ammonia, *Ammonia Liquor, Volatile Alkali.*—Ammonia liquor is a solution of the gas ammonia in water, the stronger the liquor is of this gas the lighter it is, bulk for bulk, contrary to the usual law of density; so if a Twaddle instrument constructed for liquids lighter than water be used to test it, the lower it sinks in the ammonia the better it is. The Twaddle test is a good one for ammonia liquor, as I am not aware of anything it is adulterated with in the direction of making it lighter than water. It can also be tested in the same way as soda ash as given in alkalimetry; the more oxalic acid a given quantity neutralizes the better it is. Ammonia possesses the same powers of neutralizing acids as potash or soda, and generally has similar properties to them. Ammonia is called the volatile alkali, because it flies off as gas if left exposed in an open vessel, or more quickly if heated; it is a very good solvent of several coloring matters, especially cochineal. The gas or vapor from ammonia has been sometimes used to fasten colors or mordants; it was used in some processes of the murexide color, and has been proposed as a substitute for ageing. In such cases the gas is best produced by letting the strong ammonia liquor drop in regulated quantity upon a hot steam pipe; it may likewise be produced by heating a mixture of slacked lime and sal ammoniac.

Amylaceous Matters.—All species of starches are thus designated; or substances containing or yielding starch, as flour, meal, etc.

Anchusine.—This is the name of the pure coloring matter of alkanet root, so called from the botanical name of the plant, *Anchusa Tinctoria*.

Aniline Colors.—Aniline itself is a colorless or slightly yellow oily body, made by complicated processes from coal naphtha. When acted upon by powerful chemical agents it yields several colors, the most valuable of which are the mauve or mallow, and the magenta or red; a blue coloring matter is also produced. The patented processes for making and applying these coloring matters have been so numerous these last four years that it is impossible to give any account of them here; the inquirer is referred to the specifications of patents, or to the pages of the *Chemical News*, where an abstract of them may be found.

Aniline Mauve or Lilac, is sold either in the fluid or pasty state. For silk dyeing and woollen dyeing no mordant is required; the proper proportion of clear liquor is mixed with water slightly warm, any scum that may form is cleared off, and the goods entered and worked until the required shade has been obtained; a small quantity of acetic or tartaric acid is recommended to be used in some cases. Pasty mauve is dissolved in methylated spirits before using, and great care must be taken to prevent irregularities from the tarry scum which frequently forms when the liquor is mixed with water. For printing on calico, one process consists in fixing the coloring matter with albumen or lactarine, the mauve is mixed with solution of albumen or lactarine, printed and steamed; or, the albumen alone is printed, steamed to fix it, and then dyed in a beck with the coloring matter—a quantity of soap being dissolved in the beck to prevent the whites being too much damaged. The chief processes, however, of fixing the aniline colors are with tannic acid and a metallic salt, and there are various methods of applying the materials. The cloth may be prepared with tin, as for steam colors, and a mixture of the coloring matter and tannic acid printed on and steamed with or without albumen or lactarine ; or as in the antimony process the coloring matter mixed with tannic acid is printed, steamed, and then fixed by running in a solution of tartarized antimony. Many other processes have been proposed, but these include all that have answered satisfactorily. For dyeing on cotton, the cloth or yarn is steeped in sumac or tannic acid, dyed in the color, and then may be fixed by tin, or the cloth may be sumaced and mordanted as usual with tin and then dyed. For magenta red precisely the same processes may be used as for the mauve. The blue is dyed in the same manner. Cloth prepared with oil preparations takes up the aniline blue ; for printing on calico it does not seem to be so applicable, and must be fixed by albumen or lactarine. The affinity of these new coloring matters for silk and woollen is very great, so that in piece dyeing precautions have to be taken to prevent irregularities arising from this cause. For example: in dyeing a piece of union velvet a full magenta shade, if the common "jigger" be used and the whole of the magenta liquor added at once, the first two yards will be darker than the rest, and one-half of the piece of a decidedly deeper hue than the other half, that is, the half piece first in the liquor will be darkest; it is consequently necessary to add the requisite amount of coloring matter at two or three intervals, and in such a manner with regard to the entry of the piece that the end last in at the first addition will be first in at the second addition. Notwithstanding these precautions, the ends of the piece are mostly fuller in color than the body. In most of the pasty kinds of mauve or magenta, there is a quantity of tarry matter, which being dissolved by the methylated spirits, has a bad effect on the shade; in such a case the spirits should be diluted with water as much as possible, because there is generally a strength of spirit which dissolves the coloring matter without touching the brown tarry matters; if practicable, water would dissolve the coloring matter, but as a very large quantity of water is required this plan cannot be often adopted. Aniline colors on silk are modified by sulphate of indigo to blue the mauve, and annotta to give orange or capucin shades with the magenta.

Animalization.—In the older theories of dyeing, it was

held that the animal tissues of wool and silk absorbed and re-tained colors more readily than the vegetable tissues of cotton and linen, by virtue of some peculiar animal substance they contained. As a consequence of this theory, attempts were made to communicate some animal principles to vegetable fabrics, with a view to improving their powers of receiving colors. The use of cow dung in dyeing mádder goods; the use of sheeps' dung and bullocks' blood, and urine in Turkey-red dyeing, were explained, upon the supposition that they animalized the fabric in some way or other. The present view of animalization is, that it is not possible to animalize a fabric in any other way than by actually depositing upon it the ani-mal matter in question, and that any increased facility for taking colors thus communicated, is effected by the animal matter itself held on the fabric, and not by any new property of the fabric itself. Thus, if a piece of calico is steeped in a solution of albumen, dried, and then steamed or plunged into boiling water, the albumen is fastened upon the cloth, and such cloth is then capable of receiving colors from picric acid, sul-phate of indigo, magenta, archil, and other coloring matters, which previously had no affinity for the cloth. But it is im-possible to look upon the albumen in any other light than as a kind of mordant acting as an intermediary between the color and the calico, differing, however, from ordinary mordants in some essential particulars. Besides albumen, the animal mat-ters called caseine and lactarine, possess similar properties, and have been tried on a large scale, but without any marked suc-cess as mordants or bases for some of the colors, which are not attracted by the ordinary metallic mordants. The increased affinity for colors given to calico by oil, could not correctly, under any view, be called animalization, since the oils are all vegetable oils; but in fact there appears to be a considerable analogy between this case of mordanting and that by coagulable animal matters.

Anotta; also *Annotto, Annatto, Arnotto,* etc.—This coloring matter is a pulp prepared from the seeds of a South American shrub. It is generally sold as a thick paste of the consistence of putty, but is also prepared in hard dry cakes by some London houses. In the pasty state it has a very disagreeable animal odor; it is of a reddish-brown color, does not dissolve in water, but is easily dissolved by alkalies and alkaline salts; soft soap is frequently used to dissolve it. For printing purposes anotta is used for a shade of buff orange, sometimes called salmon or nankeen color. Half a pound of good anotta dissolved with heat in a gallon of pearl-ash liquor, and half a pound of soft soap and 4 oz. borax added, thickened with tragacanth, is an old receipt giving a good result. Other receipts are similar to the following—

> Gallon of caustic potash at 14°,
> 2 lbs. anotta; dissolve and add
> 2 oz. tartaric acid,
> 8 oz. alum; thicken with gum-water.

Tin crystals are also used to modify the shade. For light shades neither alum nor tin are required, for anotta is one of those coloring matters which have an affinity for cotton of themselves. Dark anotta colors are not pretty on cotton; on account of the strongly alkaline nature of the color, it may be used as a buff discharge or resist for Prussian blues. For dyeing on cotton the anotta is dissolved in alkali, and the goods simply passed through the solution. For silk dyeing anotta is largely used, yielding bright lustrous shades; by aluming the silk is considered to take the dye better: as silk is easily acted upon by alkalies, the solution should be as little alkaline as possible. Acids and acid salts redden the shades from anotta. As solution of anotta is injured by keeping, no more should be made than is likely to be used in a couple of days or so. The pure coloring matter of anotta is called *bixine*: another coloring principle, named *orelline*, is supposed to exist in it. Orelline is a yellow principle, and bixine a red. By influence of air, moisture, and ammonia, these principles appear convertible. The great bulk of the anotta imported is consumed in coloring butter and cheese. Anotta is liable to be adulterated with colcothar, brickdust, and red ochre. (See BIXINE.)

Anti-Chlore.—Some body capable of destroying and arresting the action of chlorine. The chief substance employed is sulphite of lime, used in bleaching rags for paper, and recommended in linen bleaching, after chloride of lime treatment.

Antimony.—Antimony is a metal whose chemical properties more nearly resemble those of tin than any of the other common metals; it is sufficiently abundant to receive extended application, but up to this time has not been much used. An orange color from the sulphide of antimony was first made, I believe, by Mr. Mercer; the common black sulphide of antimony, in powder, was boiled with caustic soda and sulphur until it was dissolved; the liquor had a fetid, sickly smell, well remembered by old printers. A better preparation was made by calcining the antimony with charcoal and sulphate of soda. The result in both cases was a double compound of sulphur with soda and antimony; this was thickened and printed; containing very much sulphur, it blackened the copper rollers immediately; after drying and a short age it was passed in sours for the orange; by running it afterwards in a beck containing blue copperas, it changed to a dark olive; by passing in sugar of lead, a wood brown was produced. This color would stand washing and soaping well enough, but faded on exposure to air. The antimony orange is hardly ever made now. Tartarized antimony or tartar emetic is used in one of the processes for fixing the aniline colors; antimony as a prepare for steam colors is very inferior to tin.

Apocrenic Acid.—This is a vegetable substance, found in water, and forms one of several bodies existing in certain qualities of water, usually designated under the head of organic matter. For the tests for it and its supposed action in dyeing, see WATER.

Apricot Color.—This is a shade of buff, a little redder and browner than an iron buff. Common buff liquor is mixed with some muriate of iron and a small quantity of nitrate or sugar of lead; and after the buff has been raised in the usual way, it is rinsed in warm and very weak chloride of lime; the lead is oxidized and gives a brownish hue to the buff, which somewhat resembles the ordinary shade of an apricot. Though the name is chiefly confined to the shade so produced, a similar shade can, of course, be obtained in steam and spirit colors, and especially from catechu. (See CATECHU and ORANGE.)

Aqua Regia.—A mixture of nitric acid and muriatic acid undergoes some chemical change, producing a liquid which possesses properties different from either acid separately. It received its name from its power of dissolving gold, the king of metals.

Aqua Fortis.—An old and still common name for NITRIC ACID, which see.

Arabine.—The name of a principle extracted from gum arabic, and supposed to exist in all similar gums.

Archil; *Orchil.*—This coloring matter is a preparation from a kind of moss or dry leaf, growing on rocks and stones, called a lichen. The lichens, of which there are many varieties, have no color themselves; but, by a kind of fermentation and treatment with lime and stale urine, the coloring matter is developed. There are two kinds of archil, that in paste and that in liquor; and there are besides two colors of it called red and blue archil. Archil has a particular smell easily recognized; it mixes with water; it is turned bluer with alkalies and redder with acids. As a coloring matter it has affinity for silk and woollen, with or without mordant, but none for cotton. It is seldom used by itself for dyeing, but usually to help or top other colors; when used alone it can give very agreeable shades of violet, peach, and lilac, which colors are very loose in air, fading almost visibly in sunlight; in combination with other coloring matters it usually darkens them, giving chocolate colored shades, but archil is chiefly valued for a peculiar softness and velvet bloom it communicates to colors. Archil is used in woollen and delaine printing, chiefly for rich chocolate shades, and in combination with other coloring matters for shades of buff, chamois, wood, tan, &c. Three or four years ago a new preparation of archil, giving much faster colors, was invented and put into use. It was supplied in hard dry cakes, of a purplish color; the method of its preparation is not clearly described, but there is no doubt that a considerable improvement in fastness was obtained. It was used in calico printing to a considerable extent, until the more pleasant aniline mauve displaced it. It could only be fastened by means of albumen. It was misrepresented as being as fast as madder, while, in reality, only a loose color, so that considerable loss and disappointment was occasioned; it is very little used now. Cudbear and litmus are very similar to archil, as coloring matters. Archil may be adulterated with extracts of logwood or peachwood, a careful comparison of shades produced by dyeing silk or woollen in pure and suspected archil would indicate the adulteration. Pure archil gives no color to mordanted calico, but an adulterated archil will; pure archil mixed with water and muriate of tin, and heated, is nearly decolorized, if logwood or other extracts be present different shades will be produced. The addition of a little red prussiate to blue archil is said to give it all the properties of red archil.

Areca Nuts.—An Asiatic product, said to be capable of fixing colors by some agglutinating property.

Argols.—The crude cream of tartar goes by this name. There are red argols and gray argols used in woollen dyeing; the only valuable properties they possess are due to the bitartrate of potash they contain. (See TARTARIC ACID and TARTAR, CREAM OF.)

Arsenates, *or Arseniates,* are compounds of arsenic acid with bases; they are made by neutralizing arsenic acid with the base required. The arsenate of potash was formerly used as a resist in combination with pipeclay; the arsenate of soda has been largely used as a dung substitute; it is prepared from arsenious acid or white arsenic and nitrate of soda, heated together in a reverberatory furnace, and the product neutralized with soda. (See DUNG SUBSTITUTES.)

Arsenic, *Arsenious Acid, or White Arsenic.*—The common white arsenic is a feeble acid, and called arsenious acid in chemistry; it is a deadly poison, and should be shunned as much as possible; inhaling the dust created by moving it should be avoided. Its chief uses in connection with printing and dyeing are derived first from its weak acid properties, modifying, without neutralizing completely, the alkalies, soda, and potash; secondly, its deoxidizing powers have been used in one or two cases, as in the chrome greens; and, thirdly, it forms some colored compounds with the metals, the only ones used being the green, from copper and chromium. Arsenic is used in a good many receipts, where its action cannot be explained, and where most probably it has no useful action at all. White arsenic does not dissolve to any considerable extent in cold water, but in hot water it is more soluble; by prolonged boiling, water dissolves a considerable portion of arsenic; it dissolves to an almost unlimited extent in caustic potash and soda, forming the arsenites of those bases. When white arsenic is heated with nitric acid, it combines with more oxygen, forming arsenic acid; this acid is very soluble in water, and has strongly acid characters; it has been tried as a substitute for tartaric acid, but did not succeed. The substance called red arsenic is a compound of metallic arsenic, with sulphur; it is known also as ORPIMENT, which see.

Arsenites are compounds of arsenious acid with bases and metals.

Artichoke Green.—A patent for obtaining a green coloring matter from artichokes and thistles was taken out June 3d, 1856, but not completed. (See CHLOROPHYLL.)

Astringents.—The vegetable astringents used in dyeing and printing are represented by gall-nuts, sumac, catechu, and one or two other substances. Tannic acid may be considered as the real astringent, it possessing the astringent properties in the highest degree. It is a property of astringents to have a direct affinity for vegetable fibre, so that cotton soaked in a hot decoction of galls or sumac acquires the astringent principle, and retains it so strongly that it is difficult to remove it; it is also a pretty general character of astringents to strike a black with green copperas and other salts of iron, but this is not an essential character. In the older theories of dyeing much stress was laid upon the astringent principle as an important element of all colors, being that portion which contributed to the closeness of the adhesion of the color to the fabric. But many of the fastest coloring matters, such as indigo, madder, and cochineal, do not contain any astringent matter at all, in the ordinary meaning of the term astringent; and the supposed necessity of an astringent principle is therefore disproved. At the same time, the true astringents, as tannic acid, galls, sumac, &c., do not only themselves form very stable and intimate combinations with vegetable fibre, but also appear to confer stability to loose colors. In the great majority of cases of cotton dyeing, galls or sumac are used, and usually are the first substances employed; the astringent principle, or tannic acid, of the galls and sumac at once forms a fast and perfect combination with the fibre, and appears to enable the fibre to combine more easily and permanently with all mordants and colors than if the astringent matters were absent. The old doctrine of the importance of an astringent principle is at least partially true and worthy of attention. The method of applying the new aniline colors by means of tannic acid and salts of tin and antimony is a point in illustration, though it is not actually known what part the astringent acts in these cases. (See further, GALLS, SUMAC, &c.)

Atomic Weight.—According to the atomic theory every substance is made up of very little atoms, and each of these atoms has a regular weight of its own; that is, an atom of iron weighs so much, and an atom of lead so much more, the atom of lead being about four times as heavy as the atom of iron, and so on. The relative weights of these atoms have been very carefully ascertained by chemists, and the whole science of modern chemistry is built upon the knowledge of the laws of combination between atoms. Many chemists and philosophers do not believe in the existence of atoms at all, but allow that matter of various kinds enters into combination in certain definite proportions, which are always the same for the same substance. This is now the prevailing theory, being most in accordance with the discoveries of late years; and what were called atomic weights, are now called EQUIVALENT WEIGHTS, which see.

Awl Root.—An East Indian product said to possess some of the valuable properties of madder.

Azaleine.—Red coloring matter obtained from aniline by the action of certain metallic salts, chiefly nitrate of mercury. The shade of color not being so good as that obtained by other patented processes, azaleine has not been much used in dyeing.

Azote.—The old name of nitrogen.

Azuline.—This name is given to a blue coloring matter supposed to be derived from aniline. It is chiefly used in silk dyeing, yielding a very fine blue color; it requires the presence of a rather considerable amount of free sulphuric acid in the dyeing to secure good shades. On this account it has not yet been successfully applied to calico printing. There is more than one kind of blue coloring matter sold under this name, and they are not all of equal stability. Their discovery is so recent,that no really trustworthy information upon their manufacture can be given.

Azure.—A blue powder consisting of a glass colored with oxide of cobalt is sold under this name, also called smalts and zaffre. It has been used in finishing yarns, etc. Being quite insoluble in water, it must be suspended in some mucilaginous liquid, as starch, size, or soap, and requires considerable care to prevent unevenness.

B.

Bablah, *Babulah,* or *Neb-nab.*—This is the name of a fruit imported from Senegal and the East Indies. Upon its first introduction into Europe it was said to be endowed with the most valuable properties as an astringent, communicating permanency to all dyed colors. This was not however found to be the case, and bablah fell into disrepute, so that dyers would not buy it any price. M. Chevreul made an examination of the rinds of the fruit, and found the Senegal bablah to yield 57 per cent. of soluble matters, and the East Indian 49 per cent., while the best quality of gall-nuts give 87 per cent. Bablah contains a considerable proportion of tannic and gallic acids, and a reddish coloring matter in small quantity. With iron and alumina mordants it gives drab and fawn colors, and may be used as a substitute for sumac; but, where sumac gives a yellowish shade, bablah gives a reddish hue. Most authorities speak only of the rind of the fruit as being used in dyeing; others include the hard kernel as well.

Bandanna.—A style of work so called. It consists of a white discharge upon Turkey red; the name appears to be confined to goods produced by means of perforated lead plates, between which the Turkey red cloth in several thicknesses was tightly pressed, and the perforations being so adapted as to correspond to one another, a discharging fluid, either solution of chlorine gas in water or a mixture of bleaching liquor and acid,.was run upon the upper plate and gradually soaked through: the great pressure upon the cloth prevented the liquor from spreading beyond the pattern. This is a case of DISCHARGING, which see.

Barasat Verte, or *Green Indigo.*—A substance under this name was examined by Dr. Bancroft, who reported it to be simply blue indigo contaminated with vegetable extractive matter of a useless nature which made it appear green. It did not yield any green colors upon wool or cotton which could not withstand the action of soap. (See CHINESE GREEN and INDIGO.)

Barbary Berries, or *Seeds*, contain a coloring matter, which, according to Bancroft's experiments, in some respects resembles safflower when applied upon silk. No definite information upon these seeds was communicated to Brancroft, and he could not identify them.

Barbary Gum.—A natural gum, similar to Senegal gum and gum arabic. It is liable to contain more or less of a species of gum which does not dissolve in cold water, only swelling up and making the solution of gum-water pasty; a gum containing this kind of inferior gum does not work or keep well, and is not easy to wash off soft. By leaving a sample of gum in lumps for 48 hours in cold water, it will be easily ascertained whether there are any lumps of this fictitious gum or not, and what is their relative proportion to the bulk.

Barilla.—This is a very impure kind of soda ash imported from Spain, Sicily, and other places. It is produced by burning sea weeds and collecting and preparing the ashes. It was formerly the chief source of soda, but it is now only used in some exceptional cases. In old works upon bleaching and printing, wherever barilla is mentioned or prescribed, soda ash in perhaps one-fourth of the quantity would be found to have an equivalent effect.

Bark.—The contracted term "bark" is generally used amongst the dyers and printers of Lancashire, to designate the quercitron bark, extensively used in garancine dyeing. The barks of a few other trees are or have been used in dyeing, such as alder bark, oak bark, pomegranate bark, pine bark, willow bark, etc., for an account of which see ALDER, etc.

Barwood.—This is a dyewood obtained from Angola in Africa, and neighboring places. It is one of the red woods, and closely resembles sandal wood in its properties; it is compact, taking a good polish of an orange red color. Its coloring matter is not easily extracted by water, for boiling water only dissolves a small quantity of it, and this precipitates in great part as the water cools; there is, therefore, no barwood liquor or extract, and in dyeing with it the rasped or ground wood has to be used just as madder is used in madder dyeing. The goods take the color from the water as fast as it takes it from the wood; the coloring matter is gradually transferred until the desired shade is obtained or the wood spent. The colors it gives upon the luminous mordants are reds of a yellowish-brown shade according to Bancroft; when these are saddened by green copperas they produce a good imitation of the bandanna red. The same author states that the red from it was used as a bottom for dark indigo blues, saving indigo. At present barwood is chiefly used in yarn dyeing to produce an imitation

6

Turkey red. It is also used for a red lake or pigment employed by the paper printers. The pure coloring matter of this wood is considered to be identical with santaline extracted from sandal wood.

Barwood red is obtained by first steeping the yarn or cloth for several hours in a decoction of sumac with a little vitriol, about four pounds sumac to every twenty pounds cotton. After the sumac has had time to form an intimate combination with the cotton, the yarn is next wrought in a solution of nitro-muriate of tin, or barwood spirits standing at 8° Tw.; the tin combines plentifully with the astringent principle of the sumac and constitutes the mordant; the goods are now transferred to the boiler or beck, where about their own weight of barwood finely rasped is added, the water being nearly boiling, and the goods worked about until the required shade is obtained. The red so produced is more permanent than any of the other wood reds, and stands next, though considerably inferior, to madder red. The practical dyers say that it is more difficult to get regular and good results from barwood than from any other wood, and that a great many fail to obtain the best red.

Barwood is said not to work well with other dyewoods, if any combinations or modifications are required by the assistance of other woods, they have to be applied after the barwood by a separate operation.

Baryta, or *Barytes.*—There is a very rare metal called barium, its oxide is called baryta, and has properties nearly like quicklime. The very common mineral substance, which is sold under the names of barytes, mineral white, ground heavy spar, etc., is a sulphate of baryta, prepared by finely grinding the native heavy spar. It is used for "weighting;" that is, for giving weight and apparent body and firmness to inferior goods; it is not the only, and probably not the best substance for this purpose. China clay, pipeclay, flour, and aluminous shale are used also in this species of falsification. Beyond this use of the sulphate, the compounds of baryta have not yet been employed in printing and dyeing on a large scale.

Base.—In chemistry, a base is some body which neutralizes an acid, generally forming crystalline compounds with it, which are called salts. Thus, lime neutralizes acids, and is a base; litharge or oxide of lead, which is quite tasteless, would be found upon trial to completely neutralize acetic or nitric acids, producing a third body, which is a salt, either acetate or nitrate of lead; oxide of lead is therefore a base. Nearly all the metals are bases, and form salts with acids; and when we speak of sulphate of iron, nitrate of copper, and other similar salts, we understand that the bases iron, copper, etc., have neutralized the acidity of sulphuric, nitric, and other acids. Besides mineral bases, there are others of purely vegetable origin, and some derived from the animal kingdom. They are all distinguished by neutralizing or depriving acids of their acid characters.

Basic Salt.—A salt containing more than the usual quantity of base, as basic acetate of lead.

Bassora.—The name of a kind of gum, which is like tragacanth; it swells up in water and forms a kind of paste, but does not really dissolve. It contains a principle called bassorine, which exists also in tragacanth and salep. It seems probable that a good deal of this kind of gum comes mixed with Senegal and Barbary gum, from which it cannot be easily distinguished. It is not a good gum for calico printing, because it does not wash off well, and leaves a harshness upon the cloth.

Baume'.—This is the name of the hydrometer which is most generally in use on the continent, and fulfils the same purposes that Twaddle's hydrometer does in England. The degrees of the two instruments do not correspond, nor is there any simple relation between them; but as a guide for the translation of receipts, it may be considered that each degree of Baumé is equal to $1\frac{1}{2}$ of Twaddle, as far as the thirtieth degree of Twaddle; thus 10° Baumé is equal to $14\frac{1}{2}$° or 15° Twaddle; 20° Baumé is equal to between 30° and 31° Tw.: past the 30° of Baumé, the difference is greater, equal to about $1\frac{3}{4}$ of Twaddle for each degree Baumé; at 50° Baumé, each degree is equal to 2° Twaddle, and so on. A table is given in my "Chemistry of Calico Printing," of the exact correspondence between the degrees of these instruments.

Bear-Berry (*arbutus uva ursi*).—A substance employed in dyeing black.

Berries.—The only berries commonly used in dyeing or printing are used for the sake of their yellow coloring matter. There are as many as seven or eight different qualities, but all appear to be derived from the same kind of shrub, which in France is called the dyer's buckthorn (the botanical name being *rhamnus infectorius*), and which, besides growing extensively there, flourishes in the island of Candia, in Wallachia, and in Asia Minor. The French berries are of small size; they are generally known under the name of Avignon berries; the berries coming from Turkey are called Turkey berries, and also Persian berries. It is the Persian berry which is most generally consumed in England; it is larger than the Avignon berry, and contains, weight for weight, a larger amount of coloring matter. Its coloring principle is easily soluble in hot water, and may be concentrated to a strength of 20° or 30° Twaddle;

during boiling the berries give off a peculiar sweetish odor;
the "berry liquor," if long kept, deposits a pale yellow starchy-
looking sediment, which appears to be nearly pure coloring
matter. With alumina and tin mordants berries yield a very
pure and agreeable yellow, which, however, is deficient in stability, not resisting well either soap or exposure to air; on this
account, and because quercitron bark, fustic, and chrome oranges
and yellows are cheap and manageable, Persian berries are
hardly ever used in piece dyeing, their application being confined almost exclusively to printing. In woollen or calico printing the berry liquor is scarcely ever used for producing a yellow,
though with crystals of tin a good yellow can be obtained.
The chief consumption of berries is as the yellow part for
greens; it yields brighter and livelier greens than either bark
liquor or fustic. It is used also for olives and in chocolates; added
to cochineal red in small quantity, it brightens the color, turning it towards the orange or scarlet; it seems to be used by the
French as a sightening for alumina mordants, but I have never
seen it so employed in Lancashire. Persian berries have been
slightly used in garancine dyeing to produce an orange upon
an acetate of tin mordant. The yellow lake extensively used
by artists and in paper hangings, called "*stil de grain*," and
manufactured in Holland, is made by preparing a decoction of
berries in alum, and precipitating it by white and pure chalk.
In preparing berry liquor for yellows upon silk or wool it is
desirable not to boil too long, nor to exhaust the berry; the
coloring matter which dissolves first is purest, and should be
taken off for yellows; the liquor obtained from the second and
third boilings answers very well for greens, olives, and chocolates. The pure coloring matter of berries is called rhamnine,
but Kane distinguishes two coloring matters, which he calls
respectively *chrysorhamnine* and *xanthorhamnine*.

Bichrome, or *Chrome*.—An abbreviation of bichromate of
potash. (See CHROMATE OF POTASH.)

Bile, *Ox Gall, Gall, Biliary Fluid.*—The biliary fluid of
oxen, under the name of gall, has been employed from the
earliest times as a suitable material for cleaning colored fabrics.
Looking at its chemical constitution, which is nearly the same
as soap, we are not at a loss to explain what its properties are
owing to. It actually operates as a very mild kind of soap,
dissolving grease and oily matters without injuring even the
most delicate shades of color. It can be dried and preserved
for an indefinite length of time, and dissolved in water as required. The uses which ox-gall receives in the fine arts may
possibly be extended to dyeing and printing; they are certainly
deserving the attention of the experimentalist. Some beauti-

ful, but evanescent, shades of color are produced by the action of sulphuric acid and sugar upon ox-gall.

Cow dung contains, besides the coloring matter of the bile, very frequently the biliary fluid itself. Some attempts were made to show that it had something to do with animalizing mordants, but that theory has been relinquished. According to an anonymous writer in the *Bulletin of Mulhouse*, ox-gall has a slight deteriorating effect when mixed with the water used in madder dyeing.

Birch.—The bark of the birch, or the birch broom, has been employed in dyeing, but principally by the peasantry. I have no exact information upon the nature of the coloring matter, but it is probable that these substances were valued on account of a small amount of astringent or tannic matter, which, with copperas, would stike shades of drab, gray, or olive; and with alumina mordants would give inferior yellows.

Bismuth.—Bismuth is a metal somewhat resembling lead. In a patent granted to Emile Kopp, July 10th, 1855, a claim is made, amongst others, for the use of aceto-nitrate of bismuth as a mordant for garancine. Prior to this date I had tried various salts of bismuth as mordants, but without obtaining any good result. The specification claims the production of bright crimson shades by means of the aceto nitrate of bismuth mordant, and dark crimson and purple crimson shades when it is used in combination with a nitric solution of arseniate of iron. By following the directions given I did not succeed in obtaining anything commercially valuable, and when I had an opportunity of seeing the results obtained by the patentee I found them no better than those I had produced: although highly ingenious, and somewhat novel, as the combinations were, for practical purposes they are of but little value.

Bixa Orellana.—The botanical name of the plant from which anotta is obtained. From the first part of the name comes the word bixine, the name of the supposed red coloring matter; and from the second part is derived orelline, the name of the yellow-colored principle of anotta.

Bixine.—Name given to the supposed pure coloring matter of anotta: also the name given to an improved preparation of the seeds of the bixa orellana, by which a much more powerful coloring matter is produced, devoid of the repulsive animal smell of the crude product, and giving equally good or better shades of color. A sample of the commercial bixine I examined was about three times as powerful as average qualities of anotta.

Black.—This is probably the most important of all dyed colors, whether viewed with regard to the universal use of it,

or the peculiar difficulties attending its production. In a philosophical point of view black is not a color. It is the absence
of color, or the extinction or absorption of all the colored rays
of light, which produces black. There is no purely black body;
such a body would be perfectly invisible, since it would neither
emit nor reflect any rays of light by which it could be seen. The
best blacks have always some shade of color discernible to the
practised eye; hence we distinguish jet blacks, brown blacks, blue
blacks, purplish blacks, red blacks, etc.: it is these shades which
make black visible. Black results from a mixture of all the
elementary colors; thus the artist by mixing red, blue, and
yellow pigments produces the neutral shades, which when weak
give gray, and when concentrated give black: the same mixture, when made so as to reflect light well, produces not black
but white. The famous family of the Gobelins, whose success
in dyeing was imputed to supernatural assistance, produced
their best blacks by a mixture of the elementary colors—red,
blue, and yellow. The cloth was dyed red with madder, then
dipped in the indigo vat for blue, and lastly finished in weld
to give the yellow shade; the whole producing a very perfect
and durable but expensive black. The earliest blacks we have
account of in England were the so-called "mathered blacks,"
being a madder red dyed upon a dip blue ground; but this was
very expensive, and could only exist by legal enactments, which
forbade the use of logwood in dyeing black. This law was
either repealed or neglected towards the end of the last century, since which period the chief ingredients in black are
galls, sumac, logwood, and salts of iron. I will first give the
methods of obtaining blacks in silk by dyeing and printing;
then blacks on woollen and mixed fabrics; and lastly, blacks
by dyeing and printing on cotton goods.

Black Dye on Silk.—The common cheap silks are dyed with
logwood and fustic for coloring matters, and some iron salt as
mordant; the better class of silks are dyed with galls; the use
or non-use of galls in dyeing black on silk divides the colors
into two classes. The logwood blacks are all distinguished by
turning immediately a bright red when a drop of spirits of salts
is put in contact with them; the galled blacks, even when
topped with logwood, do not give a red immediately, and then
it is of a dull purplish color.

Blacks on Silk without Galls.—The silk properly scoured is
worked for a greater or less time in some iron salt. I believe
the common nitrate of iron is as good as any of the many mordants in use, sometimes ordinary green copperas is used, but
that will not yield a deep black; a per-sulphate of iron is also
employed, and also a mixture of sulphate and nitrate; the ace-

tate of iron or common calico printers' iron liquor is also in use and answers very well; each dyer has his favorite mordant which he considers the best for the peculiar shade of black he wants. An hour is generally sufficient in the iron, the goods washed well in cold water to remove the unattached iron, and then worked in the logwood at a very moderate heat. To obtain a jet or brownish-black there should be about one pound of fustic for every five pounds of logwood. It is customary to add a little copperas to the dye vat to raise the color just before finishing, the goods being previously lifted.

Blue Black.—Mordant in nitrate of iron, raise in logwood, to which as much white soap has been added as will make a lather; no copperas must be added.

Black on Blue Ground.—Dye a Prussian blue, mordant in iron, raise in logwood with copperas at the end.

Deep Hat Black.—Work five pounds of silk in a decoction of

2 lbs. fustic chips,

1 lb. quercitron bark; lift, and add

6 oz. verdigris,

6 oz. copperas; work for fifteen minutes, and leave overhead all night; wash and dye in a decoction of 5 lbs. logwood, with as much white soap as will make a lather—(Napier).

Union Velvets—a mixed fabric in which the pile is silk and the back cotton; they are extensively dyed in the neigborhood of Manchester by one or other of the above processes. The cotton back is a very light color compared with the silk face, showing the different affinities of the two materials; no single process at present known will enable the cotton to take the same shade as the silk.

Blacks on Silk with Galls.—The silk having been scoured is steeped in a decoction of galls made from bruised galls by boiling; for 2 lbs. of silk 1 lb. of galls is taken, after twenty-four hours the silk is rinsed in water and dipped and worked in solution of copperas, afterwards it is worked in warm decoction of logwood, then again in the iron, washed out, and if the shade is not deep enough, the same process repeated as often as necessary. This is essentially the old process by which silk was dyed, and it is the existing process except that galls are replaced by other cheaper astringent matters. In France, chestnut wood and bark is extensively used in black dyeing. An infusion of the wood and bark is prepared, the silk is steeped in it for three or four hours, during which time the astringent combines with the silk, and the latter acquires a yellow nankeen shade; it is well washed and steeped in the iron bath which is kept at near the boil. The iron bath is composed of green copperas or iron liquor with some metallic iron intended to keep down the

acidity and supply iron to the bath as the silk withdraws it; a
certain quantity of gum is dissolved in this bath, also with the
intention of making it somewhat mucilaginous, so that the black
particles of tannate of iron may be held suspended in the
liquor: a little sulphate of copper is also added by some dyers.
The silk as it comes out of the bath is reddish, but speedily goes
black on exposure to the air. It requires four or six treatments
to obtain a good black—(Dumas). The Lyons dyers are stated
to find an economy of 50 per cent. by using chestnut extract
instead of galls, and to obtain better results.

The Genoese dyers were formerly celebrated for the goodness and fastness of the black colors they produced. They
used immense dye vats, which were never emptied, composed
of water, vinegar, sour beer or cider, oatmeal, alder bark, sumac.
oak bark, gall nuts, and metallic iron, along with various other
substances, the use or application of which modern chemistry
does not explain. Bancroft examined a sample of Genoa black
velvet, and found no blue basis as was supposed. These black
vats, or *tonnes au noir*, were being continually replenished, and
the sediment occasionally cleared out.

By galling, silk increases in weight, so that by repeating
several times the steeping in galls a very considerable increase
of weight can be communicated to silk, so much so, that it has
become a species of falsification, and not only is the twenty-five
per cent. of gum which silk naturally loses in scouring made up,
but sometimes another twenty-five per cent. in weight or more
is added to it. The deposition of so much foreign matter in
the fibre of the silk injures its wearing qualities.

The use of logwood in conjunction with galls is condemned,
for though it gives a fuller and more blooming color it speedily
becomes brown by wear.

According to Dumas, the practice of giving a dip blue bottom
to black is nearly abandoned. Prussian blue is sometimes used;
but logwood and copperas, with some sulphate of copper, are
chiefly employed to give a blue shade to blacks.

Black for Printing on Silk.—These colors are comparatively
simple, being derived essentially from logwood or galls as col-
ouring matter, and some salt of iron as mordant. The other
ingredients assist to develop or modify the shade.

Black for Silk.—Blotch.

1 gallon logwood liquor at 7° Tw.
10 oz. acetate of copper at 40° Tw.
16 oz. red liquor at 18°,
18 oz. starch: boil, and when cold, add
7 oz. nitrate of iron at 80°.

For blocking, size or Carragheen moss is a suitable thickening.

Black for Silk.—Roller or Block.—(Persoz.)

1 gallon logwood liquor at 14°,
10 oz. starch,
1 lb. 10 oz. British gum ; boil, and when cool add
10 oz. crystals of nitrate copper.
8 oz. proto-per nitrate of iron.

This color should age some time before steaming.

Gall Black for Silk.

1½ gallon logwood liquor at 5½°,
5 oz. powered gall-nuts; boil until reduced to
1 gallon mixed logwood and gall liquor,
1¼ lbs. starch; boil, and when cool add
1½ oz. alum,
5 oz. sulphate of copper,
1½ oz. sulphate of iron,
3½ oz. nitrate of iron at 80°,
4 oz. melted suet or lard.

Other receipts only vary in quantities or by additions of a little oxalic or tartaric acids; it is not necessary to multiply examples.

Black upon Wool by dyeing.—For the best quality of woollen goods the process consists in, first, giving a dark blue by means of the indigo vat, and then, after the cloth has been well washed, it is passed for an hour in a boiling decoction of sumac and logwood, using about 4 lbs. of sumac to 1 lb. logwood ; the strength of the decoction depends upon the weight of the cloth. At the expiration of an hour the cloth is lifted, and aired both to cool it and let the oxygen of the air act upon it ; in the meantime green copperas is added to the dye vat, about one pound to three yards of cloth ; the vat is cooled down until the hand can be held in, and the cloth entered again and worked for an hour, the heat being kept just below the scald. These operations are repeated three times, or until the cloth is saturated ; it is then well washed and finished. Blacks so dyed are very stable, and are said to have a very characteristic greenish hue, communicated by the blue bottom and the yellow of the sumac. This plan of dyeing is too expensive for the lower qualities of woollen cloth, which are dyed as follows : For about 70 lbs. of woollen cloth—

14 lbs. of logwood,
4 lbs. galls in powder,
2 lbs. fustic ;

Boiled together for half an hour, the vat cooled down, the cloth entered and moved about for four hours, during which time the vat is brought as near the boil as possible. The cloth is then lifted and aired; 4 lbs. green copperas are dissolved in the hot liquor, which is cooled down, and the piece entered again for an hour; this process is repeated until a satisfactory color is obtained. A variation of the above consists in adding sulphate of copper or blue copperas along with the green copperas; it produces a more lustrous black, which is, however, easily faded on exposure to air and light; acetate of copper or verdigris answers the same purpose, and is open to the same objection. In other processes no gall-nuts are used, but sumac instead.

Geneva Black.—M. Dumas, in detailing this process, says that this black is very fine, does not injure the wool, possesses a brilliancy which no other has, and can have a lively blue tint.

For a piece of cloth of about 40 yards, weighing 70 lbs.:—

         6 lbs. green copperas,
         6 lbs. tartar,
         1 lb. sulphate copper,
         2 lbs. fustic,
         2 lbs. logwood.

These materials are boiled a short time, and the cloth entered and worked at the boil for three hours, then washed; afterwards entered into a fresh vat, in which 11 lbs. of logwood have been boiled, and boiled for an hour; taken out, and again entered in the same vat for half an hour, and finished. It is impossible to see anything either in the materials or management which would make this black superior to the others. Tartar may be useful in preventing the wool becoming harsh, and it may modify the color; but it is not likely to assist the fixing of the iron, nor contribute to the general stability of the color.

Napier gives the following for 10 lbs. of woollen cloth: work for one hour in a bath, with 8 oz. bichromate potash, 6 oz. alum, 4 oz. fustic; wash well, and then work for one hour in another bath, with 4 lbs. logwood, 4 oz. barwood, 4 oz. fustic; lift, and add 4 oz. solution of copperas, work half an hour in this, wash and dry.

Richardson's patented process, May 16th, 1855, consists in boiling the woollen cloth in a mixture of bichromate of potash, tartar and sulphuric acid for an hour, this forming the mordant; then entering it in a vessel containing chiefly logwood, with a little camwood, fustic, sulphate of indigo, and sulphuric acid. The use of bichromate of potash in woollen dyeing has become very general within a few years, though its action is not clearly understood. Grumel's patent, April 8th, 1859, is for a black obtained by means of chromates and logwood.

Another receipt from Napier, gives 8 oz. camwood, work in twenty minutes, lift and add 8 oz. sulphate of iron, leave the goods in all night; wash out and raise in a bath containing 5 lbs. logwood and one pint chamber lye for an hour, lift and add 4 oz. copperas, work in this half an hour longer; wash and dry.

Black dyed broad cloth is nearly all sold as "woaded," an expression which originally indicated that the black had a fast blue basis derived from woad, a variety of indigo; afterwards the indigo vat foundation, having precisely the same value as woad for practical purposes, was substituted; but as the majority of the samples of "woaded" cloth that I have tested do not indicate any blue basis, it must be presumed the term "woaded" has received some new and conventional meaning. Genuine woaded black does not turn red when a drop of muriatic acid touches it; after a time it becomes purplish, because more or less logwood is always used; a common logwood black acquires a bright red color instantaneously by contact with a drop of acid.

Black for Printing on Woollen Goods.—Logwood is the basis of all black colors for printing on wool, iron is the chief fixing agent, nitrate of copper is the oxidizing agent, alumina is employed to modify the shade, extracts of other dyewoods are added occasionally, with a view to increase the intensity of the black, or give it a shade favorable to the contiguous colors. Here is a selection of receipts, with remarks:—

Black for all Wool.—Block.

8 lbs. calcined farina,
4½ quarts logwood liquor at 20°,
12 quarts water,
1 pint sapan liquor at 20°,
4 quarts red liquor at 10°,
3 lbs. crystals nitrate of copper,
2 quarts nitrate of iron at 80°,
1 quart acetate of iron at 14°.

By altering the thickening this color would serve for roller; instead of 8 lbs. calcined farina, 6 lbs. starch should be taken; the three last ingredients to be added only when the color is cold.

Blotch Black for all Wool.

4½ quarts logwood liquor at 6°,
4⅓ quarts peachwood liquor at 6°,
18 oz. starch; boil, and whilst warm add
6 oz. sulphate of copper,
4 oz. sulphate of iron,
6 oz. pasty extract of indigo; and when cold add
12 oz. nitrate of iron at 80°.

This is also a block color, and would be found too thin for machine.

Black for Merinoes, all Wool.

6½ quarts logwood liquor at 9°,
2 quarts blue archil at 10°,
36 oz. starch,
¼ pint gall liquor at 19°; boil, and add
2 oz. copperas,
2 oz. sulphate of copper,
12 oz. pasty sulp. of indigo; and when cold
15 oz. nitrate of iron at 80°.

A variety of blacks can be made by modification of the above receipts; the addition of ammoniacal cochineal is recommended in some, oxalic acid in small quantities is prescribed in others, some contain alum. I translate some receipts from Persoz, Dumas, and Thillaye. In French receipts it will be noticed that the logwood and other liquors are at a strength never seen in England; but as water frequently enters into the receipt, a compensation can be made by leaving out the whole or part of the water, to suit the strength of liquor obtainable.

Black for Objects.—Wool or Mixed Silk and Wool.

1 gallon boiling water,
½ gallon peachwood liquor at 22°,
1 gallon logwood liquor at 48°; add gradually
½ gallon water, in which has been dissolved
¾ lb. bichromate of potash; thicken with
3¼ lbs. starch,
4 lbs. gum substitute; and while hot add
1¼ lbs. sal ammoniac,
2⅜ lbs. acetate of cop.; when cooled a little add
1¼ lbs. oxalic acid; then mix very well with
¼ pint turpentine; and when quite cold add
3¾ lbs. nitrate of iron at 90°,
3¼ lbs. refined extract of indigo.

The use of bichromate of potash will be found to present great difficulties in practice; very few color mixers can manage to obtain a workable color by this receipt.

Black for Blotch and Objects.—All Wool.

1 gallon logwood liquor at 5°,
1 lb. starch; boil, and when cold add
1 lb. nitrate of iron at 80°,
4 oz. nitrate of copper at 80°,
Pint of gall liquor at 5°.

Another.

1 gallon logwood liquor at 5½°,
1 quart gall liquor at 8°,
1 pint archil liquor,
1½ lbs. starch; boil, and add while warm
10 oz. extract of indigo; when cold add
30 oz. nitrate of iron, which has been neutralized by addition of acetate of lead.

Black upon Cotton by Dyeing.—The old fast black upon cotton was obtained by giving a blue ground with indigo, then galling and working in sulphate of iron, sometimes with addition of logwood; alder bark, and other similar substances were also employed; and the goods usually finished in an emulsion of oil, to take off the harshness which iron mordants so generally communicate. Later on, what was called the Manchester black, was obtained by first steeping in galls or sumac, then working in the copperas vat, and afterwards in logwood containing some verdigris, and repeating these operations until the desired shade was obtained. Galls are now scarcely ever used; sumac, which is cheaper, being employed in substitution; and the processes, though almost infinite in details, consist essentially of steeping in sumac, then working in an iron bath, and afterwards raising in logwood. One method said to give good results, consists in steeping in sumac for twelve hours, then working through lime-water, and exposing to the air until the light green color at first produced passes to a dull heavy shade; the goods are then passed through solution of green copperas, and exposed to the air until they appeared black while in the wet state; if dried, they would be found to be only gray or slate color. To fill up the color the goods are passed into the logwood bath (some authorities say it is advisable to pass them through lime-water first) for a sufficient time; lifted, some copperas added and the goods raised in it; for light goods this suffices to produce a black, heavier goods require a repetition of the processes. A rapid continuous method of dyeing black on light goods is practised in Lancashire; the goods are passed through a decoction of catechu, then immediately into a solution of bichromate of potash, next into decoction of logwood, then into green copperas, and lastly through a decoction of some red wood, as camwood or Brazil wood. The order of these liquids may be changed within certain limits. A simpler method of dyeing by means of bichromates is also given, which consists in steeping the goods in logwood, exposing them to the air and drying, then passing them into bichromate of potash neutralized by crystals of soda, by which the logwood is "struck" of an intense black and fixed. Velveteens are dyed black by reiterated passages in logwood and green copperas until a dark brown is produced, then passed in sumac and sulphate of copper, with sometimes addition of peachwood or Brazil wood. Fustic is an ingredient in all dyes where a brownish or jet black is desired.

Black is one of the most difficult colors to dye, and no one but a practical man understands the difficulties of obtaining regular and good results, especially when first class colors are aimed at. It is useless to give weights and quantities when these are really only inferior elements of success; a slight change in the quality of the sumac, something different in the "ageing" or "mastering" of the logwood, some slight modification in the temperature and pressure of the "stills" in which the liquors are made, and other causes not more conspicuous, have frequently in my experience put works almost to a stand still. And when I have been called in for advice, it has been evident that chemistry could only give conjectures as to what was wrong. These failures in producing satisfactory colors would not be apparent to an unpractised eye; the defects would only consist in those hues and reflections of shade being wanting which were most esteemed and usually produced: Though it is exceedingly difficult in most cases to trace the actual cause of inferior results, there have been in my practice very evident occasions in which a most trivial and apparently unimportant cause has produced very embarrassing effects; the closest attention on the part of a foreman or manager is most essential in order that these things may be avoided, or if they occur that their cause may be discovered.

Black on Cotton by Printing.—The oldest black applied topically on cotton goods, was that called "chemical black," made from gall liquor and nitrate of iron. I have a receipt for chemical black dated 1804, with a patch annexed; the color has considerably faded, but not so much as a logwood black would have done. The receipt runs as follows:—

Chemical Black, 1804.

28 lbs. gall·nuts,
16 galls. tar acid (pyroligneous acid), boil for six hours and strain the clear liquor, make up to 16 galls. and thicken with 26 lbs. of good flour, add 14 gills aquafortis killed with iron nails (nitrate of iron), "then boil it well and get it out, or else it will go thin, cool it and it is fit for work." ·

Other receipts are very similar, but generally pure water is used to make the gall liquor, sometimes vinegar is prescribed ; the nitrate of iron is usually added after the gall liquor has been boiled with the thickening, and that is undoubtedly the preferable way. This black withstands a good deal of rough treatment in the way of dunging, dipping, and dyeing after-wards, and was much used in styles that had to be dipped and dyed after printing, as indigo blues, madder reds, weld yellows, etc.

All the modern steam blacks on cotton may be reduced to logwood liquor, thickening matter and a salt of iron ; other wood extracts and drugs may fulfil useful purposes with regard to the shade of black, contiguous colors, facility of printing. washing off, etc.; but the only essentials are the three materials mentioned.

Steam Black for Calico.

1 gallon logwood liquor at 6°,
1½ lbs. starch ; boil, and while hot add
5 oz. green copperas, stir well and when nearly cold add
2 oz. olive or gallipolli oil,
10 oz. nitrate of iron, well saturated with iron, or neutralized by addition of one·third its weight of acetate of lead.

Another Black.

3 gallons logwood liquor at 12°,
1 gallon red liquor   " 16°,
1   "   iron liquor   " 28°,
1   "   acetic acid   " 8°,
7½ lbs. flour,
3 lbs. British gum ; boil for half an hour.

The following black contains a large quantity of fatty matter, this is·added for the purpose of enabling it to temporarily re-sist the penetration of chemical agents which would injure or destroy it, but which are necessary to the development of some other colors printed along with it ; for example, it is used in conjunction with fast blue to be raised in soda ; it is used on'

Turkey reds which have to pass through strong solution of chloride of lime to produce a white, etc. The black is for a time waterproof. The prussiate may be left out at discretion, but it is used for giving a blue black.

Soda Black or Spermaceti Black.

2½ gallons logwood liquor at 12°,
1 gallon red liquor at 16°,
1 " acetic acid at 8°,
8 oz. yellow prussiate,
4 lbs. starch if for blocking; 6 lbs. if for machine, boil well and add a warm mixture of
1¼ lbs. spermaceti,
10 oz. gallipolli oil,
10 oz. turpentine, and when cold add
1 quart nitrate of iron.

Black colors for delaines are similar to those for wool; logwood and nitrate of iron being the chief ingredients, with sulphate of indigo for a blue material, and red woods for browning the shade. Receipts for black on delaine frequently assume an extreme degree of complexity, and at other times are nothing but logwood and nitrate of iron. Out of a great number of receipts, I give two as illustrations.

Black for Delaines.

3 gallons logwood liquor at 12°,
3 lbs. starch; boil well, and when cooled to 90° F. add
1 quart nitrate of iron at 84°.

Another.

1 gallon logwood liquor at 8°,
1 pint wood acid at 7°,
1½ pints of bark liquor at 10°,
2½ oz. extract of indigo,
¼ oz. bichromate of potash,
2 lbs. flour,
8 oz. British gum; boil, and when nearly cool add
4 oz. sal ammoniac,
½ pint of muriate of iron at 80°,
½ pint of nitrate of iron at 80°.

For madder black, see MADDER.

Recapitulation.—Blacks may be divided into the following classes:—

1. *Compound Blacks*, produced by mixture of separate

elementary colors, or the extreme condensation of one or two colors. This includes the ancient Gobelin black, the old English "mathered blacks," and the blacks of all kinds dyed on a blue basis. They are generally very fast and permanent colors, but for no other reason than that the separate colors from which they are produced are each the fastest and best of their kind. Indigo, madder, and weld give respectively the fastest blues, reds, and yellows, their combination gives, consequently, the fastest black; so madder and indigo without weld give an extremely permanent brownish-black; and all goods dyed black with an indigo ground and gall and logwood top are fast in proportion to the depth of that blue ground. Prussian blue is sometimes used as a basis for black, and peach-wood and Brazil wood for the red part; these are true compound blacks, but as the elements have no great stability, so the compound color itself is not a permanent one.

2. *Astringent Blacks*, derived from gall-nuts, sumac, chestnut wood, and similar bodies. These blacks owe their color to the formation of a dark colored compound produced by the combination of tannic or some similar acid with oxide of iron. They are very stable, resisting extremely well ordinary wear and exposure; but on account of the low covering power of this tannate of iron, and the consequent necessity of the accumulation of large masses of it upon fibrous material in order to produce a good black, it is rarely used except in combination with logwood.

3. *Logwood Blacks*.—The black pigment, produced by combination of the coloring matter of logwood and oxide of iron, has great depth and lustre. It fades, however, very rapidly upon exposure to light and air, going brown and rusty, and if there be not some more permanent black in combination with it, or a fast blue basis, cloth dyed with such a black is speedily injured. The comparative cheapness of logwood continually incites the black dyers to use too much of it in proportion to galls and sumac.

4. *Chromate Blacks*.—Neutral chromate of potash gives a deep black precipitate with logwood liquor, and several methods have been devised of forming the black compound on cloth; but it does not appear that this combination of coloring matter and oxide of chromium possesses any greater stability or powers of resisting atmospheric influences than the corresponding iron compounds.

Bleaching.—The word "bleach" is derived from a French word, which means "to whiten," and the rough meaning of bleaching is therefore whitening, in the sense of taking away the substances which color the material being bleached. The old method of bleaching consisted in washing with water, soap, and soda, and exposure to the air; it was a very slow process. Towards the end of the last century, a French chemist, Berthollet, discovered that the gas called chlorine, then itself but recently discovered, was capable of destroying vegetable coloring matter without injuring vegetable fibre, and in a short time it was practically applied. The present process of bleaching calico consists in, first, removing from it all greasy matters, dust, etc., which it has acquired in transit or manufacture, and then submitting it to the bleaching action of chlorine combined with lime. Cotton being so nearly white in itself requires but little chlorine to bleach it. The most important steps in the bleaching process are those which are undertaken to remove the greasy substances and mechanically adhering dirt not actually belonging to cotton in its natural state.

Bleaching for Madder Dyeing.—The method now generally used for the best bleaching for madder and garancine dyeing consists of the following operations:—

1. Singeing, followed by "rot steep" or "wetting out steep."
2. Liming—boiling with milk of lime and water from twelve to sixteen hours.
3. Washing out the lime and passing in muriatic acid sours, or weak vitriol.
4. Bowking in soda ash and prepared resin, ten to sixteen hours.
5. Washing out of the bowk.
6. Passing through solution of chloride of lime.
7. Passing through weak sours chiefly muriatic acid.
8. Washing, squeezing, and drying.

The singeing is not a part of the bleaching properly considered, it is merely to remove the loosely adhering filaments and so improve the cloth in appearance and for printing.

The "rot steep" (so called because the flour or size with which the goods were impregnated were formerly allowed to enter into fermentation and putrefaction) is intended to thoroughly wet the cloth; this takes some time on account of its throwing off water in places owing to greasy matters in it; if the cloth be not thoroughly moistened there is risk of irregularity in the after processes, and attention must be paid to this point.

The liming takes place in large kiers or kettles capable of holding from 500 to 1500 pieces of cloth; the lime is very carefully slacked some days previous to being used, and brought to a smooth milk of lime, being sieved so that no small lumps of quicklime should get into the kier; it is equally distributed upon the cloth as it enters the kiers, the cloth is pressed over-head in the liquor, and the boiling commenced and continued for a period of from twelve to sixteen hours. At the end of that time the lime liquor is run off and clear water run in to cool the pieces, which are then taken out and washed. The liming is usually performed at a low pressure; but a patent process where a pressure of 40 lbs. or more is used seems to answer very well and to save time. The apparent utility of liming consists in its acting upon the greasy matters, forming a kind of insoluble soap with them which is easily taken out by the subsequent processes.

The souring after liming removes all excess of lime and breaks up the insoluble lime soap referred to in the previous paragraph, still leaves the grease upon the cloth, but in such an altered state as to be easily dissolved in the bowking which follows. Muriatic acid sours are sometimes used in this sour-ing; but it is my opinion that common vitriol sours may be safely used, for any sulphate of lime which might remain in the cloth would be converted into carbonate by the soda ash.

The bowking or boiling with alkali and soap has for its ob-ject the removal of the greasy matters; it dissolves them, and all the dirt held by them now comes out of the cloth leaving the cotton nearly pure. The kind of alkali used is soda ash, the soap is made from resin and called prepared resin. The boiling in this case need not last so long as the liming, but de-pends in great measure upon the size of the kier and the number of pieces.

The last process of passing through clear solution of bleach-ing powder is to destroy the slight tinge of color of a buff or cream shade still adhering to the cotton; the bleaching powder solution is very weak, so that probably a piece of calico of the ordinary size does not take up more than the soluble matter from a quarter of an ounce of bleaching powder. The goods are allowed to rest some time with the chloride of lime in them, and then passed through sours for the final operation. The acid has the effect of setting the chlorine free from the bleaching powder and completing the destruction of the color; at the same time it removes the lime and acts upon any traces of iron that may be on the cloth. I think there is no doubt that muriatic acid makes the best sour for the last souring, both because it obviates the danger of the sparingly soluble sulphate of lime being fixed in the fibre and giving bad whites in dyeing, and also because it leaves the goods softer and more effectually removes any iron rust that may be on the cloth.

Bleaching for dyeing self colors need not be pushed to the

extent of madder work, and where the colors to be dyed are dark shades, such as blue, black, or brown, it is not necessary to have the cloth white; all that is required is to cleanse it well from foreign matters which would tend to make the dye uneven or irregular.

On the other hand, goods sent into the market as white goods must be of a pure color, and there is no necessity for that searching treatment to which madder goods are subjected; the shortest and least expensive means of making them white are adopted; if, however, the goods are not " well bottomed," they will not remain white long when brought into domestic use.

The proportions and strengths of the substances used in bleaching are not of much value, since circumstances must influence them very considerably; however, as a kind of guide, I may give the proportions used in one or two cases coming under my observation. For 14,000 yards of nine-eighth printing cloth 66 reed, there was used 250 lbs. of quicklime in the liming; the same quantity required 110 lbs. muriatic acid for the first souring. The bowking was done with 140 lbs. of soda ash at 48 per cent. alkali and 80 lbs. prepared resin (see RESIN). The last souring was vitriol sours at 3°; the quantity of bleaching powder used was not ascertained, but the solution stood at 1° Tw. A French process communicated by a friend gives only 150 lbs. of lime to 66,000 yards of calico weighing about 13,000 lbs., the liming lasted eighteen hours; 400 to 500 lbs. muriatic acid was used in the souring, and the bowking was continued for the long space of thirty-six hours.

On the continent, caustic soda is frequently used in bowking, perhaps generally; it requires much care to prevent damage to the fibre: sometimes the sours are used warm.

Linen is not so easily bleached as cotton, and it appears to suffer considerably by boiling with lime, and by contact with chloride of lime; it is mainly bleached by continual boilings with alkali and a few sourings, with a chloride of lime treatment; or, as lime appears injurious, the chloride of potash or soda is frequently used instead.

Bleaching of Woollen.—Woollen goods are bleached by treating with very mild alkaline liquors, which remove the fatty matters; putrefied urine and soap, with crystals of soda, being the only substance usually employed. Sulphurous acid, or vapors of burning sulphur, are used to finish wool, giving it whiteness and lustre. The following is one of the processes given by Persoz, as followed in France for bleaching woollen for printing. It is for 40 pieces, each 50 yards long:—

1. Passed three times through a solution of 25 lbs. carbonate of soda and 7 lbs. of soap, at a temperature of 100° F : freshen up with ¾ lb. of soap every four pieces.
2. Wash twice in warm water.
3. Passed three times through a solution of 25 lbs. crystals of soda, at a temperature of 120°: freshen up with ¾ lb. crystals for every four pieces.
4. Sulphured in a room for twelve hours, using 25 lbs. sulphur for the 40 pieces.
5. Passed three times through crystals of soda as in No. 3.
6. Sulphured again as in No. 4.
7. Crystals of soda again as in No. 3.
8. Washed twice through warm water.
9. Sulphured a third time as in No. 4.
10. Washed twice in warm and then in cold water.
11. Blued with extract of indigo according to taste.

According as the goods are meant for dark or blotch styles, or for fancy styles, so the process may be shortened or must be adhered to.

Bleaching of Delaines.—This is carried on upon precisely the same principle as bleaching wool, but does not require so many operations; two passages through soap and soda crystals, washing in warm water and repeating the soaping, then sulphuring by Thom's patent for twenty minutes twice over, is usually sufficient for all styles.

Bleaching of Silk.—Nothing but soap and sulphur are used in silk bleaching, excepting a slight amount of soda crystals, which helps to save soap. Alkalies destroy or injure the fibre of silk very much, and must be either avoided or applied with extreme care. Bran is sometimes used along with soap in order to neutralize any excess of alkali which it might contain and the process terminated by passing in an extremely diluted sour, so weak as scarcely to be acid to the taste. Sulphuring is not necessary when the silk is to be printed or dyed dark colors, and in any case must be cautiously and sparingly applied.

Bleaching Powder, or *Chloride of Lime, Chemic ;* sometimes also *Oxygen, Hypochlorite of Lime.*—Ordinary bleaching powder is made by slacking lime to a fine powder, and exposing it to chlorine gas in properly constructed chambers ; it absorbs the chlorine in large quantity, and gives it up again when treated with acids which seize the lime. Good chloride of lime is dry and dusty, very white, and does not smell very strong: in a dry place it keeps good for a considerable period; in a damp place it absorbs moisture, becomes pasty, gives off chlorine gas, and loses strength: it is not entirely soluble in water, always leaving a sediment. The clear solution, when in quantity, has a greenish color; it is slowly injured by air, heat, and light, and should consequently be kept in a cool, shady place, and covered up.

Testing of Bleaching Powder.—The precise quality and value of a sample of bleaching powder cannot be ascertained without chemical testing; and as it is a substance liable to great variations, it is very desirable to have some means of ascertaining its value. The processes given in chemical works are quite satisfactory, but require several apparatuses only found in a laboratory. I give here a process suited to a color-shop, which will enable a practical man to tell whether the bleaching powder is below a certain standard or not. The materials required are fresh crystals of tin, spirits of salts, and a weak solution of extract of indigo, with jugs to mix them in. Weigh two ounces of fresh crystals of tin, and mix them with half a pint of water and a glassful of spirits of salts, and stir till dissolved. Weigh out two ounces of the sample of bleaching powder and mix it with a half pint of water, crushing all the lumps; when properly mixed pour it slowly into the tin solution, stirring very well until it is all added; blow off the gas from the liquor, and if it now smells very strong of chemic, and bleaches some of the extract of indigo liquor dropped in, it is a sign that the sample is not, at least, very bad; but if it does not smell of chemic, and does not bleach the blue extract, it is a sign that it is weak. Two ounces of a first-rate bleaching powder will stand mixing with two ounces and a quarter of crystals of tin and still smell strong of chemic, and bleach extract of indigo liquor.

Testing of Bleaching Liquors in the course of use.—It is frequently required to know how much strong liquor should be added to a partly spent solution of chemic to bring it up to proper strength again. The method in use in Lancashire consists in ascertaining how much of a certain solution of sulphate of indigo a given quantity of the liquor can bleach; and as the quantity of the original stock which can bleach it is known, a tolerably correct idea of how much strong liquor is to be added is arrived at. Mr. Crum devised a simple practical method, depending upon the color which chloride of lime communicates to a mixture of muriate of iron and acetic acid. Twelve white glass phials of equal size are obtained, and a mixture of equal measures of muriate of iron at 40° and acetic acid at 8° being prepared, an equal measure of it is put into each phial; if the phial be four and a half inches high, the mixture should stand only half an inch, that is one-ninth of the height. There are now prepared twelve strengths of bleaching liquor, beginning with the full strength used in bleaching, and going down, by regular weakening with water, to the weakest strength the liquor is likely to be brought to in use, and the bottles are filled up with these liquors, corked, numbered, and preserved as standards for comparison. The color of the liquor in the bottles is proportioned to the strength of the bleaching liquor, and by taking a similar phial, putting in the same amount of aceto-muriate of iron, and filling up with a sample of bleaching liquor of unknown strength, a shade of color will be produced which must be like one of the twelve standards; the strength of the liquor examined will then be the same as that with which the bottle was made up. (See CHLORINE.)

Bleaching Liquor.—This fluid is essentially the same as a solution of bleaching powder, though made somewhat differently. Instead of passing chlorine gas over dry slacked lime, it is made to traverse cisterns filled with a mixture of lime and water. It has precisely the same properties as the solid powder, although some persons seem to think it preferable. Its value cannot be correctly ascertained by the hydrometer, because common salt is frequently mixed with it to make it stand high on the glass.

Block Printing.—The difference between block printing and cylinder printing resides in the fact, that while the block not only deposits the color upon the cloth, but to a greater or lesser extent forces it in, the cloth in cylinder printing has to absorb the color mainly by capillary attraction, since the wrapping on the bowl does not generally suffice to press the cloth completely into the engraving of the roller. The same colors will not answer indifferently for block and roller. Block colors can usually be worked much thinner than machine colors, and it is possible to apply colors by block that it is very difficult to work in a machine, such as contain insoluble matters like pipe clay, sulphate of lead, etc. For dark shades upon woollen cloth the block has an undoubted advantage over the cylinder, because not only does wool demand much more coloring matter than cotton to produce a similar shade, but it does not draw it up so quickly; its fibres are not wetted so soon as those of cotton, and consequently it does not take up the color from the engraving in sufficient quantity. Dark blues, chocolates, greens, etc., on the finest class of French woollen cloth, require blocking twice or three times to apply sufficient color to give rich dark shades.

Blood.—The blood of oxen has been used for a long time in dyeing Turkey reds. It seems as if it was expected that some of the red color of the blood would be absorbed by the cloth, enhancing its shade; but there is not the slightest ground for such a belief. If blood be really of any use in the dye, it will be probably owing to the presence of the serum and fibrine, substances coagulable under certain conditions, and possessing characters somewhat resembling albumen. In the old hanging stoves of the calico printing, the hangers frequently tore their fingers with the hooks, and blood would get on the pieces and would dye up in madder of a dull brownish-red color, showing that blood acted as a mordant. In the old receipts for Turkey red about as much ox blood as madder is directed to be used, and in some cases the weight of blood would be double the weight of madder; there can be no doubt that this quantity of blood would have an influence of some kind, although it is not exactly known in what it consists.

A species of albumen called blood albumen is prepared from blood, and answers most of the purposes of the egg albumen.

Blue Colors.—Under this head I bring together the various processes in use for producing blue colors upon silk, wool, and cotton; where the explanations of the chemical actions do not seem sufficient, reference must be made to the drugs used, where their properties are more fully described.

Blues upon Silk by Dyeing.—The earliest dyed blues on silk were from the indigo vat, these are probably never produced now; they fell at once into disuse upon the discovery of the method of fixing Prussian blue upon silk, which was the next blue in chronological order. Saxony blue or sulphate of indigo blue was early in use for light shades; within these two or three years artificial blue colors prepared from aniline or similar bodies have been largely used.

For dark Prussian blues the silk is mordanted in a per-salt of iron and a salt of tin. In England, nitrate of iron is generally used as the iron mordant. In France, a species of per-sulphate of iron made by dissolving green copperas in nitric acid is used, it is known under the name of "Raymond's solution." In England, the tin salt employed is usually the common crystals of tin, but it is found useful to have the tin present as sulphate in order to allow of the tin combining easily with the silk; for this purpose sulphuric acid or sulphate of soda must be used in combination with the tin. A method yielding excellent results consists in taking the quantity of crystals of tin to be used, and pouring upon them their own weight of strong vitriol and stirring up and then dissolving the pasty mass in water; this may be considered as a solution of sulphate of tin in muriatic acid. The nitrate of iron may be mixed with this or may be added separately to the dyeing vessel. The silk is worked in the mixture of tin and iron in the cold, and then passed through clear water to remove all loose mordant. The color is raised in another vat which contains yellow prussiate of potash and made sharply acid by addition of either vitriol or spirits of salts, the silk is worked here until it has taken all the color it can, then rinsed in water and put through the same process again, even three or four times for the fullest shades. A final passage in alum and a little vitriol is thought to brighten the shade. It is necessary to wash the silk rather roughly, or else a quantity of loose uncombined Prussian blue will be dried up in the fibres, which will make the silk feel harsh and cause it to be dusty, besides injuring the color. Washing between the mordant and prussiate is recommended for obtaining regularity of shade and keeping the lustre and softness of the silk in its best condition. Some dyers, however, do not think this necessary, and merely drain the goods between the different processes. When a large quantity of tin is employed, the blue acquires a reddish shade, if the tin is deficient it has a greenish shade. Some blues are produced from red prussiate of potash, these require the protonitrate of iron for mordant.

Light sky blues are obtained by refined extract of indigo, with a little alum and sulphuric acid.

Aniline Blues.—The new blue coloring matters which yield magnificent shades are produced by working the silk, without any mordant, in the coloring matter. Most of the blues at present in use require raising in warm vitriol sours to take off a reddish hue which exists on them after dyeing; in some cases the vitriol may be added to the dye, and the operation completed at once. The best Prussian blues cannot compete with the azuline blue in softness and brilliancy; they are tolerably stable, and leave nothing to desire but a reduction in price.

Bilberries, elderberries, mulberries, whinberries, and privetberries have been used to give blue shades on silk, and are still employed on a small scale.

Blues upon Silk by Printing.—The blues obtained by printing on silk are derived from sulphate of indigo chiefly, dark blues from prussiate, some shades of blue are produced by logwood and copper salts.

Logwood and Extract Blue for Silk.

1 gallon logwood liquor at 16° Tw.,
1 gallon red liquor at 16°,
10 lbs. ground gum; stir till all dissolved, and add
10 oz. tartaric acid,
10 oz. nitrate of copper,
1 gallon extract of indigo.

This produces a violet blue on account of the red liquor and logwood modifying the extract.

Extract Blue.

1 gallon water, hot,
3¼ lbs., more or less, according to strength, extract of indigo,
½ lb. alum.
1 lb. tartaric acid,
6 lbs. gum, or less, according to thickness required.

Prussiate Blue.

3 lbs. yellow prussiate potash,
1 gallon warm water, dissolve, and add
1¼ lbs., tartaric acid; cool, and thicken the clear liquor with
7 lbs. gum in powder, and add
2½ lbs. bichloride of tin at 80°.·

The steam blues given for woollens and delaines will be found applicable to silk, but will stand bringing down with gum water.

Blue Colors by Dyeing upon Wool.—Wool is dyed blue: (1) by the indigo vat; (2) by sulphate of indigo; (3) by prussiate; (4) by logwood; (5) by the new blue colors azuline, cyanine, etc. The first method, which gives the fast and permanent but rather dull blues used in the army and navy, presents no other difficulties than occur in setting the indigo vats, for which reference must be made to INDIGO. The yarn or cloth properly cleansed and wetted out is dipped in the vat, left in for not more than a hour, and then lifted and aired, to be dipped again if deeper shades are required. The wool takes up a considerable quantity of indigo, which being a very expensive material, has induced many parties to try and save by only half dyeing with indigo and then finishing or topping with logwood. This species of adulteration is detected by putting a drop of strong acid upon the cloth: if all indigo, no change takes place; if logwood is present, a violet, purplish, or reddish color is immediately produced. Indigo blues are also topped with archil, which gives them an agreeable bloom, but which fades directly in air and light, and is immediately washed off by soap.

The sulphate of indigo blues are of very simple application; the extract is mixed or dissolved in the water, to which is added some alum and some acid, sometimes tartaric acid or cream of tartar, and sometimes sulphuric acid; occasionally, also, oxalic acid is used. Only light shades of blue can be thus dyed, and they have a greenish shade when compared with Prussian or azuline blue. Logwood is frequently combined with this kind of blue, and yields dull grayish blues.

The prussiate blues upon wool are very good colors, and when properly done possess a fair amount of stability; there are several methods of producing them, all of which will be included under one or other of the following processes. The ordinary method consisted in working the wool in nitrate of iron, and then in yellow prussiate of potash, acidified with sulphuric acid; the shades thus produced are remarkably improved by adding a salt of tin to the iron; in fact, no really good and dark blues can be obtained without a considerable portion of tin being fixed upon the wool. The salt of tin and nitrate of iron are mixed, and the cloth worked in for half an hour or more, and then taken to the prussiate bath, which is worked hot; if the shade is not deep enough the process is repeated. Very fine royal blues are obtained from first working in a mixture of muriate of iron and muriate of tin, and then in red prussiate of potash liquor; repeating the processes until the required depth of shade is obtained. Dumas recommends in all cases a little red prussiate to be used with the yellow, added towards the end; it strikes a blue with iron which has been deoxydized by the wool, and thus takes off the greenish shade of blues dyed with yellow prussiates only. Another process of obtaining blue consists in doing without iron salts altogether, and resembles almost exactly the prussiate steam blue for woollen and delaine, and depends upon the decomposition of the prussiate itself under the combined influence of acids, heat, and air. For a piece of thin woollen cloth, seventy yards long, the following materials are employed according to M. Dumas:—

12 oz. yellow prussiate of potash,
12 oz. sulphuric acid,
17 oz. alum.

The whole dissolve hot in a sufficient quantity of water to turn the piece through in an apparatus like a "jigger," from twelve to twenty gallons; the piece is worked in at a temperature of 100° F. for the first hour, at 140° for the second hour, and raised to the boil during the third hour; about half way in the last hour the piece is lifted in order to add about half an ounce of crystals of tin, and then entered again. The piece is then washed, and afterwards turned for an hour through a cold mixture of alum, sulphuric acid, and crystals of tin. This is evidently a costly process, but it is difficult otherwise to obtain regular, even, light shades of blue. I found that by first preparing the wool with stannate of soda, very good blues could be obtained by this process, with much less time than given in the above directions.

Logwood blues are so loose and deceptive as to have been at various times prohibited by law; they can be made to imitate indigo tolerably well, and are sometimes sold as indigo blues. I believe a law passed in the twenty-third of George III., imposing a fine of £20 per piece for dyeing blue from logwood and copper salts is still unrepealed; but of course, not enforced. The process of obtaining this blue consists in aluming with tartar and alum, and then dyeing in logwood to which sulphate of copper is added; or mordanting in alum, tartar and sulphate of copper, adding logwood and dyeing, finally raising with sulphate of copper. A good many blues on woollen consist of this logwood blue dyed on a light indigo blue ground.

Aniline Blues.—Aniline blues are extremely simple to work; the coloring matter is properly diffused in water with addition of acid, and the goods worked in until the color is exhausted; afterwards they are passed in warm dilute sulphuric acid to improve the shade.

Blue Colors by Printing on Wool.—Sulphate of indigo is the chief coloring matter employed for printing blues, alum and acids being used in combination; when it is desired to have the blue of a reddish hue, ammoniacal cochineal is added. For deep royal blues, prussiate of potash in combination with acids and tin salts is employed; for these blues, the cloth should be previously prepared with some preparation of tin. (See Preparation.)

Deep Blue for all Wool.

2 quarts water,
6 oz. starch; boil, and while warm incorporate
12 oz. pasty extract of indigo,
5 oz. alum,
2 oz. tartaric acid,
3 oz. oxalic acid.

Since the quality of extract or sulphate of indigo is extremely variable, it is evident that receipts in which it is a chief or important ingredient must be of a rather vague character, and merely approximative in the quantities given.

Ordinary Dark Blue.

1 gallon gum water,
6 oz. extract of indigo,
8 oz. alum,
3 oz. oxalic acid,
¼ pint cochineal liquor.

Any further receipts for this kind of blue would only differ from these two in the thickening or the quantities of material used, which are partly influenced by the shade to be produced and partly by caprice. The red part, however, may be increased to a considerably higher proportion than given in the receipt above with advantage for certain shades of color.

Dark Royal Blue—All Wool, Block.

1 gallon water,
13 oz. alum,
16 oz. oxalic acid, dissolve and thicken to style with, say
7½ lbs. gum, when cold add
¼ lb. bichloride of tin,
2½ lbs. red prussiate of potash,
13 oz. per-nitrate of iron at 80°.

There are many modifications of this receipt, but as the steam blues given below for delaine may be all applied upon wool, it is not necessary to detail them here.

Steam Blues for Delaine, applicable also to Wool.—Dark Blotch Blue.

4 lbs. starch, more or less according to requirements,
8½ gallons water,
1½ gallon red prussiate liquor at 80°,
8 pints tragacanth gum water; mix, boil, and while hot add
1½ gallon prussiate of tin (tin pulp),
4 lbs. tartaric acid,
6 oz. oxalic acid, and when cold add the clear liquor from
8 lbs. prussiate of potash,
8 lbs. tartaric acid,
2½ gallons hot water.

Another.

1 gallon water,
2 lbs. starch; boil well, and add while hot
10½ oz. muriate of ammonia,
2 lbs. 10 oz. yellow prussiate of potash,
1 lb. 5 oz. red prussiate of potash; when cold add
8 lbs. tartaric acid,
1 gallon prussiate of tin pulp.

Another Dark Blue.

Precisely the same as the last, except the addition of 5½ oz. of oxalic acid after the tartaric.

Dark Royal Blue—Delaines.

5 lbs. starch,
2 gallons water,
2 gallons chloro-prussiate liquor at 30°,
1 quart tragacanth gum water; boil, and add
6 quarts prussiate of tin,
2¼ lbs. tartaric acid,
6 oz. oxalic acid; when cold, add
8 lbs. yellow prussiate of potash,
10 lbs. tartaric acid.

Light Blue—Block Delaine.

3 quarts water,
½ lb. starch,
¾ lbs. tragacanth gum water; boil, and when cold add
1 quart red prussiate, liquor at 30°,
2 oz. tartaric acid,
3½ oz. bichloride of tin at 100°,
1 lb. prussiate of tin pulp.

It is hardly necessary to say that these are all steam colors, and require raising either in bichrome or chemic before washing off.

Blue Colors by Dyeing upon Cotton.—The chief blue upon cotton by dyeing is from indigo fixed by the vat; the skill in dyeing these colors rests principally in the preparation of the solution of indigo, which each dyer has to make for himself. The production of the indigo styles forms, therefore, a separate subject which will be treated under INDIGO.

Prussiate colors upon cotton goods are obtained by nearly the same process as upon silks; for dark shades the cloth should be prepared by steeping in stannate of soda at 14° Tw., wringing out and passing in vitriol sours at 4° Tw.; this gives a good basis of tin and shortens the time of dyeing considerably. Next the cloth is worked in nitrate of iron of a strength proportioned to the shade required, about thirty minutes will suffice to fix iron enough for a medium shade; the goods are rinsed and the color raised in yellow prussiate, sharpened with vitriol or spirits of salts; if the shade is not deep enough, the process must be repeated (but not the preparation), and crystals or muriate of tin may be mixed with the nitrate of iron bath. For sky blues no tin is required; but for deep blues it is necessary either in the preparation or mixed with the nitrate of iron. It is generally considered that the blues are brighter and softer when they are finished off in weak clear alum water than when simply washed off in common water.

Napier gives the following as a logwood blue upon cotton, the materials being for 10 lbs. cotton. A light but fast blue is first dyed in the vat from indigo, the goods are put in a decoction of 2 lbs. sumac for several hours, and then worked for fifteen minutes through water containing one pint red liquor and one pint iron liquor; wash from this in two tubs full of hot water, then work twenty minutes in a decoction of 2 lbs. logwood, lift and raise with half pint red liquor, work ten minutes longer, wash and dry. Since part of the blue color here is derived from indigo which is quite fast and another part from sumac which is tolerably fast, this blue will be of moderate stability, but of a heavy dull shade compared with Prussian blue.

Girardin gives a process used in France for obtaining a blue on cotton as follows: For 100 lbs. cotton take 5 gallons logwood liquor at 4°, 2 ounces of bichromate of potash, and 5 ounces of muriatic acid; the cotton is entered cold and gradually brought to the boil. It is not clear whether this is actually to dye cotton or merely the finishing of a dye began with acetate of copper and logwood liquor.

Blue Colors upon Calico by Printing.—Excluding those blues which are derived from indigo, and which will be found under INDIGO; the only common blues are derived from the prussiates and the receipts given for delaines will answer perfectly well for calicoes. In order to obtain good blues the cloth must be well prepared with tin in some form or other (see PREPARATION and STANNATE); for light blues this is not so essential. I give here a few receipts for blues on calico not applicable to delaines.

Steam Blue for Calico.

3 gallons water,
4 lbs. starch ; boil, and add
1 lb. muriate of ammonia,
6 lbs. crystals bisulphate of potash,
4 lbs. tartaric acid,
4 lbs. yellow prussiate potash,
8 oz. oxalic acid,
1 gallon prussiate of tin. (See TIN.)

This blue reduced with gum water of suitable thickness yields the light shades required. Whenever sulphuric acid or bisulphate of potash are used in blues, considerable care is required to prevent corrosion or burning of the cloth; the mixing must be scrupulously attended to, for if any of this acid be left free it is sure to injure or rot the cloth. For the cheaper styles of work sulphuric acid may be used with economy instead of tartaric acid, but the mixing of the colors must be carefully watched.

Another Blue for Calico.

1 gallon water,
1¼ lb. starch ; boil, and add
3½ lbs. tartaric acid,
10 oz. oxalic acid,
3½ lbs. yellow prussiate ; and when cold
½ lb. oil of vitriol.
1 pint prussiate of tin pulp.

Spirit Blue for washing off simply.

1 gallon water,
1½ lbs. starch ; boil, and cool to 110° F.,
1 qrt. Prussian blue pulp (see BLUE PRUSSIAN),
¾ pint oxymuriate of tin.

Common Blue, Standard.

2 gallous water,
4 lbs. yellow prussiate of potash,
12 oz. alum,
24 oz. oil of vitriol at 169°.

Common Steam Blue.

2 quarts gum water,
1 quart blue standard,
Extract of indigo to sighten.

All receipts for blue will resemble one or other of the receipts given in this article; the processes may be much varied in detail, but the usual method of mixing the color for machine consists in boiling the water and starch, and, while quite hot, stirring in the powdered prussiate and sal-ammoniac ; then, when the color has somewhat cooled, stirring in the ground tartaric acid (or the bisulphate); and when almost cold, the oxalic acid is added; and last of all the prussiate of tin pulp is well incorporated. There is always formation of bitartrate of potash in the best steam blues, which is disseminated through the mass in small crystals; but if the color is pretty hot when the tartaric acid and prussiate of potash are mixed together, the crystals are apt to be of some considerable size unless the color is well stirred until nearly cold ; this is objectionable for many reasons, and should be obviated by so managing the mixtures that the stirring is continued until the color is cold ; the crystals are then so small that they are not observable. The prussiate of tin pulp being added last, and cooling down the color will usually prevent large crystals forming; but if once formed they are difficult to strain out, and the color should be warmed up to about 120° F., and cooled quickly, with constant stirring. See POTASH PRUSSIATE, etc., for explanation of the chemical changes involved in the production of these colors.

Blue Azuline. (See AZULINE.)

Blue Azure, *smaltz, zaffre.* (See AZURE.)

Blue Chemic.—Name frequently given to sulphate of indigo or extract of indigo. (See INDIGO SULPHATE.)

Blue, China.—A style of blue obtained from INDIGO, which see.

Blue, Chinese.—A variety of Prussian blue is sold under this name which is soluble in oxalic acid, and which has been largely used in finishing printed calicoes. (See BLUE PRUSSIAN.)

Blue, Cyanine.—The same as QUINOLEINE BLUE, which see.

Blue, Dip.—The name of dip blue is given to the variety of styles produced by dipping cotton goods into indigo properly dissolved by means of lime and copperas. (See INDIGO.)

Blue, Distilled.—This curious name is given to a purified solution of sulphate of indigo, obtained as follows: Crude sulphate of indigo is dissolved in water nearly boiling, and a quantity of old but clean white flannel or other woollen articles worked in it until saturated with color, then washed well in cold and afterwards in warm water until the color begins to "bleed," that is, until the washing water begins to remove the blue and become tinged with it; the woollen rags or flannel are then washed sufficiently; they are then treated with hot water containing a feeble proportion of carbonate of soda, about half a pound of crystals to 10 gallons of water; this removes the blue color very rapidly from the woollen rags, leaving them of a dull brown color. The blue thus dissolved is considered as being purified on the one hand from hurtful substances soluble in water, which are removed by washing the wool, and from a reddish coloring matter which is retained by the wool and its shade improved. A little acid being added to the extracted blue enables it to dye up a good clear blue upon silk or woollen. (See INDIGO SULPHATE.)

Blue, Fast.—The conventional name for one of the loosest colors obtained from INDIGO, which see.

Blue, Finishing.—The use of blue in finishing is to counteract the cream color which most bleached goods possess; this cream color may be considered as a very pale orange and compounded of red and yellow; the addition of blue with a strong

8

reflecting white surface beneath neutralizes the shade and produces what passes for white. But this point is practically impossible to hit, and all blued goods have always an excess of blue. Each market has its own peculiar prejudice as to shade, and so in accordance various finishing blues have to be used. This apparently trivial matter is frequently a source of the greatest perplexity to the bleacher and finisher, so that a great number of blues for finishing are in the market. These consist chiefly of indigo in paste, being simply indigo very finely ground; sulphate of indigo in a more or less imperfect state, various kinds of Prussian blue in solution or suspension, and also preparations of smalts and ultramarine.

Blue, Opaline.—A new product of chemical art has been so called from its yielding a shade of color like the blue opal. Its color upon delaine is of nearly the same shade as China blue upon calico, but infinitely more lustrous and beautiful. The process of obtaining this coloring matter is kept secret, but there is no doubt that it is obtained from aniline or some similar body.

Blue, Parisian, or *Bleu de Paris.*—Name given to a blue compound produced by the action of bichloride of tin upon aniline at a high temperature and under pressure. The process was published in 1861 by Messrs. Persoz, de Luynes, and Salvetat.

Blue, Paste.—This name is usually intended for sulphate of indigo, it may sometimes mean Prussian blue in a pasty state, the context will show which blue is intended.

Blue, Pencil.—A particular kind of blue obtained from indigo, and so called because formerly applied by means of a modification of an artist's pencil. (See INDIGO.)

Blue, Prussian.—This color, which was one of the earliest contributions of chemistry to the list of artificial coloring matters, was obtained by accident in the capital of Prussia in 1710.; but it was nearly one hundred years afterwards before any good process was discovered for fixing it upon textile fabrics; and it is hardly twenty years since the present means of fixing it as a steam color was discovered and put into practice. Accepting prussiate of potash as the correct name for the salt so known, then Prussian blue is a prussiate of iron, and the readiest way of producing it is to mix together a solution of iron and prussiate of potash, when it forms as an insoluble pulp which can be drained, washed, and dried. There is more than one kind of Prussian blue, and there are several methods of preparing it. I give receipts of some methods used on print and dye works when Prussian blue is required to be made either for finishing or color mixing.

Prussian Blue for Finishing.

6 lbs. green copperas,
1½ gallons water, dissolve;
6 lbs. yellow prussiate of potash,
1½ gallons water; dissolve separately and mix
   with agitation, add to the whole
1 lb. oil of vitriol,
24 lbs. spirits of salts, stir up well and let stand some hours; the sediment will have a very pale blue color, to bring it up to full shade it must be oxidized, which is most conveniently accomplished by clear solution of bleaching powder of chemic. Take a rather weak solution of chemic and add it gradually to the liquor, stirring all the time until it begins to smell decidedly of chlorine; it is then time to stop putting in the chemic. The blue which is now of an intense dark color is left to settle; the clear drawn off and fresh water poured upon the blue to wash it; this repeated several times until all the acid is removed, leaves the blue fit for use. If warmed with a small quantity of oxalic acid it partially dissolves and forms a clearer color.

Prussian Blue for Spirit Colors.

4 lbs. prussiate of potash,
1 gal. water; dissolve, and separately dissolve
8 lbs. green copperas in
1 gal. water; mix the two solutions, and add
1 quart nitric acid.

Leave some hours, then wash three times by decantation, and drain on a filter to a paste. The nitric acid here acts the same part that the bleaching powder did in the previous receipt. Prussian blues are made immediately by mixing pernitrate of iron and yellow prussiate, but the product does not answer so well because it does not dissolve in oxalic acid or tin salts so easily as that prepared by one of the above methods. The reason of the methods of dyeing blue with pernitrate of iron and yellow prussiate will be now intelligible; the cloth takes iron from the nitrate, and then when brought to the prussiate it acts upon it, producing the blue; but this would not take place unless the prussiate was acid, because then the iron and it would never come into actual contact. The insoluble blue powder being formed in the pores of the cloth is fast, but if the cloth has been worked in the blue ready formed, the color would only have been on the surface and easily washed off.

Red prussiate of potash and green copperas give at once a

fine dark blue; red prussiate and per-nitrate of iron give a dark olive color, which becomes a splendid blue upon addition of muriate of tin.

The mere exposure of prussiate of potash mixed with an acid to heat and air produces a kind of Prussian blue without addition of any iron, and it is from this reaction that our finest blues are obtained. The chemical changes which take place are not clearly understood; but it is known that prussic acid is evolved, and probably some of the iron which naturally exists in prussiate of potash forms the basis for the blue.

Blue, Quinoline, or *Cyanine.*—This was an artificial blue color, discovered by Greville Williams, made from a refuse product obtained in the manufacture of quinine; its production was the result of exquisite chemical knowledge, it yielded very fine colors on silk; but they were so susceptible to the action of strong light as to be entirely useless. I have seen a magnificent blue velvet become a plain drab color in less than four hours' exposure in a window.

Blue, Royal.—That shade of Prussian blue which has a reddish or purplish reflection; the existence of tin seems absolutely necessary for the production of this shade. (See BLUE COLORS.)

Blue, Saxony.—Old name for sulphate of indigo. (See INDIGO.)

Blue, Soluble.—Also a name for sulphate of indigo, but lately also applied to a modified Prussian blue. Dry Prussian blue treated for forty-eight hours with strong mineral acids and then washed, is said to lose iron and dissolve easily upon addition of a minute quantity of oxalic acid.

Bluestone.—Common name for sulphate of copper, called also blue vitriol and blue copperas. (See COPPER SULPHATE.)

Blue, Ultramarine. (See ULTRAMARINE and PIGMENT COLORS.)

Borax.—This substance is a salt composed of boracic acid and soda, and because boracic acid is a very feeble acid, the soda retains some of its alkaline properties in this salt. Borax can be used as a weak alkali; it is milder than crystals of soda, it has cleansing or detergent properties, it dissolves resin, shellac, anotta, and some other coloring matters; it is but little used at present in printing or dyeing.

Bowking or *Bucking.*—One of the operations in BLEACHING, which see.

Bran.—Bran has some detergent powers, and is frequently recommended to clean fabrics of very delicate colors. It is now sparingly used to clear some styles of goods, as logwood blacks, garancine pinks, etc.; it was formerly very much used in calico printing and dyeing. Before soap was applied to clearing the whites of printed goods, boiling in bran and exposure to air were the only means used. Bran added to a dye has the effect of causing lighter and clearer shades to be produced. Growses' pink was produced by mixing madder with a large excess of scalded bran and dyeing mordanted cloth in the mixture; it is long since abandoned in favor of better methods, but is an illustration of the effects of bran upon dyeing matters.

Brauna Wood.—This wood is mentioned in a patent dated April 25th, 1857; it is said to grow in the Brazils, and its coloring matter to have great affinity for cotton, with or without mordants, producing shades of brown, drab, slate, fawn, and black.

Brazil Wood, or *Brasil Wood.*—This is one of the class of red woods whose coloring matter is largely soluble in water. It is from the same kind of tree and nearly identical with peachwood, Lima wood, and sapan wood. The richest variety is from Pernambuco, and is sometimes called Fernambuc wood. The real Brazil wood is said to be one-half less rich than the Fernambuc variety, while peach, sapan, and Lima woods are still more inferior. They all, however, contain the same kind of coloring matter, and present the same kind of chemical reactions. Brazil wood when freshly rasped communicates a bright red color to water in a few minutes; by this test it can be distinguished from logwood, which does not sensibly color the water, while inferior qualities of red wood give a reddish brown color. Santal wood and barwood do not, under similar circumstances, color water. Decoction of Brazil wood gives a bright red with alum and crystals of tin, which distinguish it from logwood, which give purplish precipitates.

Brazil wood is usually kept some weeks after rasping in a moist state before being made into liquor. Though this does not appear so necessary for Brazil wood as for logwood, it is very generally thought to be beneficial. It is considered that a decoction of Brazil wood improves greatly by age, both with regard to the depth and purity of the colors it gives, so that it is frequently kept several months in vats; a fermentation appears to go on, and tarry and other matters are deposited, the absence of which improve the shade. Several methods of improving Brazil wood liquors have also been given, but they seem rather impracticable. One method consists in adding skimmed milk to the liquor, and raising to the boil; the caseine of the milk coagulates, and carries with it some substances injurious to the color. Another consists in sprinkling the wood, before extracting, with water containing a small quantity of glue or bone size, and leaving it for a few days.

Applications.—Brazil wood is used in dyeing for common qualities of reds and crimsons, and as a constituent in other shades where a red element is required. In calico printing it is also used for the cheaper kinds of reds and crimsons, and as a component of many of the more complex shades, as brown and chocolate.

The pure coloring matter of Brazil wood is called Bresiline. As fixed upon textile fabrics it is one of the loose fugitive colors, and only acquires a moderate degree of permanency when combined with relatively large amounts of astringent matter.

Braziletto or *Brasiletto.*—An inferior kind of Brazil wood, said to come from Jamaica, and sometimes called Jamaica red wood.

British Gum. (See GUM SUBSTITUTES.)

Bromine.—The name of one of the elementary bodies. Excepting mercury, it is the only one existing in a liquid state at natural temperatures; it is comparatively rare, and has received no application as yet.

Bronze Colors.—A bronze color is a kind of brown, usually with a greenish reflection, or, perhaps, rather with some kind of a shade which reminds the observer of a metallic reflection. There are many shades of bronze. I select a few examples of methods for producing what are called bronze shades.

Manganese Bronze.—This color was at one time very popular, but is now scarcely ever required. It can be produced of various shades, from a brown so dark as to appear black, down to a light nut shade, according to the strength of the liquor used. The bronze liquor was generally muriate of manganese, but sometimes also sulphate of acetate; this was simply thickened according to the style, printed and aged for a short time, preferably in a hot stove, then raised in a hot solution of caustic soda, and winced in clear water until the shade was developed. For dark grounds the pieces were finally winced in weak solution of bleaching powder, to raise the full shade of color.

The bulk of manganese bronzes or browns are self colors, and produced by padding the cloth in bronze liquor at about 28°, slightly thickened with gum, drying, and raising or fixing in a hot and strong solution of caustic soda, the caustic standing as high as 30° for the darkest shades. The oxidation is finished by a passage in weak chloride of lime. Designs can be produced upon these grounds by printing a discharge of crystals of tin. (See DISCHARGE.) The color is due to the deposition of oxide of manganese upon the cloth, which is oxidized by exposure to the air, and by the chloride of lime into the peroxide of manganese. (See MANGANESE.)

Bronze upon Wool.

100 lbs. of wool,
10 lbs. fustic,
20 lbs. alum,
5 lbs. tartar;

boiled for three hours in this mixture with sufficient water, then boiled with 20 lbs. of madder, and afterwards dipped in the blue vat until the required shade is obtained. (Dumas.) The bronze in this case is a mixed color produced from yellow, red, and blue, in which the yellow predominates, or it is a green browned by orange. Another cheaper bronze on wool is given as follows:—

60 lbs. fustic,
40 lbs. quercitron bark,
5 lbs. logwood, are boiled together for an hour; then is added
24 lbs. alum,
4 lbs. madder, and the cloth entered and boiled for four hours. The cloth lifted, 2 lbs. green copperas added, and the cloth worked in again hot. A greenish bronze is also obtained by boiling the wool for an hour in a mixture of $2\frac{1}{2}$ lbs. bichromate potash and $1\frac{1}{4}$ lbs. tartar, then dyeing in a mixture of 20 lbs. fustic, 3 lbs. logwood, 3 lbs. santal wood, 6 lbs. madder, 2 lbs. turmeric, and $1\frac{1}{2}$ lbs. alum.

A bronze brown upon silk may be obtained by working for half an hour in fustic and archil and raising in copperas.

See further BROWN COLORS, of which bronze is actually one.

Broom.—A kind of broom, called "Dyer's broom" (*genista tinctoria*), is locally used to obtain inferior yellow colors upon woollen, by means of an alum and tartar mordant.

Brown Colors.—Brown is produced by the reflection of mixed rays of red, blue, and yellow in unequal proportions; when reflected in equal or chromatic proportions they produce so-called blacks or whites, and when the reflection is imperfect the class of gray colors result. It is the predominance of the orange over the blue which characterizes brown; and there are an infinite number of shades of it. Instead of attempting to collect under this head the methods and processes employed for all kinds of brown colors, it will be found more advantageous to confine the remarks to general principles, with a few processes

of a characteristic nature to illustrate them; and to refer to
the body of the book for most special shades of brown. The
popular names of the brown colors assist this arrangement and
permit them to be described under distinctive heads, such as
BRONZE, FAWN, CHOCOLATE, NUT, WOOD, &c.

If we consider chestnut brown as the middle type of a brown
color, the gradations of the shade darker and lighter may be
considered as due in the first case to the increase of the blue
element, and in the latter to the increase of yellow or red
parts. Thus, if blue be added to chestnut brown it becomes a
chocolate; if mixed yellow and red be added it becomes nut
color; if an excessive amount of blue is added the brown
passes into black, or an extremely dark chocolate; and, on
the other hand, if a large quantity of orange is added it
passes to fawn and buff. As the greatest number of brown
shades are produced directly by combining yellow, red, and
blue woods or dyes, this hint should be a sufficient guide
as to how the shades may be modified at will. The only difficulty consists in the want of a distinct comprehension as to
what colors certain ingredients contribute to a mixture; about
indigo, weld, and madder, with alum mordant, there is no difficulty, because it is known they are distintly, blue, yellow, and
red. But logwood does not yield a pure elementary color;
with alum it gives a color which is a mixture of blue and red,
the blue predominating; with iron it gives a blue so dark and
absorbent as to appear black or gray—it may be considered
as a blue part in brown colors. Anotta gives a color which is
a mixture of red and yellow, and only requires blue to produce
light browns. Sumac and gall-nuts are blue and darkening in
their action. Catechu and other substances give a brown
without any combination. These simple natural browns will
be treated under the head of their coloring matter.

Brown on Silk by Dyeing.—The largest class of browns on
silk are obtainable by first dyeing an orange or yellow ground
with anotta, and then superadding a blue or black pigment, as
in the following illustrations:—

Red Brown.—Dye the silk first in anotta, and then work it
in a mixture of logwood and nitromuriate of tin or plum
spirits. (See SPIRITS.) Here the lilac of the plum spirits,
composed of blue and red, adding itself to the yellowish-orange
of the anotta gives a light shade of brown.

Dark Brown.—Dye a deep orange in anotta, work in copperas
liquor, wash and work in fustic, logwood, and archil, or peachwood may be substituted for archil; finish in alum water.

Quantities.—10 lbs. silk dyed with anotta, 1 lb. green copperas, 20 minutes; 6 lbs. fustic, 1 lb. logwood, 1 quart archil,

or 1 lb. peachwood, 30 minutes; one pint of alum liquor, 15 minutes.

There is no limit to the depth and quality of shade to be obtained by varying the quantity of woods; the archil contributes greatly to the fulness and richness of the color, but may, nevertheless, be replaced by the red woods. The production of brown from the above materials may be explained by the basis containing yellow and red; a further amount of red and yellow is added by the fustic and archil or peachwood, the logwood adds the blue, the alum forming a basis for the woods. The copperas darkens the whole by its forming the black-blue color with logwood.

Other Browns.—Anotta, though yielding the brightest browns, is not necessary as a basis; for a variety of browns are obtained by first aluming the silk and then working it in a decoction of logwood for the blue part, peachwood or brasil wood for the red part, and fustic for the yellow part.

Deep Chocolate Brown. Quantities.—10 lbs. silk, steep 60 minutes in alum at 1 lb. to the gallon; wash, 6 lbs. peachwood, 2 lbs. logwood, 8 oz. fustic, 30 minutes; 1 quart alum solution, 15 minutes.

Brown on Silk by Printing.—The same general principles apply as in silk dyeing, and nearly the same materials are employed, as will be seen by the receipts following:—

Chestnut Brown on Silk.

1 lb. logwood liquor at 3°,
1 pint berry liquor at 6°,
3 quarts Brazil or sapan wood liquor at 3°,
1 lb. starch, or 2 lbs. gum if for block,
8 oz. alum,
4 oz. nitrate of copper at 80°,
8 oz. oxymuriate of tin at 80°.

Exactly the same ingredients, but in different relative quantities, may be used for obtaining a dark chocolate or a light nut brown. For chocolates the logwood or blue part must be in greater quantity; for the nut shades the berries or yellow part must be increased.

Another Chestnut Brown on Silk.

1 gallon berry liquor at 11°,
3 quarts brasil wood or peachwood liquor at 7°,
3 pints logwood liquor at 7°,
1¼ lbs. alum.
10 oz. sulphate of copper, thickened with
8 lbs. gum, more or less, to pattern.

In a few receipts the red part consists of ammoniacal cochineal, but it is questionable whether this expensive liquor is any better than a decoction of one of the red woods in such a color. In all the cases where copper salts are used with woods the addition of muriate of ammonia will be found beneficial.

Brown on Wool by Dyeing.—The following is an example of a fast and durable, but expensive brown :—

Chestnut Brown.—The wool is first dyed yellow in a decoction of weld and fustic, or else in quercitron bark and fustic; alumed and dyed in madder, then dipped in an indigo vat until the right shade is obtained.

Quantities.—100 lbs. of wool, 50 lbs. yellow woods, 60 minutes at boil; 25 lbs. alum and 5 lbs. tartar, boil for three hours; three days, 60 lbs. madder; two hours, indigo vat at discretion.

In this illustration the fastest known yellow, red, and blue elements are combined, and the product is a fast color. This example serves very well to show the effects of the mixture of the elementary colors, the disappearance of each particular shade, and the blending of the whole in a complex hue. For cheaper woollen cloths cheaper dyeing materials are used; for example, instead of dipping in indigo, the blue part is given by sumac, logwood, and copperas, or by sumac and copperas without logwood. The fast but expensive red from madder is substituted by similar color from santal wood or brasil wood, and the yellow obtained from fustic. There are many methods of combining the elementary colors on wool to obtain brown, a few examples of which will suffice.

Brown on Wool, No. 1.—Mordant in bichromate of potash and alum for half an hour, wash and work in a decoction of fustic, madder, cudbear, logwood, and cream of tartar. The quantities of those woods must depend upon the shade desired.

Brown on Wool, No. 2.—Work the wool in a decoction of fustic, madder, peachwood, and logwood, and raise in copperas.

Brown on Wool, No. 3.—The wool is boiled in a mixed decoction of galls, santal wood, madder, brasil wood, and fustic; then raised in a mixture of logwood and green copperas.

Quantities.—These quantities are only suggestive, and admit of great latitude. For 100 lbs. wool, first receipt, 3 lbs. bichromate, 3 lbs. alum, 3 lbs. tartar, 20 lbs. fustic, 10 lbs. madder, 5 lbs. peachwood, 3 lbs. logwood.

No. 2 Brown—100 lbs. wool, 20 lbs. fustic, 20 lbs. madder 10 lbs. peachwood, 2½ lbs. logwood, 1¼ lbs. copperas.

No. 3 Brown—100 lbs. wool, 6 lbs. gall nuts, 12 lbs. santal wood, 6 lbs. madder, 4 lbs. brasil wood, 5½ lbs. fustic, three hours; 3 lbs. logwood, 2 lbs. green copperas, 45 minutes.

Brown on Wool by Printing.—The following receipts for brown will serve to show the method of obtaining this color on wool by printing:—

> *Chestnut Brown—all Wool.*
>
> 4 pints bark liquor at 18°,
> 4 pints cochineal liquor at 4½°,
> 2 lbs. gum,
> 8 oz. oxalic acid,
> 6 oz. alum,
> ½ pint bichloride of tin, at 100°,
> 3 oz. extract of indigo.

Archil enters largely into all the dark or chocolate shades of brown for wool, and may be used, but with less advantage, for the more yellow shades, as in the following receipt:—

> *Wood Brown—all Wool, Block.*
>
> 7 quarts bark liquor at 18°,
> 3 quarts archil liquor at 10°,
> 7 quarts cochineal liquor at 6°,
> 3 lbs. starch; boil, and add
> 9 oz. alum,
> 6 oz. oxalic acid,
> ¾ pint bichloride of tin at 100°,
> 3 oz. extract of indigo.

Or, as again, in the following receipt for a similar shade o color, obtained by rather different means:—

> 1 gallon berry liquor at 18°,
> 1 gallon archil at 18°,
> 2 lbs. starch: boil and add
> 1 lb. alum,
> ½ lb. tartaric acid,
> ½ lb. green copperas.

Brown on Calico by Dyeing.—The following methods will serve to illustrate the compound brown on calico:—

Spirit Brown.—Dye first a yellow from bark, by mordanting with sumac and tin—(see YELLOW)—then pass into peachwood or brasil wood mixed with logwood for half an hour, lift, and add alum water to raise the colors. The peachwood here gives the red, and the logwood the blue or purple constituent. The shades may be modified to wish, by altering the quantities of the materials.

Quantities.—10 lbs. cotton dyed yellow, 2 lbs. peachwood, 1 lb. logwood, 3 oz. alum; time, half an hour.

Brown with a Chrome Yellow Basis.—The cloth or yarn is dyed chrome yellow. (See CHROME COLORS.) The remaining process is exactly the same as the above.

Brown with Anotta Basis.—Dye in anotta liquor (anotta dissolves in pearlash), wash out, and work in decoction of fustic and sumac; lift, and add green copperas liquor, and work in again; wash, and work for twenty minutes in a mixture of red wood, fustic, and logwood; lift, and again raise with alum. This produces a fawn, or yellowish-brown, on account of an excess of yellow. The anotta color may be considered as yellow with a little red, and then fustic being again twice used, the yellow accumulates and gives a tone to the brown. The sumac darkens the color.

Quantities.—10 lbs. of cotton dyed with anotta, 2 lbs. fustic, and 1 lb. sumac, twenty minutes; 3 oz. copperas, twenty minutes; ¼ lb. logwood, ½ lb. each of fustic and peachwood, twenty minutes; 1 oz. alum, ten minutes.

The great majority of brown colors upon cotton are obtained from catechu, which is a distinct brown coloring matter itself. In calico printing many shades of brown and chocolate are obtained from madder and garancine with mixed mordants. For information upon those colors the articles CATECHU, GARANCINE, and MADDER may be consulted.

Brown Colors on Calico by Printing.—Catechu cannot be used to advantage in steam browns, and the mixture of elementary colors is necessary. Steam brown on calico is very seldom required; it differs from chocolate by containing more red and yellow and less blue. Frequently the color is obtained by mixing steam orange and steam lilac together, the blue part of the latter turning the orange to brown. I give a couple of receipts as sufficiently indicating the nature of the mixture used for brown.

Steam Brown for Calico.

3 quarts bark liquor at 12°,
2 quarts sapan liquor at 10°,
3 quarts berry liquor at 12°,
2 quarts logwood liquor at 12°,
12 lbs. British gum; boil, and add
12 oz. alum,
8 oz. sal ammoniac,
8 oz. sulphate of copper,
¼ pint nitrate of copper at 80°,
3 quarts of lilac standard. (See below.)

Lilac Standard for Brown.

1 gallon logwood liquor at 6°, heat to 180°, and dissolve in it
4 lbs. gum senegal,
8 oz. red prussiate of potash,
12 oz. alum,
1 oz. oxalic acid,
2 oz. binoxalate of potash.

Wood Brown on Calico.

1 gallon berry liquor at 3°,
2 quarts peachwood liquor at 8°,
¼ pint logwood liquor at 8°,
1½ lbs. crystals nitrate of copper,
1½ lbs. alum; thicken with gum water, according to shade
  required.

Browns on Delaine by Printing.—These are nearly the same
as upon wool. I give one or two examples:—

Dark Brown for Delaines.

5 pints berry liquor at 8°,
12 oz. alum,
1 pint of archil at 8°,
½ pint sapan liquor at 8°,
¼ pint logwood liquor at 11°,
1 lb. starch; boil and add
2 oz. oxalic acid.

Wood Brown for Delaines.

1 gallon peachwood liquor at 9°,
1 gallon berry liquor at 18°,
2 quarts archil (strong),
2 lbs. starch; boil, and add
1½ lbs. alum,
4 oz. sal ammoniac,
2 oz. acetate of copper.

The absence of logwood or sulphate of indigo in the latter
receipt would cause a yellowish or buff brown; when the blue
part predominates, as before stated, the brown passes into chocolate. For delaines, sulphate of indigo may be used as the blue
part, but not exclusively, since only the wool takes blue from
this coloring matter. (See CHOCOLATE, CATECHU, etc.)

Browning.—Neutral colors of the gray and dove species
upon cotton goods are darkened by passing them through a
weak solution of green copperas alone, or mixed with a small

portion of logwood liquor or decoction of galls. This process is the one sometimes called "browning," but in Lancashire the more usual term is "saddening," and colors so modified are known as "saddened" colors. All the wood colors are turned darker by copperas; and even red colors, from garancine and madder, are turned to a chocolate shade.

Buccinum Lapillus.—A species of shellfish or whelk, obtainable on the English coast, which contains a viscid white matter that acquires a purple color when applied on calico. It passes through several shades before it is wholly changed into purple; becoming, first, pale yellowish-green; secondly, an emerald green; thirdly, a dark bluish-green; fourthly, a blue beginning to purple; and, finally, a purple. In strong sunshine these changes take place in less than five minutes; in the dark the color does not get beyond the second or emerald green stage. This peculiar and interesting liquid is mentioned by the most ancient writers, as Aristotle and Pliny; and there seems to be no doubt that the coloring matter was formerly employed for dyeing. There is also strong reason for supposing that the famous Tyrian purple of the ancients was derived from this or some similar shellfish.

Bubuline.—A supposed constituent of cow dung, to which some chemists desired to attribute its useful actions in dyeing and printing. (See Cow Dung.)

Buckthorn, Dyers'.—A plant called "*nerprun*" in French seems the same as the dyers' buckthorn. Some attempts were made recently to obtain green dyes from it by M. Michel, who was led to the experiment from ascertaining that the Chinese extracted a green color from the same species of plant— (*Rhamnus utilis* and *R. chloroforus*.) The results so far prove that there exists a colorless substance in the French indigenous buckthorns, which upon exposure to light becomes green; but it has not yet been extensively used, probably because the color is neither pretty, durable, nor cheap. (See ARTICHOKE, CHINESE GREEN.)

Buff Color.—A color so named because resembling the shade of leather prepared from the buffalo skin, called buff or buffalo leather. The continental colorists, probably more familiar with the dressed skin of the chamois than of the buffalo, gave the name "chamois" to this color. It is yellow mixed with a little red, or, according to Chevreul's nomenclature, yellow with some orange of a low tone.

The chief buff color, or the one distinctively so called, is from iron, and prepared as follows:—

Buff Liquor, Ordinary.

4 gallons water,
20 lbs. sulphate of iron (green copperas),
5 lbs. brown sugar of lead,
2½ lbs. white sugar of lead.

Another Buff Liquor.

10 gallons water,
48 lbs. green copperas,
20 lbs. brown sugar of lead.

Both these are proto-acetates of iron with undecomposed sulphate. The following liquor contains an excess of lead, found to work well with chromed styles, or where a somewhat yellower or softer buff was required.

Lead Buff Liquor.

10 gallons water,
25 lbs. acetate of lead,
20 lbs. green copperas,
½ gallon acetic acid.

In all cases the sulphate of lead is allowed to precipitate, and the clear field only used. It is thickened either with gum, flour, or starch. Old receipts give nitrate of potash along with the other ingredients, and direct the liquor to be kept for six months before using. Modern French receipts give the nitrosulphate of iron as a buff liquor; but for printing on calico there can be no question of the superiority of simple acetate of iron. The addition of white arsenic and salts of copper, found in some receipts, seems more likely to injure than assist the color.

The buff color being printed, is aged for a night, and then fixed or raised in an alkaline bath, consisting of well slacked lime with a small quantity of soda ash. The pieces are entered carefully, and, when evenly wetted, are winced in the lime for ten or twenty minutes, then winced in clear water until the shade is raised, which may take half an hour or more; washed, dried, and finished.

The lime and soda take the acetic acid or sulphuric acid from the oxide of iron which is then retained by the fibres; but it is in the state of protoxide, and has a greenish color. By wincing in water the iron absorbs oxygen and becomes peroxide, which is the coloring body.

To obtain regular and even shades requires a good deal of care and attention. The cloth must be well bottomed in the bleaching; the gum used for thickening must be one that washes off well and easily; and in the raising it is highly important that the pieces be kept moving—any stopping in the process is injurious.

Steam Buff for Calico.

3 gallons madder liquor,
1 gallon bark liquor at 10°,
2 gallons red liquor at 14°,
7 lbs. starch; boil, and add
2 oz. crystals of tin.

Steam Buff for Wool.

1 quart bark liquor 4°,
1 pint archil at 40°,
6 oz. alum,
1½ oz. tartaric acid,
1 gallon gum water.

Steam Buff or Chamois on Delaine.

5 quarts catechu liquor, at ½ lb. to the gallon,
8 oz. alum dissolved in
1 quart hot water,
3 oz. acetate of copper,
10 oz. nitrate of copper,
6 quarts thick gum water.

The above buff is a simple color; but those most in use are compounded of red and yellow. Anotta gives a species of buff on calico. On silk and woollen the yellow part from Persian berries, and the red part from cochineal, yield all shades required. (See NANKEEN, RED, and YELLOW.)

Buffaloes' Milk.—According to the accounts of the missionaries in India, at the end of the last century, buffaloes' milk was rather largely used by the natives in dyeing fast madder colors. It was applied at the same time as the astringent matters, and appeared to partly answer the same purpose that oil does in Turkey-red dyeing.

Butternut Tree.—The common name for a tree growing in the New England States—a species of walnut tree (*Juglans oblonga Alba*), so called because the fruit it yields is very oily. The bark is stated to be capable of communicating a lasting black color to fibrous matters prepared with iron mordants; with alumina mordants it gives a tobacco brown color. The rinds of the nut have also the same dyeing powers as the bark of the tree. (See WALNUT.)

C.

Cactin.—Vogel extracted a carmine red coloring matter from the blossoms of the *cactus speciosus;* the leaves yielded also a quantity of a scarlet red substance, soluble in water. It would be interesting to know whether these colored matters were similar in their composition to the colors from cochineal —for this plant is one of the species of shrubs upon which the cochineal feeds. Wittstein examined the sap of the branches, and the ripe fruit of another species of cactus (*c. opuntia*), but was of opinion that the coloring matter of the cochineal did not exist in the tree, and that what was extractable by solvents was something different, and quite useless in the arts.

Cactus Cochenillifer.—The botanical name of the tree or shrub upon which the cochineal insect is nourished; is a native of America, and there are several species of it, producing fruits of various colors, as yellow, red, violet, etc. It is observed that the crimson-colored fruit contains a mucilaginous juice, which strongly colors the urine of those who eat it. It seems probable, that if the cochineal insect is merely an extractor of the coloring matter of the plant, that the fruit, etc., might be more directly and economically applied as a dyeing substance, than as food for insects.

Calcined Alum, *Alumen Ustum.*—In some old receipts alum is directed to be dried in an earthen pot, and made red hot before being applied in dyeing. Although good modern alum cannot be improved or changed beneficially by such a process, it is quite possible that inferior and impure alum would be better for a moderate calcination. The heat would have a tendency to render the iron in an impure alum insoluble in water by expelling a portion of the acid with which it was combined : if the alum was also of a very acid nature, some of the excess of acid would also, be removed and its quality improved.

Calcined Copperas.—When sulphate of iron, or green copperas, is raised to a low, red heat, in an earthenware or iron basin, it loses water and some acid, and gains a little oxygen. Provided the heat be not forced too high, there is no doubt that copperas thus treated is improved for several of its applications in dyeing and color mixing; but the goodness and variety of the iron mordants, at present obtainable in trade, obviate the necessity of such treatments to obtain suitable solutions in iron.

Calcined Farina.—A kind of thickening matter largely used in calico printing, and made by exposing the starch or

farina of potatoes to a roasting heat; it is one of the Gum Substitutes, which see.

Calcium.—This is the name of the metal which exists in
lime, chalk, etc., and from which a good many chemical names
are derived; thus, in strict chemical nomenclature, lime is the
oxide of calcium, chalk is the carbonate of calcium, muriate of
lime is the chloride of calcium, and so on.

Camwood.—This is one of the red woods obtained from the
Gaboon, in Africa, and from Sierra Leone, where it is called
by the natives *Kambe*, whence, by abbreviation, *Kam* or *Cam*.
It has the same properties as brasil wood, but dyers are not
agreed as to their relative value. Some say it is inferior both
in richness and durability to Brazil wood, whilst the contrary
is also maintained. It appears to yield more scarlet shades
than peachwood, having some portion of yellow in its composition, and may generally be employed in all cases where
peachwood, sapan wood, or brasil wood are prescribed. It is
evident from the contradictory nature of the statements made
with regard to this wood, that it is either very variable in
quality or that the methods of its application are not generally
understood.

Caoutchouc or *India Rubber.*—Attempts have at various
times been made to use a solution of India rubber as a vehicle
for pigment colors, but, so far as is known, without success.
In a few cases solution of India rubber has been applied to
fabrics by block, as a means of fixing flock and metallic designs, but it is unsuitable to mix with pigments. It is soluble in coal, naphtha, turpentine, oils, and bisulphide of carbon.

Capucine Color.—The color called capucine is a deep
toned reddish orange. In Chevreul's nomenclature, it is called
3 red orange of 11 or 12 tone; it has some resemblance to a
deep chrome orange on cotton. Upon wool and silk it is obtained by a proper mixture or combination of red and yellow,
having the red in excess, as the following receipt for dyeing
50 lbs. of wool.

Yellow Part.—3½ lbs. fustic,
3 lbs. oxymuriate of tin,
1 lb. cream of tartar.
Red Part.—2 lbs. oxymuriate of tin,
¼ lb. cochineal.

The yellow is first dyed, and then the cochineal and tin added.
In printing it suffices to mix at once a little made scarlet color
with orange, as for example :—

Cupucine for Wool and Shawls.

4 quarts orange for wool,
4 pints scarlet for wool.

Carbazotic Acid.—The same as PICRIC ACID, which see.

Carbonate.—In chemical language a carbonate is a compound of carbonic acid with a base. The carbonates are all insoluble in water, except those of potash, soda, and ammonia. The soluble ones have all an alkaline reaction, and can neutralize acids. All carbonates are known by giving off the carbonic acid as a gas when a strong acid is poured over them; thus, when muriatic acid is poured on chalk, which is a carbonate, a strong effervescence or bubbling takes place, owing to the carbonic acid gas forcing its way out of the liquor, being set free by the muriatic acid taking the lime or calcium which previously held the gas in a solid state.

Carbonic Acid.—This acid is a gas under ordinary circumstances, it is one of the weakest acids in chemistry, never completely neutralizing the alkalies. It exists in small quantities in the air, is the cause of exposed lime water being covered with a skin of solid matter, but has no direct influence in printing or dyeing.

Carmelite Color.—The color so called is a yellowish orange mixed with brown, darker than the colors called wood colors. In Chevreul's nomenclature—3 orange, 15 tone. The following receipt is given by Dumas:—

Carmelite.

1 quart sapan wood liquor at 6°,
1 pint berry liquor at 6°,
1 pint logwood liquor at 6°,
10 oz. starch ; boil, and add
12 oz. oxymuriate of tin.

In woollen dyeing and in cotton dyeing carmelite is obtained by saddening orange or using logwood to brown it.

Carmelite shades are also obtained upon calico by printing or padding in a mixture of equal parts of bronze liquor and buff liquor, and raising in lime.

Carmine.—This name is understood in England as indicating a red pigment used by artists, prepared from madder or cochineal by secret processes. French writers, and from them English, however, speak of *carmine of* "*indigo*," meaning a refined sulphate of indigo ; also "*purple carmine*," or "carmin de pourpre," meaning murexide, using this term generally for some preparation yielding fine colors without regard to what kind of color. This term frequently occurs in specifications of patents and translations from French.

Carragheen Moss, *Iceland Moss, Irish Moss.*—This substance has been frequently proposed as a suitable thickening agent for colors, and was probably the first gum substitute tried in this country. Towards the end of the last century it was put into use, but has never made any progress; the mucilaginous jelly it yields is deficient in nearly every quality of a good thickening; it is watery, has no solidity, and is glairy. It is a little employed in block printing on silk and in finishing.

Cartamus, *Carthamus* (?)—An empirical mixture of cochineal, tin salts, and safflower, was patented under this name, January 22d, 1853, for dyeing tissues or stuffs of silk and cotton.

Carthamine.—The name of a pure coloring matter extracted from safflower, so called from the botanical name of the plant—*Carthamus tinctorius.*

Caseine.—The name is given in chemistry to the pure curd of milk, obtained by acting upon milk with weak acids and purifying the curdy precipitate from fatty matters attached to it. It is the same substance which is extensively used in this country under the name of LACTARINE, which see.

Catechu, *Terra japonica, Cachou, Cashew.*—Catechu is the dried up juice of certain trees, from which it is obtained either by natural exudation or through cuts made for the purpose. It is a resinous looking body, dark on the exterior of the lumps, but light colored within. Its quality varies very much, not only from differences in its origin and method of collection and drying, but also because it is susceptible of alteration by age, and especially by moisture. Soft and uniformly dark colored, catechu is reckoned inferior; it should be brittle enough to break upon the stroke of a hammer, and the interior should not be pitchy colored or soft, but rather of a buff or a cream color, somewhat fibrous, and capable of being scraped into powder with a knife, without adhering to it. Though these are the external characters of a good quality of catechu, there are good samples which vary in appearance from this; the color may be darker, and the consistency of the mass less brittle, so that a knife does not scrape it. That depends upon several circumstances of the carriage and storing of this drug, and must be decided upon according to the judgment and knowledge of the examiner. The chemical characteristics of a good catechu are, unfortunately, not very well defined. It ought to be all soluble in hot water, and then have a brown, not blackish color: it ought not to be all soluble in cold water, for that is an indication of heating and partial decomposition of the catechu; and the hot water solution should, upon cooling, deposit a portion of the catechu in a fine granular state. The only actual and reliable test for the quality of catechu is to make some color from it, or to dye up samples from it, in comparison with a known quality.

This substance was formerly supposed to be of mineral origin, and went under the name of Japan earth. It was long known in medicine before it became cheap enough to be applied in dyeing and printing. Its first successful applications in calico printing were about 1830; it was used in combination with madder colors, and as its application was kept secret for a while by the one or two houses who used it, much skill and ingenuity were wasted by others in endeavoring to discover the new *mordant,* which it was thought had been used to obtain this brown shade from madder. Its applications in madder and garancine styles have been of the greatest service to the trade; it has allowed a scope of design and variety of coloring which has done much to extend the use of printed goods. In dyeing, it is largely used to give various shades of brown, and the lighter colors which spring from it.

Catechu is one of the astringent or tannic substances, but not of the same kind as gall-nuts. Its acid is called japonic acid, and possesses different properties and characteristics from the tannic acid. The method of application of catechu in calico printing shows it to be very different from most other coloring matters.

For the purpose of obtaining browns in printing, it is mixed with sal ammoniac and nitrate of copper; sometimes the acetate being used instead. From this it would appear that copper is its proper mordant; but the copper is not so much an actual mordant as it is an agent for effecting a chemical change in the catechu. The copper salts by themselves are oxidizers of coloring matters, and when mixed with sal ammoniac, their oxidizing powers are greatly strengthened, in a proportion, indeed, far beyond the amount of oxygen which is present in the whole of the copper salt used. It acts as a medium for obtaining oxygen from the air, and transferring it to the catechu, which of itself absorbs oxygen in a very slow manner. The effect of this oxidizing upon catechu is to change its properties, to give it a great hold and affinity for the fibre of the cloth, and to render it insoluble and unacted upon by water. Some of the copper remains combined with the color; but the greater part is removable without injuring the fastness or shade of the catechu brown—a fact which seems to point out that catechu can fix itself without a mordant. Such is really the case, but the length of age necessary is excessive and imprac-

ticable. Even with copper salts, when it is desired to get dark
shades, several days' ageing is required. Lighter shades take
less time.

Iron and alumina mordants do not give agreeable colors
with catechu in printing; but several shades can be obtained
by dyeing a mixture of such mordants and catechu in madder
and garancine, the resulting color being a mixture of the
catechu shade itself and that which has been produced by the
mordant and dying material used.

Muriate of iron and catechu give shades of drab, stone, and
gray, when dyed up in garancine. Acetate of alumina, red
liquor, and catechu give shades of red brown, varying, for the
different amounts of red liquor, in a certain quantity of color.
Mixtures of catechu with salts of manganese, and other mineral
matters, are in use.

In dyeing with catechu alum mordants are mostly employed;
iron and tin salts can be used to obtain various shades—copper
being but little used in these cases. The bichromate of potash
has a powerful oxidizing action upon catechu. It cannot be
applied mixed with it, because it combines with the catechu,
rendering it insoluble and curdy; but a solution of bichrome
can be used to pass catechu colors in. It fixes them and makes
them darker. Soda and potash are also used to raise catechu
colors in dyeing; their action seems to be an oxidizing one—
enabling the catechu to absorb oxygen with so much the
greater rapidity from the air, and become fixed upon the cloth.
The affinity of catechu, in its altered or oxidized state, for the
fibre of cotton, is very great. It is one of the most difficult of
all colors to discharge from the cloth. It is valuable in calico
printing, because it is fast enough to stand the dunging and
dye beck, and all the subsequent clearing operations. It
suffers, of course, in passing through all the various operations,
but still sufficient is left to form good colors. It is usual to
add a little bichrome to the dunging to help the catechu; but
this is rather dangerous for the other colors, and should not be
used except the circumstances have compelled the goods to go
to the dye with a deficient age. If time enough can be given
to catechu colors, the addition of bichrome to the dunging is
not necessary; but in case of heavy browns, which have only
had two or three days' age, it may be used with advantage.
These shades of drab, etc., in which the proportion of catechu
to a gallon of water is only small, require no longer age than
the other colors they go with, and no chrome in the dung.
Bleaching powder solution should be very sparingly and
cautiously applied to catechu styles—it soon takes the "top"
or bloom of the color.

The use of large quantities of nitrate of copper in colors is very disadvantageous in printing, and it is much to be desired that some milder oxidizing agent could be discovered for catechu. It compels the use of composition doctors, and even acts upon them. A resist brown, containing lime juice or citric acid, is very difficult to work on account of its corrosive action upon the doctor, and consequent scratching of the roller. Without any acid, the regular brown will resist light covers, but not heavier ones; and I believe the difficulties in the way of printing such an acid mixture have caused the abandonment of this style almost altogether. I have tried many chemical mixtures instead of copper, but none of them gave any results worth following up. The applications of catechu are still limited, and its chemical properties but little known. It offers a good field for the exertions of those who have leisure and knowledge of coloring matters, for it is doubtless capable of many more valuable applications than it has yet received.

Catechu Colors.—The following receipts and remarks will sufficiently illustrate the various methods adopted for using this coloring matter :—

Catechu Colors on Woollen.—I am not aware that catechu is employed in woollen dyeing in any other way than as an assistant in some kinds of black dyeing; the browns which it yields are not so desirable as those which can be obtained from a mixture of coloring matters and mordants. For obtaining dark shades of drab, of a red or brownish tinge, it is largely used in woollen printing, in combination with other coloring matters, to modify the shade.

Tea-Drab Color—all Wool.

1 gallon catechu liquor at 24°,
2 quarts sapan wood at 11°,
4 oz. extract of indigo,
1 pint cochineal crimson,
6 oz. alum,
6 oz. oxalic acid,
4 lbs. gum if for block ; 8 lbs. if for machine.

Drab—all Wool.

1 gallon catechu liquor at 24°,
3 oz. extract of indigo,
½ pint cochineal crimson,
2 lbs. gum, more or less at discretion,
8 oz. oxalic acid.

A great variety of shades are obtained by varying the strength of the catechu and the quantities of the red and blue parts. The cochineal crimson is ammoniacal cochineal with alum.

Wood Ground for Delaine.

1 gallon of water,
2 oz. catechu; dissolve, strain, and add
2 quarts substitute gum water,
1½ oz. green copperas,
1½ oz. acetate of copper crystals.

Catechu Fawn, on Cotton.—Work the goods to be dyed in catechu liquor containing a little sulphate or nitrate of copper; wring from this, and work in a weak solution of bichromate of potash. The fixing by bichromate is not necessary; modified shades can be obtained by passing the goods which have been worked in catechu into caustic soda or into lime water, or into water in which green copperas has been dissolved. By using other dyewoods after the catechu a vast variety of shades may be obtained. (See DRAB, STONE, etc.)

Catechu Fawn or Brown.

1 gallon water,
2½ lbs. catechu,
1 pint acetic acid; dissolve hot, and add
3 lbs. gum senegal,
10 oz. nitrate of copper.

This color was for raising in lime or soda, in conjunction with fast blue (indigo); it required at least three days' ageing, and was found to age best in a cool, moist place.

Catechu Brown for Garancine.

6 lbs. catechu,
1 gallon water,
1¼ lb. sal ammoniac; boil, strain, and add
2 gallons gum water,
2½ pints nitrate of copper at 80°,
1½ pints acetate of copper (p. 45).

By adding red liquor to this color the resulting brown is modified towards the red side. This color must be aged for not less than three days; dyed in garancine as usual.

Catechu Brown for Madder.

4 lbs. catechu,
2 quarts ordinary vinegar,
2 quarts acetic acid at 8°,
1 lb. sal ammoniac,
1 quart acetate of copper.

The clear liquor from the above only was used, and served as a standard or stock liquor. To make a dark brown take as follows :—

Dark Brown for Madder.

1 gallon catechu liquor above,
4 oz. sal ammoniac,
½ pint acetate of copper,
17 oz. starch ; boil and strain.

This color must be aged as long as possible before dunging; by mixture with red liquor red browns are obtained.

Catechu Brown.

9 gallons water,
15 lbs. powdered catechu ; boil and dissolve, take the clear liquor only, and add
11 lbs. flour,
1 lb. gum tragacanth ; mix, and boil well ; when cold, add
8 gallons caustic soda at 6°.

After printing, steam; then pass, for a minute or two only, in chrome liquor, neutralized with carbonate of soda, and heated up to 170° F.

The following receipt is only inserted to show the possible modifications of this color :—

Catechu Brown for Madder (French).

1½ gallon water,
1 pint acetic acid at 8°,
2¾ lbs. catechu ; boil, dissolve, and add
3½ lbs. sal ammoniac,
1 pint acetate of lime liquor (see below),
7 lbs. gum senegal ; when cold, add
10 oz. nitrate of copper at 90°.

Acetate of Lime Liquor.

2½ lbs. lime, slacked,
3 quarts acetic acid ; use the clear.

The acetate of lime would undoubtedly resolve itself by contact with the nitrate of copper into nitrate of lime and acetate of copper, as far as the quantities would permit; the nitrate of lime being a deliquescent salt, would tend to keep the color soft, and so help its ageing. (See further, DRAB, FAWN, MADDER COLORS, etc.)

Chamois Color.—This color is as nearly as possible the same as buff. It is a mixture of yellow and red of a low tone. In the systematic nomenclature of Chevreul, it is 2 orange, from 2 to 6 tones. (See COLORS.) The word chamois is very little used among British colorists as yet. Analogous colors are STRAW, SALMON, FLESH, NANKEEN, etc., which see.

Charcoal.—The charcoal or carbon obtained from the imperfect combustion of vegetable substances constitutes the basis of all printing inks, but up to the present time it has not been used in calico printing for a black. For a shade of gray obtainable from a species of charcoal see LAMP BLACK. Very finely ground charcoal has been successfully employed as a sightening in block printing.

Chayaver or *Chay root.*—An East Indian product, belonging to the same natural order of plants as madder, and capable of dyeing good and permanent reds. It is said to be the only coloring matter used by the natives of Malabar, and the Coromandel coast, in producing the well-known durable red colors of those localities. The accounts and descriptions of it which I have been enabled to consult are not clear or satisfactory, but it appears to be similar to the MUNJEET, employed to some extent in Lancashire.

Chemic.—Name commonly given to bleaching powder or bleaching liquor in Lancashire. Chemic blue is the vulgar name for sulphate or extract of indigo.

Chestnut Bark and Wood.—As stated under BLACK, the bark and wood of the chestnut tree are successfully employed in France as cheap and effective substitutes for gall-nuts and sumac. I am not informed as to their being generally used in this country. The right to use it in Great Britain was secured by a patent dated Nov. 8, 1825, in which it was proposed to form a solid extract from it by decoction. This extract received the uncouth name of "Damajavag."

Chestnut Colors.—These are browns, and treated of under that head. Marron is the French word equivalent, which, under the English form "Maroon," expresses the same meaning. In Chevreul's nomenclature chestnut is 4 orange, of 16, 17, or 18 tone, which is equivalent to saying it is black lightened down by a mixture of rather yellowish orange.

Chica, *Crajura* or *Carajura.*—A pigment used by the Indians, in Central America, for ornamenting their persons, applied by means of the fat of the cayman or alligator. It has also been used in painting to a limited extent, but not, that I can ascertain, in dyeing or printing. It is extracted, by maceration and heat, from the leaves of the *bignonia chica,* a plant growing in equinoctial America. A sample, which I examined

some years ago, to ascertain its value as a dyestuff, is in fragments and powder of a bright red raddle appearance. The
lumps are softer than starch, and acquire a metallic appearance
when rubbed by the thumb nail or any polished body. It is
scarcely soluble in water, alcohol, or ether. It does not dye
cotton mordanted with iron or alumina, nor communicate anything beyond a faint color to wool or silk. This inertness I
found was owing to the coloring matter being in combination
with some earthy base, for, upon treating the chica with acid,
or with muriate of tin, I found I could dye up very deep colors
upon wool; and by treating it with acid, and washing the free
acid away, I succeeded in dyeing up full and deep colors upon
mordanted calico. The colors it yields with the alumina and
tin mordants are very similar to those yielded by lac dye; they
are not nearly so bright as cochineal, but they appear more
permanent, resisting washing and exposure somewhat better. I
have no doubt that, if the product was imported at a reasonable price, it would find a permanent place in the dyeing arts.

Chicory.—According to a statement in the specification of
a patent dated Nov. 21, 1844, the leaves of the chicory plant
are, being treated as the leaves of the woad plant are usually
treated, capable of dyeing blue and other colors.

China Blue, *Crockeryware Blue, Bleu de Faience.*—One of
the colors obtained by a peculiar method of fixing indigo upon
calico. (See INDIGO.)

Chinese Green, *Chinese Green Indigo.*—This is a coloring
matter introduced into Europe about 1857 from China. It is
in thin scaly pieces, of an olive green color, scarcely soluble in
water or alcohol, but brought into solution, or a fine state of
suspension, by means of alkalies. It is applied to dyeing silk,
as follows: 80 grains of the color (lo kao) is steeped three days
in 500 grains of alum liquor at 7° ground up with it, and
mixed with a pint of the same alum solution, the mixture
stirred frequently; the liquor becomes dark green, almost black;
it is made up to a quart by water, the sediment is kept for a
fresh treatment; the quart of liquor is mixed with five gallons
water, preferably a hard spring water, and 1 lb. of silk, previously bleached and wetted out, is entered and worked in it
for thirty minutes; to obtain dark shades, four or five such
treatments are necessary. It would appear that the presence
of lime in the water facilitates the fixing of the color, and lime
is considered as being its appropriate mordant. The color so
produced is an agreeable green, whose most esteemed property
is that of keeping all its lustre, and nearly the same shade, by
gas or candlelight as in daylight. It is tolerably stable for a
fancy color. The Chinese dye common cotton cloths from a

similar green coloring matter, and obtain a deep, but somewhat dull, grass green. It appears that the green dye sent to Europe, and which sells at an enormous price, is obtained by extracting the excess of color from pieces originally dyed by a process that very much resembles indigo dipping. It is evident, therefore, that the green color used in Europe is a kind of lake, and that the Chinese retain themselves the original materials for dyeing a cheap green. M. Michel has shown that certain plants, indigenous to France, contain a colorable matter, which, upon exposure to air and light, give a green similar to the Chinese green (see BUCKTHORN), and has endeavored to apply them practically as dyeing materials; but it does not appear that these natural self-colored greens have as pleasing an effect as the greens compounded by blue and yellow.

Chinese green is interesting, as being the first known green not compounded of blue and yellow. The chemists and colorists who examined the samples of green-dyed cloth, obtained by the agents of the French Government, were surprised at not being able to split the color into its supposed elements, and were forced to conclude that it was a natural simple green; further inquiry showed their surmises to be correct. Since that period (1851) many attempts have been made to extract and apply the green color of grass and leaves, but up to this time, with no particular success.

Chlorate of Potash.—This salt is used in calico printing as an oxidizing agent, or, perhaps, as a chloridizing material, for it is difficult to see how it can directly yield oxygen in any of the forms in which it is applied. It is found in receipts for sapan wood reds, for chocolates, for greens and olives, and as a prepare for hastening the ageing of madder and garancine mordants. Its purity is ascertained by testing it with nitrate of silver and chloride of barium, with neither of which should it yield any precipitate. If the crystals are large and clear, they may be depended upon as pure.

Chlorides.—The class of bodies called chlorides are so called because they contain the element chlorine, in combination with some metal or basic substance. They can, for the most part, be produced directly from chlorine gas and the metal, but are more usually prepared from hydrochloric acid and the metal. Thus, hydrochloric acid, which is a compound of hydrogen and chlorine, when put into contact with zinc dissolves it with effervescence and escape of a gas. The gas is hydrogen, and the chemical action consists in the metal zinc taking the place of the gas hydrogen, and the result is chloride of zinc. The term muriate, frequently used in this book, and in general use in trade, is equivalent to chloride. Thus muriate of manganese and chloride are the same. So also muriate and chloride of iron, etc. The chlorides have few general properties as a class, each one owes its character more to the base than to the chlorine. All the chlorides are soluble in water, except those of lead, mercury, and silver; and all the soluble ones are sufficiently characterized by giving a white precipitate with nitrate of silver, which is not dissolved by pure nitric acid. The individual chlorides or muriates are described under the head of their metallic base.

Chlorine.—This is the element spoken of in the above article. It is a gas of a yellowish-green color, and has a most suffocating and irritating effect when inspired. It is produced when bleaching powder and vitriol are mixed together, and is the real bleaching agent in all cases where chloride of lime or bleaching powder is employed in conjunction with an acid. When chlorine gas was first proposed as a bleaching agent, by the celebrated Berthollet, it was used much in the same way as sulphur is now in bleaching woollen goods; afterwards a solution of this gas in water was employed, but the final improvement was Tennant's patent, of combining the gas with lime, to form chloride of lime. In more recent times it has been proposed to return to the older methods, but it is difficult to see where the advantage lies.

Theory of Chlorine Bleaching.—Chlorine is supposed to destroy colors, by combining with the hydrogen of them to form hydrochloric or muriatic acid. But as dry colors are not destroyed by dry chlorine, some theorists consider that water acts an essential part, and that the oxygen of the water is the real bleaching material, it being set free by the chlorine removing its hydrogen. Either hypothesis presumes that a coloring matter is such on account of a nicely-balanced arrangement of the atoms of carbon, oxygen, and hydrogen composing it, and if this arrangement be disturbed by the removal of a portion of the hydrogen a new arrangement ensues. There is nothing which would enable us to predicate of this new arrangement that it must produce a colorless body; but in the vast majority of cases this is evidently the fact. But it is an accidental fact; and it would be incorrect to assume this as an essential property of chlorine. There are some substances at least to which chlorine communicates a color, such as colorless solution of aniline, which it turns purple, or white silk or woollen, which it turns yellow.

Chlorine acts upon cotton in a very energetic manner when warm and concentrated, and it acts injuriously even in the cold and dilute if contact be prolonged. Care must be taken, therefore, in all chlorine treatments to get the chlorine well out of the cloth as soon as it has performed its part.

Chlorophyll.—This is the name given to the green coloring matter of leaves, grass, etc. Until the introduction of the Chinese green color, no idea was entertained of applying this coloring principle to dyeing; but since that time several attempts have been made to extract it in a state fit to dye with, but as far as my information extends, with but little practical effect. By treating common grass with boiling water, so as to remove all soluble principles, and then digesting it in carbonate of soda, at 4° Tw., the green coloring matter will be dissolved in a tolerable state of purity. The chlorophyll being of a resinous or waxy nature, is not sensibly acted upon by water only, but yields to alkaline solutions; this property enables us to separate the various soluble matters in grass from it. By neutralizing the alkaline solution with muriatic acid the coloring matter is then thrown down in an insoluble flocculent precipitate. This precipitate being washed a little, dissolved in an alkaline solution, thickened and mixed with salt of tin, yields green colors upon wool and silk by printing. Upon tin mordanted cloth the shades can be obtained by dyeing. It is not probable that chlorophyll will ever be an important dyeing material, on account both of its instability, its cost, and the comparative dulness of the shades it yields.

Chocolate Colors.—A chocolate is a dark brown, in which the blue part overbalances the red and yellow. There are many shades of chocolate, such as red chocolate, black chocolate, purple chocolate, green chocolate, etc. In M. Chevreul's nomenclature the color of chocolate in cakes is defined as 5 orange 18½ tone; but *puce* and *grenat*, which are French equivalents for some of our shades of chocolate, are classed very differently, *puce* being 4 blue violet 13 tone, and *grenat* 3 violet red 16 tone. I give some receipts for chocolate colors, with observations how to proceed in modifying the shade.

Chocolate Colors by Dyeing.—There is little to add on this head to the information given under browns. The general rule is to increase the blue or violet part of the brown, to produce the chocolate shade. As an illustration, let us take a catechu brown upon cotton, dyed by immersion in catechu, and fixed in bichromate of potash. To convert this into a chocolate, work the cotton in logwood, and raise in alum. This adds blue to the neutral or rather yellowish-brown, and converts it into a chocolate, the shade and depth of which depend upon the proportions of materials used, and may be varied at will.

In the cases of brown dyeing where the red, yellow, and blue parts are distinctly defined, it is entirely a question of quantity of blue as to what shade of chocolate is produced. For example, several kinds of browns are obtained by first dyeing a yellow with a tin mordant and bark, and then dyeing again in a mixture of some red wood (like peachwood) and logwood. If now, the peachwood be left out altogether, and alum and logwood alone used, the resulting color will be a chocolate instead of a brown, because the logwood dyes a blue with little red in it, which, mixing with the yellow already formed, gives the chocolate shade.

Chocolate Colors by Printing.—The colors for silk, wool, and delaine are very similar, so that what answers for one will frequently answer for the other.

Chocolate for Silk.

1 gallon sapan wood liquor at 20°,
1 quart logwood liquor at 20°,
10 oz. alum,
2 oz. sal ammoniac,
8 oz. acetate of copper,
5 lbs. gum.

In this color we only have the red and blue elementaty colors, but the blue is of a very dark nature, and the mixture will produce a bluish or purplish chocolate; the want of a yellow makes it heavy and dark.

Chocolate for Silk.

1 gallon peachwood or sapan wood liquor at 5°,
1 quart berry liquor at 11°,
1 gallon logwood liquor at 5°,
2½ lbs. starch,
4 lbs. gum substitute; boil, and while warm add
1¼ lbs. alum; when cold,
10 oz. sulphate of copper.

In this color the three elementary colors are present, and the quality of the chocolate may be varied at will by increasing the strength or quantity of the red, yellow, or blue part. The alum is the real mordant; the sulphate of copper is useful in developing the colors. As with other copper salts, its action is supposed to be of an oxidizing nature.

Dark Chocolate for Silk.

1 gallon sapan or peachwood at 20°,
5 quarts berry liquor at 17°,
7 pints of logwood liquor at 20°,
15 lbs. gum in powder,
2¼ lbs. alum,
8 oz. sal ammoniac,
14 oz. acetate of copper.

Archil Chocolate for Wool.

1 gallon archil liquor at 17°,
1 gallon gum water,
2 oz. sulphate of indigo.

Dark Archil Chocolate for Wool.

5 quarts blue archil at 17°,
10 oz. alum,
1½ oz. sal ammoniac,
1½ oz. oxalic acid, agitate until the effervescence has subsided,
    then thicken with
14 oz. starch, and
14 oz. calcined farina, and add
5 oz. sulphate of indigo.

In order to make this chocolate darker it may be mixed with a small quantity of black color for wool. (See BLACK.) To obtain a redder chocolate cochineal liquor is the best addition, and the ammoniacal cochineal is to be preferred.

Red Chocolate on Wool from Archil and Cochineal.

1 gallon archil liquor at 17°,
7 oz. prepared ammoniacal cochineal,
4 oz. alum,
1 oz. oxalic acid,
4 oz. sal ammoniac; stir until the effervescence has subsided,
    strain, and thicken with
1 lb. starch, and then add
3 oz. sulphate of indigo.

The colors from archil have a peculiar softness and lustre upon wool, which causes it to be largely used; but, as this coloring matter has no affinity for cotton, it cannot be employed upon cotton goods, and only sparingly upon delaines, where, in fact, it is useful for the woollen part of the fabric only. Chocolates may be also obtained upon wool by the process given below for delaine.

Dark Chocolate for Delaine.

1 gallon sapan wood liquor at 11°,
1 pint bark liquor at 14°,
1½ pint logwood liquor at 14°,
6 lbs. gum,
1½ lb. alum,
1 lb. muriate of copper crystals,
5 oz. sal ammoniac.

Dissolve the last three ingredients in one gallon warm gum water, and mix all together.

Another Dark Chocolate for Delaines.

2½ gallons sapan liquor at 8°,
3 quarts red liquor at 16°,
3 quarts nitrate of alumina,
1 gallon logwood liquor at 10°,
3 pints bark liquor at 18°,
7½ lbs. starch, boil; and when nearly cool add
4 oz. red prussiate of potash,
8 oz. yellow prussiate of potash,
8 oz. chlorate of potash.

The blue part in this chocolate is partly composed of Prussian blue. This color can be made darker by addition of a small quantity of black color, of the following or similar composition :—

Black for Darkening Chocolate.

1½ gallon logwood liquor at 12°,
1 quart iron liquor at 24°,
1 quart red liquor at 16°,
2 lbs. starch.

Another Chocolate for Delaines.

1 gallon sapan wood liquor at 30°,
1 gallon red liquor at 24°,
½ pint vinegar,
3 lbs. starch,
4 lbs. gum substitute; boil, and add
¼ lb. alum, dissolved in
1½ pint logwood liquor at 30°, and
1 pint bark liquor at 30°, then add
1 lb. chlorate of potash, dissolved in
3 quarts of tragacanth gum water, and
8 oz. muriate of ammonia,
3 oz. sulphate of copper.

10

This is a French receipt. The liquors being very concentrated, it would yield a good chocolate for small objects. As a blotch it is too strong, and this would be found upon washing off, on account of the large amount of color which would come out. It might be worked with rollers engraved for calico. In the above and following receipts, the oxidizing powers of chlorate of potash are employed, and the quantity of copper salts greatly reduced. This is an advantage in some respects, as for example, in the printing it cleans better, because the doctor is not so liable to corrosion; but in steaming the color is liable to evolve chlorine, and attack pale colors from woods, but it is not to be feared in conjunction with masses of blue, green, or olive.

Gum Chocolate for Delaines.

1 gallon sapan wood liquor at 20°,
1 gallon nitrate of alumina,
1 quart bark liquor at 30°,
3 pints logwood liquor at 30°,
14 lbs. gum in powder.
10 oz. chlorate of potash, dissolved in
8 quarts of boiling water; and, lastly,
3 oz. sulphate of copper.

The nitrate of alumina for the above chocolate is obtained as follows :—

2 gallons hot water,
10 lbs. alum,
13 lbs. nitrate of lead.

The nitrate of alumina supplies alumina as a mordant, and at the same time nitric acid as an oxidizing agent. I conclude the chocolate receipts on delaine by one on which a portion of the blue is, as in a previous case, composed of Prussian blue, derived from an impure solution of red prussiate of potash, called chloro-prussiate liquor.

Chocolate for Delaine.

1 gallon sapan wood liquor at 16°,
3 lbs. starch ; beat up, and add
3 quarts red liquor at 16°,
2 quarts logwood liquor at 12° ; boil, and add
3 oz. tartaric acid,
6 oz. sal ammoniac,
5 pints chloro-prussiate liquor at 30°,
¼ pint oil.

The chocolate colors for calico have a great resemblance to those for delaines, but are less concentrated, on account of the lesser capacity of pure cotton for absorbing coloring matters. One or two examples will suffice as specimens, although the modifications are numberless.

Paste Brown Chocolate for Calico.

9 quarts sapan wood liquor at 8°,
9 quarts water,
6 quarts logwood liquor at 12°,
3 quarts red liquor at 16°,
12 oz. muriate of ammonia,
12 oz. sulphate of copper,
12 oz. alum,
13½ lbs. flour,
3½ lbs. British gum.

The liquors are thickened, and the salts stirred in as usual.

Spirit Chocolate for Calico.

3 quarts sapan wood liquor at 8°,
2 quarts logwood liquor at 10°,
1 quart bark liquor at 14°,
2 lbs. starch; boil, cool to 110°, and add
1 pint oxymuriate of tin,
½ pint nitrate of copper,
1 pint oil.

To be aged three days in a cool place, and washed off.

Red Chocolate for Calico.

3 gallons sapan wood liquor at 9°,
3 quarts nitrate of alumina (see below),
1½ gallon logwood liquor at 12°,
6 oz. yellow prussiate,
6 oz. red prussiate,
6 oz. chlorate of potash,
9 lbs. starch.   To be boiled well.

Nitrate of Alumina for Red Chocolate.

1 gallon boiling water,
3 lbs. nitrate of lead,
3 lbs. alum ; dissolve, and add
3 oz. crystals soda.   Use the clear only.

Another Chocolate for Calico.

5 quarts sapan wood at 8°,
3 pints red liquor at 16°,
3 pints nitrate of alumina.
4 pints logwood liquor at 18°,
1½ pint bark liquor at 18°,
4 lbs. starch; boil, and, when cool, add
2 oz. red prussiate of potash,
4 oz. yellow prussiate of potash,
4 oz. chlorate of potash.

For the chocolate colors produced in madder and garancine styles see GARANCINE and MADDER respectively.

Chromates.—The chromates are a class of salts formed by the union of chromic acid with a metal or base. They are all colored, without exception; but the dyer has only been able to avail himself of two of them as coloring matters, viz., the red and yellow chromates of lead.

The only commercial chromates are those of potash and soda, and there are two of each. The bichromate, or red chromate of potash, is a salt containing two atoms of chromic acid to one of potash; the yellow chromate of potash contains but one atom of chromic acid to one of potash, and is, consequently, less rich in chromic acid. There is a corresponding bichromate and chromate of soda, but they are not commercial articles. The salt sold as chromate of soda is of variable composition, and cannot be correctly represented by any chemical name or formula.

Bichromate of Potash, Bichrome, Red Chrome, etc.—This is the chief and only trustworthy chromate for the use of the dyer and printer. If it be in clean, well-defined crystals, and of a uniform color, without admixture of white or yellow crystals, it may be considered as pure. It does not lose weight by drying, being an anhydrous salt. A gallon of cold water will dissolve about one pound of bichromate; in hot water it is much more soluble, but the excess over one pound crystallizes out on cooling.

Yellow Chromate of Potash.—This salt is rarely met with in trade; it can be prepared by adding caustic potash to solution of red chromate until it becomes slightly alkaline; the red chromate loses its characteristic color, and becomes of an intense yellow. In practical receipts it is sometimes directed to neutralize bichromate with soda crystals; this is practically making neutral or yellow chromate. The yellow chromate is much more soluble in water than the bichromate, and is often used in printing or padding on that account.

Chromate of Soda or Chrome Salts.—The samples of yellow chrome salts that I have had occasion to test or examine have varied so much in quality and actual value, and are so easily adulterated, that I would advise having nothing to do with them unless their quality can be satisfactorily ascertained. If they contain a proportionate quantity of chromic acid, they can be applied to all the purposes of bichromate of potash with equal results.

Chromate and Dichromate of Lead.—These substances are trade pigments, but not applied in printing or dyeing, They form the yellow and orange chrome colors on cotton; but as they are produced by a dyeing process in the fibres of the cloth they are considered under CHROME COLORS.

The yellow chromate of lead is sometimes employed in printing as a sightening for chrome mordants. It is prepared by mixing nitrate of lead and bichromate of potash, both in solution, washing the precipitate and draining to a pulp.

Applications of the Chromates.—Bichromate of potash is applied in printing and dyeing in a great variety of ways, the whole of which may be classified in three divisions. (1) *Cases where the chromic acid acts as a coloring matter :* these are all included in the article on CHROME COLORS. (2) *Cases in which the chromic acid acts as an oxidizing agent.* (3) *Cases in which the application depends upon the oxide of chromium ;* the third class of cases are included in the article on CHROMIUM COLORS. It only remains, therefore, to indicate here the cases in which the oxidizing powers of bichromate are brought into play. The raising or development of the steam blue, green, and olive colors, depends upon the oxidating effect of bichromate of potash; so also the fixing of catechu colors, and those few cases in which bichromate of potash is used in color mixing. The chromate discharge upon indigo blue is another illustration of its oxidizing powers, in this case exerted, not to develop, but to destroy a color. (See DISCHARGE.) Whether the use of bichromate in woollen dyeing belongs to the second or third class of applications is a point upon which there appears to be no satisfactory information. Some experimenters incline to believe that the salt, as a whole, is taken up by the wool, and that when worked in the dye-wood it is deoxidized by the organic matters, reduced to the state of green oxide of chromium, in which condition it acts as a mordant. It seems proved that nearly the same colors are obtained whether the chromed wool enter the dye in its yellow state, or whether it be brought to the green state by deoxidizing agents, but that the color is more quickly dyed when it is in the yellow state. Other experimenters think its principal action is in oxidizing the wool.

In,cases where the wool has been sulphured before dyeing, I should attribute part of the useful action to the oxidation of the sulphurous acid retained in the wool.

In steam blues, and catechu colors, raised by means of bichromate of potash, a small quantity of chromium remains attached to the cloth or color; this has been looked upon as constituting an essential part of the color; but, as it is very evident, that other oxidizing agents can procure the same effect, and it is at least probable, that in the very act of oxidation, some oxide of chromium falls on the fibre and is retained, it is more reasonable to look upon the presence of chromium as accidental rather than as essential to the color.

Chrome Colors.—As before mentioned, the only colored chromates which adapt themselves to the wants of the dyer and printer are those of lead, and the chrome colors are consequently limited to the orange and yellow shades yielded by the lead basis. The chromate of lead, which contains single atoms of the acid and base, is yellow; it is produced by simply mixing solution of bichromate and any salt of lead. The orange-colored chromate of lead is produced by acting upon the yellow with dilute alkalies, such as lime water or weak caustic soda. The alkali abstracts one-half of the chromic acid from the yellow chromate of lead, leaving a compound which contains two atoms of lead to one of chromic acid, and which has a deep orange color. If the alkali is strong, or its action long continued, it abstracts the whole of the chromic acid from the lead, leaving it colorless, and ends by dissolving up the lead and all. Hence, in turning the chrome yellow into orange, much caution has to be exercised not to pass the right point.

In dyeing chrome colors the cloth is first mordanted in a salt of lead, and then passed into bichromate. The insoluble chromate of lead is formed by double decomposition in the fibre, where it is retained. This is the yellow salt. It is converted into the orange by alkalies. The following practical process will illustrate the various methods of proceeding to obtain the yellow and orange shades:—

Chrome Yellow and Orange by Printing.—The mordant is very simple, either acetate of lead or nitrate of lead, or a mixture of the two salts.

Dark Paste Orange.

2 lbs. brown sugar of lead,
1 gallon water; dissolve, thicken with
1½ lb. flour;

sighten with precipitated chromate of lead pulp. The amount of sugar of lead may be increased to as high as 8 lbs. to a gallon of water when dark shades are wanted, or when the design includes small objects.

Pale Paste Orange.

1 gallon water,
¾ lb. nitrate of lead,
1½ lb. of flour.

There is apparently some advantage in employing nitrate of lead for the paler shades, but the acetate could be equally as well used. After printing, the goods are aged to soften them, and then passed in warm dilute sulphuric acid or sulphate of soda. This fixes the lead upon the cloth as sulphate. The cloth is washed to remove any loosely adhering lead salts, and then entered into the bichromate. From two ounces to half a pound of bichromate are added to a beck of warm water for each piece of calico, and the goods run in for about ten minutes, when they will have acquired a full yellow, and, if to remain yellow, must be taken out then. To convert them into orange, the readiest method is to lift the pieces when they have acquired a full yellow, and add half a pint of caustic soda, at 30° Tw., for every five pounds of bichromate employed; stir up well, and run the pieces again for ten minutes. If the full orange shade is not developed, a little more soda may be cautiously added, taking care to stop the pieces, and turn them into clear water, upon the slightest appearance of their losing color. Another process consists in taking the pieces at the yellow and wincing them in water, and then raising the orange by means of boiling lime water, or very weak caustic liquor. Or, again, instead of passing the pieces in bichromate, they may be passed into neutral chromate, made by adding crystals of soda, or caustic soda, to bichromate. The method given first leaves nothing to desire if carried out with care and intelligence.

Sulphate of lead, though an insoluble salt, when printed upon calico and dipped in lime, forms an intimate combination with the fibre, and may serve very well as a mordant for chrome orange. One method of applying it consists in melting brown acetate of lead, and adding to it strong sulphuric acid; a great portion of the acetic acid is expelled, and the lead wholly or partially converted into sulphate. This, slightly thickened, printed, dipped in lime, and raised in chrome as above, gives very good oranges. This method of obtaining the chrome orange is subject to irregularities, and is not to be recommended.

The yellow from chrome is scarcely ever produced in printing. The orange is very often worked in combination with iron buff. When orange alone is produced sulphuric acid is the best fixing agent, but if accompanied by buff, sulphate of soda must be used, and, if necessary, some carbonate of soda added to it.

Chrome Colors by Dyeing on Calico.—The colors produced by the chromates of lead upon calico by dyeing receive various names, according to their depth. A very light shade, obtained by using from 2 lbs. to 3 lbs. of acetate of lead for 100 lbs. of cotton, and raising in 2 lbs. bichromate, may be called straw color. The process consists in working the goods for twenty minutes through the acetate of lead liquor; drain, and work through the chrome ten minutes, and finish in the lead. The shades are sometimes modified by using anotta along with the chrome.

Lemon Yellow.—10 lbs. sugar of lead, 4 lbs. bichromate of potash, 100 lbs. cotton. Work as in the previous case.

By increasing the weight of drugs, or by repeating the dips, any desired depth of yellow can be obtained. Nitrate of lead may be used instead of the acetate.

Plombate of Soda Yellow.—In this method advantage is taken of the power which alkalies possess of dissolving hydrated oxide of lead in the following manner: For 100 lbs. of cotton, 5 lbs. bichromate and 10 lbs. of acetate of lead are taken. The acetate of lead is dissolved in water, and strong caustic soda is gradually added; it produces a bulky white precipitate at first, but continued addition of the soda dissolves this precipitate. In order not to have an excess present, it is well to leave a little of the precipitate undissolved. The goods are worked in the clear liquor, and then passed into the chromate as before.

Chrome Orange.—The orange is obtained by working the yellow in boiling lime-water, or weak caustic, as in calico printing.

Subacetate of Lead Orange.—A subacetate of lead is prepared by boiling 20 lbs. brown sugar of lead and 10 lbs. litharge in a sufficient quantity of rain water, working the goods in this solution, which yields up its lead more easily than common acetate, and fixing in lime-water. As it is necessary to have a considerable quantity of lead on the cotton for deep orange, this process must be repeated two or three times, then worked in the chrome 10 lbs., and the orange color raised by boiling lime-water.

Sulphate of Lead Orange.—By working the cotton in acetate of lead, wringing, and passing in sulphuric acid sours, at 4° Tw., then washing, and passing in warm bichrome and boiling lime-water, a full deep orange can be obtained without any repetitions, and with great certainty and regularity.

Plombate of Lime Orange.—Similar to the plombate of soda yellow, only that lime-water is employed, instead of soda, in excess, to dissolve the oxide of lead precipitated in the first instance. It requires three or four repetitions to obtain the darkest shade of orange.

The chromate of lead colors have a moderate degree of stability; they are weakened by soap and friction, and also by washing soda. Like all colors containing lead, they are blackened by sulphuretted hydrogen, so that they are entirely unfitted for hangings for dwelling-houses, where sulphurous emanations are always more or less present.

Chrome yellow forms the yellow basis of some green styles in which indigo is the blue, and of some shades brown and chocolate. (See BROWN and GREEN.)

Chromium.—Chromium is the name of the metallic basis of the chromates; it combines with oxygen in two proportions, the one compound being called oxide of chromium, or green oxide of chromium, and the other chromic acid. The pure metal chromium is almost unknown, and the oxide is always obtained from the chromic acid compounds or chromates by one of the methods given below. A class of colors is obtained from the oxide of chromium, which may be distinguished as CHROMIUM COLORS. A pigment green, which is an oxide of chromium, prepared by a peculiar process, has been used in calico printing, fixed by means of albumen or lactarine. This preparation, known as Guignets' green, is obtained by making an intimate mixture of about three parts of boracic acid with one part of bichromate, and calcining it, at a low temperature, in a reverberatory furnace; the chromic acid loses oxygen, and becomes changed into green oxide of chromium. There are many ways of preparing the green oxide different from this, but that which is produced by this particular method has a beauty of color entirely peculiar to itself. It is sold in the pasty state, at a price about equivalent to 10s. per lb. of the dry oxide; this high price has considerably limited its applications in printing.

Chromium Colors.—The chromium colors are those which have oxide of chromium as a basis. The first process is to make some salt of chromium, which is printed, and then raised or fixed in lime or soda, precisely the same as an iron buff.

Sulphate of Chromium Standard.

1 gallon of boiling water,
4 lbs. bichromate of potash,
2½ lbs. sulphuric acid,
1 lb. brown sugar;

dissolve the bichromate in the water, add the acid with care, and then the sugar, by small portions. A violent action follows each addition of sugar, accompanied by escape of gas and swelling of the liquid; in order to prevent loss, it is necessary, therefore, to have large vessels, and to add the sugar with care.

Muriate of Chrome Standard.

2 quarts boiling water,
2 lbs. bichromate potash,
4 lbs. muriatic acid,
14 oz. sugar;

proceed as in the making of sulphate of chromium, observing the same precautions.

The sulphate of chromium is the more frequently employed of these two solutions: when properly prepared, it is a viscous dark green fluid; sometimes, owing to a deficiency of acid, or the heat not being high enough, it sets into a jelly, or is full of curdy lumps. Such a color will be very unsafe; it should be made quite fluid either by heating it, or if that is not sufficient by adding more acid. It should stand about 90° Tw. It is difficult to thicken properly, having a great tendency to coagulate starch and bad gums. Three quarts of the standard to one quart of thick tragacanth gum water will usually work well; sometimes the liquid has a sufficient amount of thickness to work without thickening.

When raised in lime and soda it produces a grayish-green, of a shade similar to green tea leaves, hence sometimes called *tea color;* it has also been called *Victoria green.*

Arseniate of Chromium Standard.

9 gallons of hot water,
9 lbs. bichromate of potash,
11 lbs. white arsenic;

boil this mixture for fifteen minutes; an action takes place, and a precipitate forms, which is collected upon a filler and drained to a pulp; the pulp is scraped into a mug and mixed with 3 quarts of nitric acid, and kept at a boiling heat until the pulp is dissolved, or nearly so; when cold, mixed with 3 quarts of acetic acid at 8°. The clear liquor should stand at from 80° to 90° Tw., and may be thickened with tragacanth gum water.

A much simpler receipt is as follows:—

1 gallon of water,
5 lbs. bichromate of potash,
7 lbs. white arsenic,
10 lbs. muriatic acid;

heat until the bichromate is entirely deoxidized : if the acid is not sufficient, a little more must be added ; but, in order to avoid an excess of acid, it is well to boil the liquor down to 90° or 100° Tw., by which the free acid is mostly expelled.

In both these processes the white arsenic, or arsenious acid, takes oxygen from the chromic acid, and becomes arsenic acid, which combines with the oxide of chromium produced to form arseniate of chromium, which is kept in solution or dissolved by excess of acid.

Raised in lime, these liquors produce the same grayish-green shades as the sulphate and chloride of chromium.

By passing the chromium green shades through weak solution of blue copperas, a somewhat livelier tint is obtained.

Modifications of the Chromium Shades.—By mixing other mineral colors, or vegetable extracts, with the chromium standards, a variety of shades, all of a dull character, may be produced. Thus, buff liquor, bronze liquor, etc., may be mixed with them. A shade of color of a soft gray or drab may be produced as follows :—

Green Dove Color.

1 gallon of red liquor at 8°,
1 lb. of ground madder; steep 24 hours, and strain
2 quarts of the above,
3 pints of sulphate of chromium ;

thicken with soluble gum substitute. Raised in lime or weak soda, it gives a pleasant greenish dove color.

Oxide of chromium appears to be of no use as a mordant; it has but a very slight affinity for coloring matters, and the shades it gives are dull and dry.

Chrome Alum is a double sulphate of chromium and potash, and may be used for the same purposes as sulphate of chromium. It is employed on the continent, but is not, as far as I am aware, an article of commerce in this country.

Chrysammic Acid.—A golden yellow colored substance, obtained by the action of nitric acid upon aloes. It gives highly colored salts, and hopes were entertained of making it applicable to dyeing, but so far it has not been practically applied.

Cinnamon Color, *Cannelle.*—The color of Cinnamon, as exposed for sale in this country, may be defined as a yellowish brown, of a rather low tone. Chevreul defines it as a yellowish orange of a deep tone, or orange darkened with black ; he gives. it two formulæ, 3 orange 14 tone, and 2 orange $\frac{5}{10}$ black of 9 to 12 tone. In dyeing, it may be considered as a yellow brown, and its production depends upon the excess of the yellow in the composite color.

Madder and bark cinnamon shades are produced by mordanting calico in red liquor, dyeing in bark first, and then in madder, until the desired shade is obtained. The bark and madder may be mixed in the dye-beck, but the shade is more under control when they are used separately, because the madder is a stronger dye stuff than the bark, and the brownish-red of the madder can displace or drive out more or less of the bark yellow. For producing rather darker shades of cinnamon, small quantities of iron liquor may be mixed with the red; but, if the proportions amount to more than one part of iron liquor to ten parts red liquor, the color becomes nearer a chocolate than cinnamon.

For common calico shades, the processes for brown may be followed, increasing the red wood and decreasing the logwood; the yellow part being preserved of the same intensity.

Calico mordanted in copper and raised in prussiate of potash, as in dyeing Prussian blues, gives a cinnamon shade.

Upon woollen, cinnamon shades are obtained by aluming as usual, and then dyeing in a mixture of madder and some yellow dye stuff, which may be either weld, bark, or fustic. In low class work the cheaper red woods may be used instead of madder, but in that case, a little iron will have to be employed to sadden down the shade.

The class of cinnamon colors obtained by printing are illustrated by the following receipts:—

Cinnamon for Wool or Delaine.   Block.

2 gallons of bark liquor at 18°,
2¼ lbs. of ground cochineal; keep hot for some
      time, then pass through a straining cloth,
      and thicken with
7 lbs. gum senegal, and add
6 oz. sulphate of indigo,
18 oz. crystals of tin.

This color contains a large quantity of red and yellow to a small quantity of blue, and consequently yields an orange-colored brown, tending to the yellow side. Cinnamon colors upon calico can be obtained by modifying the receipts given under brown.

Citric Acid.—This acid is obtained from the juice of lemons, limes, and similar fruit. When pure, it is in colorless crystals, of a strong, pleasant acid taste; dissolving easily in hot or cold water. The pure acid is not much used either in dyeing or printing, on account of its high price; but in some places it is employed as a resist on madder work, and for throwing down the coloring matter of safflower after it has been dissolved by alkali, but for either of these purposes ordinary lime juice of good quality will answer nearly if not quite as well. Citric acid is occasionally adulterated by admixture with tartaric acid, which, besides being a fraud, is liable to cause much injury to the printer. It is possible to discover five per cent. of tartaric acid, and any greater quantity in citric acid, by means of caustic potash in the following manner; dissolve a couple of ounces of the acid to be tested in as little water as possible, from three to four ounces at the most, then take one half of the solution in a glass tumbler, and add strong caustic potash (soda will not do) drop by drop until the acid is neutralized, then mix with it the other half of the dissolved acid, and stir well up together. If a heavy white crystalline powder falls to the bottom of the glass, it indicates tartaric acid; if the quantity of tartaric acid is small, it will not appear directly, and the glass should be left for about six hours in a cool place, when, if the liquor is quite clear, or only a little gelatinous substance in it, the citric acid may be considered as being free from tartaric. ·

Citric acid, whether in pure crystals or in lime juice, is the best resistant for iron and alumina mordants. Its power does not rest simply in its acidity, although that is very great, being able to neutralize or dissolve nearly as much of an oxide as an equal weight of oil of vitriol, but partly in a power, like that of tartaric acid, of masking the usual properties of metals, and putting them beyond the action of the agents usually influenc- ing them. That it is not simply the acid is evident from the fact that when the acid is quite neutralized with either potash or soda, the citrate of potash or soda is, for the quantity. of citric acid present in any given bulk, nearly as good as before for resisting, though it cannot act as a discharge. When an iron mordant is printed over a citric acid resist, citrate of iron is formed, which, unlike the majority of the salts of iron, does not oxidize on the cloth, even when exposed for a very long time to the air, and which can be removed from it by simply washing in water; the usual fixing agents having no action upon the salt. When citrate of potash is employed, its first action is mechanical, to receive the superimposed mordant, and then to effect a decomposition by which citrate of iron is formed, and so kept from fixing itself upon the fibre of the cloth. That citric acid has the power of changing the bearings of metals to other bodies, is capable of being proved easily: if caustic potash be added to a weak solution of copperas, it precipitates all the iron as oxide; but if a sufficient quantity of citric acid be mixed with copperas, and then potash added in any quantity, there will be no precipitate formed, owing to some agency of

the citric acid; the same or a similar agency keeping the iron
or alumina of the mordant from precipitating upon the cloth,
whenever it meets the citric acid.

The combinations which citric acid makes with the alkalies
and metals are not employed in dyeing or printing, except the
citrate of soda, mentioned above, which is used to resist catechu
brown, and in one or two other cases.   :

Lime Juice, or *Lemon Juice.*—The acidity of this juice is
owing to citric acid, and its value depends upon the quantity
of this acid which is present in it. Lime juice is very variable as to quality, depending upon the method of extraction,
the quality of the fruit, and the honesty of the shipper. The
best kind in English market, for the use of printers, is a dark
treacly-looking fluid, marking from 48° to 54° Twaddle, and
containing from 30 to 36 per cent. of pure citric acid. There
are many inferior qualities which, though standing nearly as
high on the hydrometer, contain not more than two-thirds or
one-half of that quantity of real acid. The sellers in this
country charge the acid so much per gallon per degree, as indicated by an arbitrarily graduated instrument, supposed to
show the percentage of pure citric acid in the lime juice, but
which gives unreliable results. At certain times all lime juice
is bad, containing a great deal of sediment and some peculiar
substance which prevents it giving good whites as a resist for
madder and garancine work. This is understood to be owing
to a bad harvest of limes; the quantity of fruit being much
less than usual, the manufacturers abroad appear to press it
more completely to make up the quantity, besides using
damaged fruit and substances, whose only resemblance to limes
or lemons consists in their giving some kind of juice; but the
citric acid—the real, active element of lime juice—is not there
in the proper quantity, or is injured by a mass of other matters.
It may be generally observed that as lime juice becomes dearer,
its quality deteriorates; as it becomes cheaper, it gets clearer,
less sediment in the cask, and sharper and more agreeable to
the taste. The strength of lime juice cannot be well ascertained by Twaddle; the real test is to ascertain how much
alkali it will neutralize, and that no cheap acid has been added
to make it apparently strong. I have found lime juice adulterated with potashes, which raises the density without (for a
certain quantity) materially injuring its resisting power; for
citrate of potash itself is known to resist tolerably well, and
especially when mixed with free citric acid. The adulteration
can be detected by mixing very strong solution of tartaric acid
with the suspected lime juice; if potash be present there will
be formed in a short time a crystalline deposit of bitartarte of
potash at the bottom and on the sides of the vessel. It must

be observed that some potash is naturally present in lime juice, and care must be made to distinguish between what may be allowed and what is evidently a falsification. Citric acid is made from lime juice by adding ground chalk to the acid mixed in water, washing the citrate of lime from the impurities which are dissolved in the water, and afterwards decomposing it with sulphuric acid; if the citric acid is not white enough animal black is used to decolorize it. It is possible for the calico printer thus to purify his lime juice, and to bring it into citric acid if he thinks proper.

Analysis.—Citric acid or lime juice may be tested in just the same manner as acetic acid and other acids. (See ACIDIMETRY.) I have found citric acid to contain as much as ten per cent. of tartaric; this was in a sample of brown acid sent as of the first crystallization—without doubt it was purposely adulterated. I have never found the finished white crystals adulterated.

Lime juice is only valuable on account of the citric acid which it contains, and which varies considerably. A good quality of lime juice, marking from 46° to 50° Tw., will neutralize from 70 to 76 grains of pure crystallized carbonate of soda. For commercial purposes each grain of carbonate neutralized may represent a half grain of crystallized citric acid (equal to 0.38 grain of dry acid), and the value of the lime juice be calculated in proportion. Upon evaporation and calcination at a red heat the citrate of soda will be converted into carbonate, and being tested by the alkalimeter should indicate the same amount of alkali as was at first added; sometimes it will be found to indicate more, which arises from potash present in the lime juice; to correct this a quantity of the lime juice should be evaporated and incinerated separately. I have had occasion to test samples of lime juice not containing more than 18 per cent. of citric acid although marking full strength on the hydrometer; they were overloaded with vegetable extractive matter.

Citron, or *Lemon Color.*—The lemon yellow or citron color may be considered as pure yellow with a little red or orange in its composition; in Chevreul's nomenclature 4 yellow orange 6 tone. The chrome yellows upon cotton, and picric acid yellows upon wool and silk, may be called lemon yellows. Weld and alum give very pure lemon yellows. The following receipt is for a lemon yellow upon wool or delaine—

> 1 gallon berry liquor at $4\frac{1}{2}$° Tw.,
> 3 lbs. gum,
> 8 oz. alum,
> $1\frac{1}{2}$ oz. oxalic acid,
> 8 oz. oxymuriate of tin.

By mixing with a small quantity of red or scarlet the shade can be modified, as in the following receipt for machine:—

1 gallon berry liquor at 7°,
1 lb. starch; boil, and when cooled add
8 oz. oxalic acid,
10 oz. bichloride of tin, at 100°,
½ pint cochineal scarlet.

Cleansing.—In the process of calico printing, after the thickened mordants have been applied to the cloth, and exposed a sufficient length of time, the goods are ready for dyeing, in so far as the mordant has become fixed, and not removable by water. But there has been in all cases a great deal of mordant applied, which either never comes into actual contact with the cloth, or is more than it can absorb and retain. This removable by water, and, if the pieces were to be placed in the dye vessel, would dissolve, and seriously interfere with the dyeing. If the pieces were washed in water simply before dyeing, the object would be partially attained, that is, the loose mordant would be removed; but another difficulty would occur, the excess of mordant liberated at one point would be absorbed by the cloth at another where the design required a white or colorless part; or, in case of different mordants being on the same piece of cloth, they would intermix and spoil one another; the red would turn the purple into chocolate, and the blacks would give a purplish color to the reds. It was necessary for the calico printers to find some fluid in which the pieces could be washed from the excess of mordant, and the now useless thickening matter, which at the same time should prevent the loose mordant from being at liberty to fix itself upon any part of the fabric. Such a fluid was first found in a mixture of hot water and cow dung; but now various manufactured substances are successfully used for that purpose. Although cow dung left nothing to desire as far the cleansing was concerned, the supply was not regular—in certain localities it was scarce—and the animals have had to be kept near to the print works, actually for the sake of their dung. In other places the supply fails in the summer months, when the cows are grazing, and their dung is spread over the pastures; add to this the unpleasantness of working in such a material, and it will be easily understood why substitutes have been called for, and are now very extensively adopted.

The term "dunging" was naturally enough applied to this process, because nothing but the excrement of cows was formerly used for the purpose; but now that it seems probable that the use of cow dung will be given up the continuance of the name is neither desirable nor exact. The process is frequently called "cleansing," which is an appropriate name, since the real use of the process is to clean the cloth from loose matters which would interfere with the dyeing. This name is sufficiently distinct from "clearing," by which is understood processes subsequent to the dyeing. M. Persoz proposes to change the name of dunging into "fixing of mordants;" this, besides being a somewhat clumsy expression, is expressive of a theory which may not be true, since the fixing of the mordants takes place before dunging, if not wholly, at least in great part. I shall adopt the word "cleansing," as sufficiently characteristic, and on the whole preferable to the term dunging, which is in many cases obviously incorrect at this day.

The analysis of cow dung does not point out any particular principle in it which can be said to be the active agent in cleansing. Its power has been at various times attributed to each of several substances it contains: its albumen, and its peculiar animal matters, were supposed at one time to be the active elements; again, the mineral matters it contained were said to be the real principles which were useful in it, and so on in turn every single element has been at some time or by some person considered the essential matter in it. That cow dung does not possess any principle peculiar to itself, which enables it to be used for cleansing, is plainly evident from the fact of the successful employment of substitutes which have no resemblance to it in any way. But that it possesses some principles that fit it for such a duty is evident; but it does not seem necessary to fix upon any single one as the essential one, but rather to view the action exercised as one resulting from the combined influence of two or more of its constituents. My observations and experiments have led me to conclude that cow dung owes its efficacy to two things, namely, the finely ground up or chewed organic matter, the remains of the hay, grass, or other food of the animal, and to a species of greenish olive coloring matter which is present. The effect of a passage in dung appears to me in great part to be mechanical, and to be an illustration of the power of surface in attracting chemical matters. The undigested fibrous parts in the dung fix upon themselves the excess of mordant as soon as it leaves the piece, and so prevent it spreading either on the whiter or the neighboring colors. There is no difficulty in considering that this would be a sufficient explanation if it were allowed that the insoluble parts of the cow dung could exercise this affinity for the mordant which is set at liberty by the liquor. When it is considered that the chemical composition of the fibrous matter of cow dung, and its physical constitution also, resemble very

11

closely that of cotton fibre, it is not difficult to imagine it as
having at least as great an affinity for the mordant which is
loosened as the surrounding cotton fibres themselves; but,
because it is in a finer state of division, and in contact with the
actual particles of mordant, it has an advantage over the cotton
fibres which are at some distance, and could only receive the
excess of mordant through the medium of the bath; consequently, the superfluous mordant is retained by the insoluble
floating particles of the cow dung. This action of the cow
dung I consider all that is essential to it; it has other actions
upon the mordant and cloth, but they are either of no
importance in their result or only of secondary importance.
The coloring matter referred to seems to act a useful part, but
it is not clear what it is. The mordants take it up and retain
it through the washings; but in the dyeing it is driven out by
the stronger coloring matter of the madder or garancine. If a
fent, mordanted for black and purple, be dipped in hot caustic
soda, at one or two degrees of Twaddle, it will come out with
the mordants of a light buff shade; in this state they do not
dye well in madder—the colors produced are poor, and it takes
a longer time and higher temperature to dye them. If the
fent be taken from the caustic soda to a regular passage in
dung, it soon changes its color, coming to nearly the same
shade as if it had been at once passed into dung; in this state
it dyes up well and quickly. The change of color may be attributed to the absorption of coloring matter from the cow
dung; and the better colors produced upon dyeing, and the
shorter time required, must be attributed to some action of the
cow dung—not necessarily to the coloring matter, but the
weight of probability is in favor of it. Some actual chemical
change in the mordant is possible; if any, it must be in the
way of deoxidation. I have found a decoction of cow dung to
act powerfully as a deoxidizing agent.

It may not, perhaps, be accepted as a fact that the presence
of coloring matter of the cow dung would have any useful
action in dyeing, but I am convinced it has, and especially in
madder colors. Experience proves that, in the case of the
best dung substitutes, or cleansing compounds, a final wince in
cow dung before dyeing is advantageous. It is better for the
mordanting oxide that it should go into the beck in a partly
saturated state than in a state of the highest activity. In a
majority of cases the colors will be more solid, brighter, and
faster when the combination between the mordant and the
coloring matter is slow and gradual than when, on the contrary,
it is rapid and completely effected in a short time. In the case
of madder, this might be explained upon the assumption of a

variable displacing power of the true and false coloring principles present in it. It may be supposed that the dun coloring
principle cannot displace the coloring matter of the cow dung,
but that the alizarine can do so, and does so by degrees in the
course of the dyeing; and it may be supposed that, by this
means, a pure color is produced, which suffers less in the
clearing operations than if the mordant was partly filled with
the easily removable secondary coloring matters of the madder.
In other coloring matters it may be considered that they cannot
combine so rapidly with the mordant, because it is partly combined already; but this combination is a weak one, and gives
way to the power of a stronger agent, and, being formed
slowly, seems to be more stable; just as in painting, a thick
layer of paint, equal to the whole quantity required, might be
applied at once, but it would not be so good as if applied in
two or three separate coats.

The dung substitutes at present in use are chiefly the arsenite
and arseniate of soda, the silicate of soda, mixtures of the two,
phosphate of lime, and various compounds containing other
substances in addition to those named, but whose utility is of a
very questionable nature. As mentioned just now, caustic
soda would act as a dung substitute for black and purple mordants, but not for red mordants, because the alumina forming
a soluble combination with the alkali would be entirely removed. The silicate, arsenite, and arseniate of soda may be
looked upon as caustic soda, the more energetic properties of
which are modified by the arsenical and silicious acids. The
alkalinity is modified, not destroyed. The same may be said
of the carbonates, bi-carbonates, and phosphates of the alkalies,
which have been sparingly used as dung substitutes. The
chemical action of these substitutes differs from that of cow
dung; for, while I do not look upon cow dung as possessing
any notable fixing powers, it is certain that the arsenites and
silicates do enjoy such a power. They actually neutralize the
acid remaining in the mordant, and precipitate the oxide upon
the cloth in greater quantity than dung, and under different
circumstances. At the strength at which the substitutes are
employed in the cleansing apparatus, very little of this precipitation of oxide upon the cloth actually takes place; the action
of the substitute is confined to rendering the loosened mordant
insoluble at the moment it quits the cloth, and so preventing
its combining with other parts of the fabric. But, if the
strength be increased, the precipitation upon the cloth takes
place, and darker colors are produced—of course at the expense
of the dyeing material, and frequently also at the expense of
goodness of shade. It is not well that the cleansing liquor

should be also a fixing liquor; in ordinary cases of calico dyeing the two processes should be distinct.

Dunging or cleansing is one of the most important parts in dyeing for calico printing; it deserves the utmost attention of the superintendent, for upon it depends, in a remarkable manner, the success of the dyeing. The heat of the cleansing liquor and its strength must vary with the styles, and be skilfully adapted to them. The rule is to use such a temperature and strength as will cleanse effectually, and not to go beyond that strength and temperature. The nature of the thickening must be attended to; if it is a gum thickening, and one which is easily soluble in water, the temperature may be kept low, but the strength must be rather greater than for paste thickenings; if a mixture of paste and gum thickenings, the heat must be higher and the strength medium; if all paste, the heat must be high and the strength of the liquor entirely in proportion to the kind of printing—being the strongest for blotch and heavy covered patterns, and weaker in proportion as the pattern is lighter. As a general rule the heat should be kept at the lowest point, this may be as low as 100° F. for light plate patterns without black; for black and acids for madder work a temperature of from 140° to 160° is the best; for garancines the temperature must be high, but need not exceed 180° F. in the first cleansing. For light reds and purples, especially for the former, a low temperature is necessary to success. In styles which contain much acid, as also in those which have a large quantity of red in them containing crystals of tin, the addition of chalk to the cleansing dolly is very useful, and in fact necessary. The cleansing is usually divided into two parts; the open passage of the pieces by means of rollers through the "dolly," and the fly wincing afterwards. About two minutes is all the time necessary in the first liquor; that is sufficient to fix the loose mordant, and to loosen, but not to remove, the thickening. The second cleansing requires a longer time and a rougher motion of the pieces, in order to remove all the thickening matter from the cloth; it may take from fifteen to twenty minutes, or even longer, if the nature of the thickening is opposed to easy solution. There must be no stinting of time; the cloth must be well cleansed, or there will be nothing but confusion in the dyeing. The pasty thickenings sometimes get into a state very difficult to deal with; they swell out but do not dissolve from the fibre, adhering to it like a jelly; rapid motion is necessary to detach them, or some mechanical appliance. The clear substitute liquors do not act so effectively in this case as cow dung, its insoluble matter acting mechanically upon the cloth, scours it by attrition; probably if some light substances like sawdust were added to the substitutes their energy would be increased. It is not usual to wash the pieces between the cleansings, except by simply passing them through clear water; it is very necessary that they should receive a perfect washing after the last cleansing—about fifteen minutes in a dashwheel, or three to six passages through a washing machine, depending upon the style and nature of the thickening. I have tried the effect of a brush revolving against the open piece in the first cleansing to help in detaching the paste thickening; it acted very well, and if applied near the end of the dolly cannot do any harm. This or some similar mechanical appliance may be beneficially used when the thickening does not come off well; but under ordinary circumstances it is not necessary, the motion of the pieces in the beck, and a good washing with plenty of water, will generally suffice to remove all adventitious matter, and leave the cloth clean for dyeing. (See COW DUNG, DUNG SUBSTITUTES, and PHOSPHATES.)

Clearing.—By this term calico printers understand the operations of obtaining the undyed parts of their work of a pure clear white. In all kinds of dyeing upon mordanted cloth, where a portion of the fabric is left unmordanted, or from which the mordant has been removed previous to dyeing for the purpose of leaving some parts white, the process of clearing is necessary in order to give those whites their highest degree of purity. Thus in madder dyeing, in logwood dyeing, or in garancine dyeing, though the coloring matters employed are said to have no affinity for calico, unless this is mordanted, yet it is always found that the unmordanted parts are stained and discolored to a degree which would much injure the appearance of the print if finished in that state. It is necessary, therefore, to treat the cloth in such a manner as to bleach these parts, at the same time having due regard to the integrity of the colored parts. The oldest method of clearing was a modification of the old method of bleaching, namely, exposure of the printed goods to the action of the air and moisture upon grass; this was done at night to avoid the injurious influence of strong daylight or sunshine. Although this is spoken of as an obsolete method, it is yet practised where there is convenience for it.

Bran is sparingly used for clearing some qualities of dyed goods, as logwood blacks and garancine reds; the bran is scalded, mixed with water, and the goods passed through at about 140° F.

Cow dung has been used in cleaning garancines, but its employment for this purpose is not to be recommended.

Chloride of lime or bleaching powder is the most generally

used clearing agent; it is applied in several ways. For clearing madders, the goods are run through a very dilute hot solution of the bleaching powder, until the dyer or clearer considers the whites sufficiently clear; for garancines, the same
process may be adopted, but experience has shown that the
higher. temperature employed in garancine dyeing tinges the
whites more deeply than madder, while the less stability of the
colors produced render it dangerous to let the pieces run long
enough to clear the whites. In consequence, two other
methods of clearing garancines are adopted: the first consists
in padding the piece to be cleared in solution of bleaching
powder, standing at $1\frac{1}{2}°$ or $2°$ Tw,: the padding is done by an
engraved roller, and so arranged that only a small quantity of
the bleaching liquor is transferred to the piece, the piece is
then passed over the steam chest or drums and either washed
off or finished without washing. The other and better method
consists in padding the pieces in a bleaching solution not
quite so strong as the last, and then passing them through a
steam box in which. they remain two or three minutes exposed
to low pressure steam, and from this pass through three or four
boxes of water to wash them. The only delicate point in clearing is to hit such a strength of clearing liquor as shall effectually
clear the whites without acting too much upon the colors. In
dark work there is a considerable margin, because the heavy
colors do not show a little punishment by excess of clearing,
but in light colors the point is not easy to hit; and, in fact, it
is preferable to clear them in the beck, where they can be
watched.

It appears from my experiments that nearly all dyed colors
can better withstand the action of strong clearing solution for
a short time than of weak clearing solution for a longer time,
and that, in consequence, it is better to clear quickly with
strong liquors than slowly with weak liquors. In madder and
garancine styles, where the dyeing material is in very fine powder, the clearing will not be satisfactory unless preceded by a
very good washing to expel the mechanically enclosed particles of spent dyewood. (For the use of soap in clearing, see
SOAP and MADDER.)

Coccus Polonicus, *Polish Berries or European Cochineal.*—
This is a small insect very similar to Kermes, used in Southern
Russia, Turkey, and Armenia, as a red dye stuff. It seems to
belong to the same kind of insect as cochineal, and before the
introduction of the rich American cochineal into Europe, it
was one of the chief dyeing materials for producing crimson
colors upon silk and wool. It is now only employed locally
in remote districts of the tract of country where it flourishes.

Cocoa Nut Tree.—It is stated that the whole of the cocoa-nut tree, but more especially the husk which encloses the nut, and the foot stalks of the leaves, may be used as a dyeing material. From the claims made in a patent, bearing date March 29, 1825, it would appear to have astringent properties, but I am not aware that it has ever been practically applied.

Cochineal.—This important dyeing material consists of an insect, feeding on a species of cactus called *Nopal* by the Mexicans. It was formerly considered to be a berry, and even years after travellers had brought home true accounts of the real nature and origin of cochineal, European observers of some eminence persisted in the belief of its vegetable nature, not being able to detect the usual external marks of an insect. It is, however, an insect having six legs. no wings, and a very small head; the legs are disproportionately small when compared with the body of the insect; they, in fact, appear only to be used by the insect in its earliest state, for after it has settled upon the branch of a nopal it employs them no more, and they are not developed like the rest of the body; this refers to the female insect, the male insects having better defined limbs, but there is only one male to about two hundred females. The females remain fixed to the branch like berries, feeding upon the sap until they are brushed off and killed. The method of killing the insect practised by the extensive cultivators consists in putting a basket full of them into a hot stove; the smaller cultivators kill them by immersing the baskets in hot water. Cochineal insects killed by the latter process are said to be mostly burst open and acquire a reddish or foxy appearance, and to be somewhat inferior, while those killed by the stoving process are gray and most esteemed.

There are two kinds of cochineal, known commercially as silver and black. The silver cochineal is said to be the impregnated female just before laying eggs, while the black cochineal is the female after laying and hatching the eggs. The black cochineal appears to be the most valuable, but the bulk of that imported is the silver, the black only being what has been kept for breeding purposes. Formerly there was a kind of cochineal called "English black cochineal," which was the Mexican silver grain, dyed by immersion in decoction of cochineal.

The value of cochineal can be very closely ascertained by the simple inspection of a practised observer; being one of the most expensive dye drugs, there are considerable inducements to fraud, and some kinds of cochineal are actually known in the market as "doctored cochineal." The usual adulteration consists in adding some powdery matter to the cochineal, to

increase its weight. French chalk, white lead, and ground talc
are said to be chiefly used, and as much as ten per cent. of this
mineral matter may be added without the appearance of the
cochineal attracting the attention of an inexpert observer. A
fine metallic powder is also used to weight cochineal, and it is
said that as much as 30 per cent. of weight can be thus added
to it. These adulterations are easily detected upon analysis,
for by calcining the mixture at a red heat, the true matter of
the cochineal is burned away, leaving the added mineral adulteration, the quantity and nature of which can be then ascertained. Practical methods of testing cochineal consist in dyeing
up silk and woollen in comparison with a known good quality,
also in ascertaining how much chlorine is required to bleach
the decoction from a known weight of the insects, and lastly
by comparing the intensity of a colored solution obtained by
treating a certain weight of cochineal with strong ammonia.
The dyeing test is the only really trustworthy test, and when
applied with due care gives very exact results.

Cochineal is a very rich coloring substance, yielding about
half its weight of real coloring matter. This coloring matter
is very soluble, and easily extracted from the insect by boiling
it in water. The extract contains, besides pure color, a quantity of animal matter, and it is liable to putrefy in the concentrated state when kept in warm places. It acquires a very
disagreeable smell, but even then does not appear to be much
injured in its powers of dyeing or yielding colors in printing,
though it would not be safe to allow this putrefaction to continue to any considerable degree. The extract of cochineal
may be boiled down to a syrupy consistence, but for general
purposes it is not used stronger than about 9° Tw., and is reduced from that for most shades. The pure coloring matter
of cochineal has received the name of *carmine ;* it is red, soluble in alcohol and water, and besides being itself of a magnificent color, readily yields all those shades which the cochineal
itself does to mordanted cloth.

The colors which are derived from cochineal are red, and its
modifications of pink, scarlet, and crimson. The alumina mordants give the crimson colors ; they are very fine, deep, and
solid shades, but have not any great brilliancy to distinguish
them. The scarlet is obtained by means of a tin mordant, and
no color can compare with a good cochineal scarlet for brilliancy, fastness, and fire. The pink is obtained from a modification of the cochineal color, obtained by boiling the insect
with ammonia and water instead of water alone. Alumina is
the proper mordant for it. It is a difficult color to obtain in
its best state, and as a pink has not so much beauty as safflower

pink, but is much more stable. Other shades besides these can be obtained from cochineal, but they are seldom worked, being expensive and capable of close imitation by cheaper coloring matters. The greatest affinity of cochineal is for wool; with it it yields its deepest, brightest and fastest colors. It is applicable on silk and cotton, but the same depth of shade cannot be obtained. The greatest consumption of cochineal is in dyeing woollen cloth scarlet, but large quantities are also used in calico and delaine printing, for obtaining reds of the scarlet nature, and also pinks and crimsons.

Cochineal Colors. *Scarlet on Wool.*—The principal color for which cochineal is employed in dyeing is that of scarlet upon wool. The process consists in first mordanting the woollen cloth in solution of tin, and then dyeing in the cochineal, with a further addition of tin solution. The dyers generally prepare their own solution of tin; and most dyers consider that their own method has some advantages over that of others, and guard their processes as secret. There can be no doubt that very much depends upon the tin solution; but, at the same time, it is certain that almost any tin solution may be made to yield good colors. The explanation of many differences among dyers upon this matter would be found to consist in points of practical detail: for example, one solution of tin yields the metal quickly at a low temperature; another yields it but slowly or not at all at a low temperature, but well enough at the boil. It is easy to see that in the case of such differences of composition, differences of subsequent treatment are essential. The solution in use is generally a mixture of per and protochloride of tin, prepared by dissolving the metal in nitric acid mixed with common salt or muriate of ammonia. It appears from practical authorities that the solution is best when made slowly, and when the solution of the tin is not accompanied by evolution of gas, or with but little evolution; such a state of affairs indicates the formation of ammonia in the solution, and seems to point to the production of some double salt of tin. The proportions used in one case are

> 10 lbs. nitric acid (aqua fortis),
> 5 lbs. water,
> 10 to 20 oz. common salt,
> 20 oz. feathered tin.

The tin is added bit by bit, waiting for one portion to dissolve before another is added, until the whole is dissolved. Whatever the composition of the tin solution may be, and a variety of receipts will be found under the article TIN, the woollen cloth, well wetted, is boiled with it. In the above tin solution there will be rather more than an ounce of the metal in each pound of the liquid; and for a quantity of woollen cloth weighing 100 lbs., and intended to be dyed a full scarlet, about 20 lbs. of the composition are taken and divided into two portions of 13 and 7 lbs. respectively. The cloth is boiled in water, to which the larger portion has been added, along with 8 lbs. of crude tartar; a little cochineal, about 6 oz., is also added to tint the cloth and serves to indicate the progress of the mordanting. The whole is kept at the boil for about two hours, and then winced in clean water; it is next entered into the dyeing vessel, either copper or tin, or part tin and part copper, with from five to six pounds of finely ground cochineal, and the remainder of the tin solution; the whole kept at the boil for about half an hour, or until the cochineal is spent and the dye liquor nearly colorless. In some cases tartar is used in the dyeing as well as in the mordanting, its effect being to give a more fiery or orange scarlet.

Bancroft appears to have been the first to draw attention to the fact that scarlet differed from crimson by the addition of a yellow part derived from some source; but where cochineal alone is used as the coloring matter it is evident that the yellow could only be derived from the change of its red coloring matter. This change was considered at that time as being accomplished by means of the tartar used in mordanting and dyeing, and the more yellow or fiery the scarlet was desired the more tartar was employed. I consider that it was the nitrous acid or hyponitric acid contained in the tin solution which produced this effect; but however produced, it was evident that red color was destroyed, and Bancroft had the idea of economizing cochineal by first dyeing a yellow basis upon the woollen cloth. He made many experiments with quercitron bark, but the results do not appear to have pleased the dyers, the shades being deficient in brilliancy, since quercitron yellow does not withstand very well the strongly acid solutions used in dyeing scarlet. At this day the yellow colors derived from fustic and turmeric are employed to economize cochineal in obtaining the yellower kind of scarlets, but the best scarlets are still obtained from cochineal alone as coloring matter, and these alone resist wear and exposure, and are not stained by liquids or dirt as the compound scarlets are.

The quality of the water has a great influence in scarlet dyeing; it should be as pure as can be possibly obtained: all earthy matters contained in it either tend to dull the color or to bring it towards the rose or crimson shade, which is fatal to a bright scarlet. Additions are frequently made to the water with a view of correcting impurities. Sometimes bran, flour, or starch are added ; and in France it is customary to boil some coarse wool in the solution of. tin and tartar before entering the cloth to be dyed; this is supposed to attract the dreaded impurities and so remove them from the dye-bath.   To obtain the very finest shade of scarlet requires great experience and a multitude of precautions, which can only be learned by assi-duous attention to practical operations.

One ounce of cochineal is generally reckoned as sufficient for one pound of wool : inferior colors are obtained by diminishing the quantity.

The cochineal crimson on wool is obtained by mordanting in alum, tin, and tartar, and dyeing in cochineal, mixed with ammoniacal cochineal.   It is not much in request.

Crimson upon Silk.—This color is obtained by working the silk in a warm, weak bichloride of tin for a sufficient length of time, then dyeing in cochineal and water.

Cochineal Pink on Silk.—This color is obtained by the use of the prepared or ammoniacal solution of cochineal (the process of which is given further on) upon a tin mordant.   By simply diminishing the quantity of cochineal used in dyeing crimson a pink can be obtained, but not of a good shade.   The cochineal pink is not so much in vogue as that from safflower.   A scarlet is obtained upon silk by first giving it a yellow or orange ground with anotta, and then passing it through the process for crimson.

The cochineal dyed styles upon calico or cotton goods are quite unimportant.   If calico, printed with the usual mordants for madder, be dyed in cochineal, it will be found that the dark red takes a violet-hued crimson, the light reds dye up bluish-pinks, the light iron mordants dye up shades of violet gray, and the strong iron mordants dye up a grayish-black.   There is a style of work occasionally produced from cochineal by printing a red liquor mordant, dunging or fixing as usual, and dyeing in cochineal, or a mixture of cochineal and gall-nuts. On account of some peculiar action which spent cochineal liquor has upon alumina mordants, injuring the colors as if it were acid, it is found desirable to divide the dyeing into two portions, so as not to have the liquor too concentrated.

Cochineal Colors in Printing.—For the application of cochineal in printing, the coloring matter may be applied in three states : The first being the insect, ground as fine as possible, and simply mixed up with the thickening and salts employed. This method may probably insure the extraction of the largest amount of coloring matter, but it is evidently unfitted for any kind of good printing.   The second, and most usual condition, is a solution of cochineal, or *cochineal liquor*, made by boiling the insect with water until exhausted as much as possible of coloring matter, and then concentrating the solutions to a strength of 10° or 12° of the hydrometer. The third state of cochineal is where the solution of coloring matter is made by means of ammonia, and the preparation is frequently spoken of as ammoniacal cochineal, and cochineal in cake or paste. The following are some of the methods followed in obtaining this preparation, which is, without doubt, the best form in which cochineal can be employed for crimson and pink colors upon woollen and delaine:—

Ammoniacal Cochineal.—Persoz gives as follows for the dry preparation:—

> 10 lbs. ground cochineal
> 30 lbs. ammonia,
> 4 lbs. gelatinous alumina.

The cochineal is placed in a stoppered bottle or carboy, and mixed well with the ammonia; the whole is left closely corked for a month, then mixed with the alumina, prepared, I presume, by precipitating alum with crystals of soda, put into a tinned copper and kept hot until the ammonia is dissipated; when the paste has acquired a sufficient consistence it is spread upon a cloth, and the drying completed in a stove. A pasty ammoniacal cochineal is prepared by digesting equal weights of ground cochineal and ammonia for a week, and dissipating the ammonia by evaporation.

In England the compound is mostly prepared by putting the unground cochineal into a close tin or tinned copper boiler, mixing with the ammonia, and keeping at the boiling point for eight or ten hours; straining and keeping the liquor hot in an open vessel until the ammonia is dissipated. If steam, at eight or ten pounds pressure, can be used in the operation it appears advantageous.

Dumas gives directions as follows: 15 lbs. of ground cochineal are mixed with 17½ lbs. of ammonia in a close vessel, and left to digest cold for six or eight days, then heated gently for about eleven hours, with constant stirring, until the smell of ammonia has disappeared: this leaves about 27 or 28 lbs. For dark colors this answers, he says, very well, but for light and delicate shades, a dry cochineal lake is prepared by mixing the gelatinous alumina obtained from 1½ lb. of alum with the paste produced as above, and washing the lake several times upon a filter, then drying until the product weighs only 16 or 17 lbs.

I am of opinion that the method of simply keeping the ammonia and cochineal together in a heated state for some hours, gives the best and most economical preparation. It is true that the whole of the coloring matter is not exhausted, but the residuum may be extracted and used for darker and scarlet shades.

It will be observed that the ammonia is driven away before the solution is used in color mixing, therefore any action it has had upon the cochineal must have taken place during the steeping, and remained permanent; in fact, the investigations of chemists seem to prove that the ammonia, or some of its elements, have entered into intimate combination with the coloring matter of the cochineal, and then modified its properties. The following receipts for printing, in which cochineal, is the only or predominating coloring matter, are selected as characteristic specimens from a vast number in the author's possession :—

Deep Scarlet on Silk or Wool.

3½ gallons cochineal liquor, containing
1¼ lb. cochineal per gallon,
36 oz. starch; boil, and add
½ pint mixed berry and fustic liquor at 15°,   .
1 lb. binoxalate of potash,
3 oz. crystals of tin,
12 oz. bichloride of tin at 100°.

The berry and fustic liquor in this receipt, giving a portion of yellow, make the shade more lively, turning it towards a flame color; the binoxalate or, as it is sometimes called, the superoxalate of potash is useful as a foil upon which the excess of acid in the tin solution may spend itself, liberating oxalic acid, the presence of which is favorable in cochineal colors.

Crimson or Mallows red.   Silk and Chalis.

1 lb. ammoniacal cochineal, dry,
2 quarts water, boiled and strained,
1 lb. gum, in powder,
1 oz. alum,
· 1 oz. acetate of lead,
½ oz. Tartaric acid,
2 oz. oxymuriate of tin.

This color is for block; if required for machine the thickening must be altered. I have not tried this receipt, but I think · the amount of alum is decidedly too small.

Blotch Red for Delaine.

6 gallons cochineal liquor, at 6°,
1 gallon berry liquor, at 10°,
10 lbs. starch; boil and add
2¼ lbs. oxalic acid,
2¼ lbs. crystals of tin.

A strong red for objects may be made exactly as above, but the cochineal liquor must be stronger, and may go as high as 12° with advantage, especially upon the poorer qualities of delaine.

Another Red for Delaines.

In this receipt the cochineal liquor is prepared in a peculiar manner, caustic potash being used with a view to extract the coloring matter. It has yielded good colors in my hands, but I do not think there is much advantage in using potash for extracting the coloring matter.

30 lbs. cochineal,
25 gallons water,
1 gallon caustic potash, at 9°,

leave in contact twelve hours; boil, strain, and treat the residue twice over with water and potash, then boil the whole down to fifteen gallons. The liquor stands at about 8°, and is somewhat pulpy or gelatinous.

1 gallon of above liquor,
1¼ lb. starch,
6 oz. crystals of tin; when cold, add,
6 oz. oxalic acid,
¼ pint of orange color. (See ORANGE.)

The above receipts are sufficiently illustrative of the scarlet red shades from cochineal. To obtain the crimson and pink shades the ammoniacal cochineal is employed, and it will be noticed that, while tin is the mordant for the scarlet reds alumina is the mordant for the crimson and pink shades. The fact that tin mordants gave a shade more inclining to the orange, and alum mordants one inclining to the purple, has been long known, but the modification of the hue by the action of ammonia is of comparatively recent discovery. Although alum and the ammoniacal cochineal usually go together, yet some receipts will be found in which tin salts are used in conjunction with this preparation of cochineal; the result is then a dark red, which holds a position intermediate between the scarlet and the crimson. For this shade of red, seldom used in the designs of this country, no yellow part is admissible, it is consequently a somewhat dull though a rich and solid color. The following receipt illustrates one of these colors:—

Amaranth Red for Delaine.

7½ lbs. of dry ammoniacal cochineal,
2 gallons hot water,
1 pint acetic acid,
½ lb. alum,
9 lbs. gum; dissolve, and add
1 pint white muriatic acid,
2 oz. bichloride of tin, at 100°.

There is a free amount of acid employed in this color, partly for the purpose of neutralizing any ammonia left in the cochineal, partly to dissolve the alumina used in preparing the cochineal, and partly to facilitate the fixing of the color on the wool. The quantity of acid which can be advantageously used, depends upon the alkalinity of the ammoniacal cochineal. If the color be not distinctly acid the wool does not take the color; if it be excessively acid the cotton does not take color; and, in either case, "threadiness" ensues. If the utmost amount of acid be employed that can be safely used, the effect is to produce a simple red; but, as the object is usually to obtain a crimson on the purplish side, the color is worked as neutral as possible, and generally with a faint excess of ammonia

Crimson for Delaine.

10 lbs. cochineal,
20 lbs. strong ammonia; steeped for 24 hours, then boiled and strained,
1 lb. cream of tartar,

the whole kept hot until the smell of ammonia is only faint; thickened with powdered gum senegal, and while warm a half pound of alum per gallon of color is dissolved in. Just before the color is going to be worked, it is considered advisable to add a small quantity of ammonia to it, until it acquires the purplish shade which indicates a slight excess. If, however, any considerable excess of ammonia is present, tartaric acid or cream of tartar must be added.

Pink colors are obtained by reducing crimson or amaranth with gum water; or by direct preparation from a weaker solution of ammoniacal cochineal.

There are peculiar difficulties in obtaining the best crimsons and pinks on wool and delaine, which are only to be mastered by a very close attention to the colors. It is considered by some printers that the contact of copper is prejudicial to this class of colors, and they are particular to avoid copper vessels,

and have the color worked out of wooden boxes. It is true
ammonia acts upon copper, but it is doubtful if copper is injurious to the color, and certainly the finished color should not
contain any such excess of ammonia as would be likely to act
rapidly upon metallic copper.

African Cochineal.—A substance so called is imported in
small quantities from Algeria; it yields red colors with alumina
mordants, but possesses scarcely any other of the properties of
the cochineal insect. It is not employed in Europe.

Colors.—Under this head I bring together several items of
a general character, bearing upon all colors as such.

Nature of Color.—At first sight we conclude that colors are
properties of bodies of the same nature as weight, size, hardness, etc.; but a consideration of even ordinary phenomena
would lead to doubt upon this point. We find colors of natural objects varying at different times, and even showing different
shades according as we look at them from one or another point
of view. In soap bubbles, and thin films of tar upon water,
we perceive the most beautiful colors, although certain that
there is no coloring matter in them. In common spectacles
and telescopes we find that the glasses have a tendency to give
a colored fringe, like a rainbow, to objects seen through them;
fine spray of water, as in a waterfall, gives a rainbow exactly
like the great natural rainbow, with all its gorgeous colors;
and in numerous other cases we find that color must be traced
directly to some particular action of light.

Sir Isaac Newton caused a ray of sunlight to pass through a
three-corned glass prism in such a way that the ray was twice
bent from its direct path before it emerged into the air again.
Instead of coming out as it entered the glass, this great philoso-
pher found the light entirely changed; the one united ray of
natural light was broken up into several rays of colored light,
which diverged from the prism like a vertical fan, in the
following order :—

Higher or more refrangible rays.
Violet.
Indigo.
Blue.
Green.
Yellow.
Orange.
Red.
Lower and less refrangible rays.

From this experiment, and others, Newton came to the conclusion that the pure light of heaven was not a simple body but actually a compound of the seven colors given above, and that ordinary eyes were not conscious of this fact, because the colors were so balanced in the mixture that their individuality was lost in producing the general effect. Many experiments might be cited to prove the reasonableness of this opinion, but they do not properly come within the scope of this work, and must be sought for in the elementary treatises on natural philosophy. In the meantime the reader may accept this as a fact, and then the theory of colors will be comprehensible.

A body is said to be white when it receives the white rays of light and reflects them with moderate strength, and unaltered as to quality, to the eye of the observer. Thus, a sheet of white paper or white calico possesses moderate reflecting power, and sends off the light which falls from it without decomposing it, and produces the effect of whiteness, or distinct vision, with absence of color.

A body is said to be black when it absorbs and quenches all the rays of white light falling upon it. There is no substance, at least we do not know any substance, which has this perfect blackness; for the consequence of a perfect absorption of light would be the invisibility of the object, and its form could only be perceived by its intercepting the reflected rays from contiguous bodies. In a dim light there are, however, many substances so black as to be quite invisible against a dark ground.

Black and white, not being produced by the decomposition or splitting up of light, are, therefore, commonly said not to be colors; however, for the printer and dyer, they are the most important colors, and have as much right to the name as red or blue.

Colors proper may be conceived to be actually due to light, upon the following suppositions: First, that it is possible to separate a portion of the seven colored rays from the rest; secondly, that it is possible to so prepare a surface that one portion of the rays shall be quenched or destroyed, and the others left to reflect or shine out ; and thirdly, that the undestroyed rays will shine or reflect with their own color or colors.

Suppose a surface of wood, calico, or silk, so prepared that all the colored rays, except the red, are absorbed or quenched by it, it will then appear to the eye just as if no light but red light illuminated it, that is, it will be red ; again, suppose the surface to be of such a nature as to absorb all the colored rays except blue, then the color of the surface will be blue. The surface may be of such a nature that it will absorb five of the

12

colored rays and reflect two, then the color of the body will be
that produced by the mixture of the rays so reflected; in like
manner four rays may be reflected and three absorbed, or even
six rays reflected and only one absorbed, thus producing all
the mixed and compound shades.

If this theory be correct, all the operations of dyeing and
printing are merely directed to deposit such substances on the
fibre as shall possess the properties of decomposing the light
falling on the fibre, or, rather, of intercepting it, and causing
one portion to be quenched and another to be reflected.

The natural coloring matters, such as indigo, cochineal, anotta, etc., are themselves capable of so decomposing light, and
when deposited upon cloth, even in very thin layers, produce
the effect of coloring the cloth, so called; in fact, intercepting,
more or less, the proper reflection of the cloth, and substituting
their own.   This theoretical view, which seems something
superfluous with the ordinary vegetable dyeing matters (which
are themselves erroneously looked upon as not only colored
but coloring substances), may be more acceptable as explaining
the production of strong and well-defined shades from substances in themselves nearly, or altogether colorless.   Such
cases are Prussian blue from iron and yellow prussiate; chrome
orange from chrome yellow by means of lime water; the scarlet iodide of mercury from bichloride of mercury and iodide of
potassium; and numerous other cases familiar to chemists and
colorists.   It is true, that a change of color, or production of
color, is accompanied by a change of chemical composition in
the majority of cases known; but, there are well-known
instances of change of color without change of composition,
and of the same chemical body having several different colors.

Upon metals, bone, glass, and other surfaces, the mere ruling
or impressing of very fine lines close together gives the surface
the property of reflecting colored light; and this is actually a
regular and well-understood method of obtaining colored
effects, as in Barton's buttons, Nobert's bands, and De La Rue's
prismatic films.

Names of Colors.—One of the greatest difficulties which a
writer upon colors experiences resides in the fact, that there is
no generally accepted and well-defined language in which he
can express or indicate the shades of color he may be treating
of. A practised eye can distinguish say ten thousand shades
of color; but it is a question whether there are fifty names of
colors which would convey the same idea of shade to any ten
colorists in the world; the consequent hindrance to the communication of knowledge can be easily conceived. The natural
and sensible plan of comparing a color to some common sub-

stance, flower, plant, animal, or mineral, of similar shade, has been of great service, and, with a few exceptions, constitutes our nomenclature up to this date. Of the seven prismatic colors of Newton, the violet, indigo, and orange are named from the vegetable world; the other four names require some etymological research to find their originals, if even they can be found; for, being simple colors, their names must be of great antiquity. Of the other more common colors, *pink, lilac, peach, mallow, chestnut, cherry*, etc., are from fruits and flowers; *flesh, fawn, salmon, buff, chamois*, etc., are from animals; *stone, amber, emerald, aventurine*, etc., are from minerals. These, the least exceptionable names of colors, are far from being satisfactory; they acquire a conventional meaning in certain countries and districts which is not the same in others. Who, for example, can dogmatically define the color of a cherry? Who is to know that *flesh color* does not mean the color of muscle but the color of skin, and what is the color of skin throughout the world? *Salmon color* does not mean the color of the fish as caught, but of its flesh, and generally in the cooked state. Again, what is *stone color?* But, beside these, there are a number of names of colors so unfit, or so absurd, as to show but little taste or little invention on the part of their originators and adopters. Capucin, or Capuchin, and Carmelite, are from the dresses worn by the religious orders so called; but in Protestant countries this indicates very little since the orders are not in existence. Puce color has been boldly adopted by some English writers as *flea color*, and there cannot be much doubt that this unpleasant insect is the cause and origin of the name. But for a specimen of the resources to which colorists have been put for names, take that of *Isabelle*, which is a kind of buff orange. Clair Eugénie Isabelle was daughter of Philip II. of Spain, and wife to the Archduke Albert. She was present when her husband was besieging Ostend, and made a vow, most piously intended, not to change her linen until the town was taken. The siege was protracted for upwards of three years. The consequences to the Archduchess's linen may be easily imagined; but the ladies of her court applied to art to effect for them what natural causes had done for their mistress; and the color so produced was thenceforward known as Isabelle.

If we take any single color, such as blue, we become immediately cognizant of our defective nomenclature; for we find in French and English books a vast number of names which can have no meaning without the eye has actually seen the color. For example, there is Chinese blue, Prussian blue, French blue, Haytian blue, China blue, and Saxony blue; again there is king's blue, queen's blue, prince's blue, and royal blue; towards the sky we have celestial blue, cerulean blue, sky blue, azure blue; and in the opposite direction, *blue d'enfer*, which name is not yet translated into English. There is also torquoise blue, ultramarine blue, and opal blue, from the mineral kingdom; we find also bluebottle blue, pigeon blue, damson blue, chemic blue, cassimer blue, and a number of others, omitting all those which are characterized by the name of the dyestuff from which they are derived.

In the midst of this confusion of names any system which promises to bring some degree of order or regularity in the naming of colors deserves respectful attention. Such a system, moreover, coming stamped with the name of so eminent an authority as M. E. Chevreul, and the result of many years of patient investigation on his part, needs no apology for the space to devote to a very brief but sufficient abstract of his ponderous memoir in the 33d volume of the "Mémoires de l'Académie."

Chevreul's System of Naming Colors.—M. Chevreul has only six fundamental names of colors, which are the three elementary colors, Red, Yellow, and Blue; and the three secondary colors, Orange, Green, and Violet. He arranges them in a circle, like the spokes of a carriage wheel. Commencing with red, and going to the right, the next spoke is orange, then yellow; after yellow comes green, which passes into blue, and then violet, which is upon the left hand of the red. There is thus produced a wheel of six spokes, at equal distances from one another, in the following order:—

Red,	Green,
Orange,	Blue,
Yellow,	Violet.

It may be observed that in such a wheel or circle, the secondary colors are placed between the primaries which compose them; and that the colors which are lineable on each side of the centre are complementaries. Another spoke is placed between each of the six already in position, and a circle of twelve colors produced, which are as follows:—

Chromatic Circle of Twelve Colors.

Red,	Green,
Red-orange,	Blue-green,
Orange,	Blue,
Yellow-orange,	Blue-violet,
Yellow,	Violet,
Yellow-green,	Red-violet.

These colors gradually blend one into another; but even an ordinary eye can perceive that between the red and the red-orange there is room for four or five shades of color; and so also between each of the other couples. These are then placed in, and the circle or wheel may now be considered as filled up without interstice; and the colors present a gradual passing or shading into one another; the red becoming yellower until it is an orange, and this still yellower until it is a pure yellow. This yellow meets the blue until the middle point of green is reached, when it passes into pure blue, which in its turn passes through violet until it meets the red. A circle of seventy-two colors was thus constructed by M. Chevreul, assisted by persons skilled in discerning between slight differences of shade. This constitutes his complete chromatic circle, and the colors are named as follows:—

Complete Chromatic Circle of Seventy-two Colors.

Red, six shades, called—Red, 1 red, 2 red, 3 red, 4 red, 5 red.

Red-orange, six shades, called—Red-orange, 1 red-orange, 2 red-orange, 3 red-orange, 4 red-orange, 5 red-orange.

Orange, six shades, called—Orange, 1 orange, 2 orange, 3 orange, 4 orange, 5 orange.

Yellow-orange, six shades, called—Yellow-orange, 1 yellow-orange, 2 yellow-orange, 3 yellow-orange, 4 yellow-orange, 5 yellow-orange.

Yellow, six shades, called—Yellow, 1 yellow, 2 yellow, 3 yellow, 4 yellow, 5 yellow.

Yellow-green, six shades, called—Yellow-green, 1 yellow-green, 2 yellow-green, 3 yellow-green, 4 yellow-green, 5 yellow-green.

Green, six shades, called—Green, 1 green, 2 green, 3 green, 4 green, 5 green.

Blue-green, six shades, called—Blue-green, 1 blue-green, 2 blue-green, 3 blue-green, 4 blue-green, 5 blue-green.

Blue, six shades, called—Blue, 1 blue, 2 blue, 3 blue, 4 blue, 5 blue.

Blue-violet, six shades, called—Blue-violet, 1 blue-violet, 2 blue-violet, 3 blue-violet, 4 blue-violet, 5 blue-violet.

Violet, six shades, called—Violet, 1 violet, 2 violet, 3 violet, 4 violet, 5 violet.

Red-violet, six shades, called—Red-violet, 1 red-violet, 2 red-violet, 3 red-violet, 4 red-violet, 5 red-violet.

This chromatic circle is not imaginary, but actually exists, composed of dyed wools, while a good many copies, executed in chromo-lithography, are published more or less correct.

Light and Deep Shades.—So far as the seventy-two shades or hues of color go the circle is complete, but it is necessary to indicate the tone or depth of a color, or, in other words, the lightness or darkness of it. To define this M. Chevreul constructed a circle, in which each of the seventy-two shades was dyed of twenty different degrees of depth, from the lightest which could be discerned from pure white to the most intense depth, approaching to brown black. These he calls tones; and the lowest or lighest tones are near the centre of the circle, and the higher or darker tones near the outside or ·circumference. This circle, then, has seventy-two colors, each of twenty tones, or tints, as I prefer to call them, making a total of 1440 shades. To express which tint of a color is meant the number is written after, as, for example, 3 blue-violet 13 tint. As the number of shades and tints above are by no means so numerous as the colors of all natural or artificial substances, it will frequently happen that the color of a substance does not coincide exactly with any of the 1440 shades, but falls between two of them, and not equally distant from each; this relation is expressed by fractions.

Darkened or Broken Shades.—The 1440 shades, defined above, are so many modifications of six colors, mixed two and two, of different depths. When these colors are mixed with gray or black they are modified, being darkened or·broken. Between the lightest gray and the deepest black Chevreul counts twenty equal tints, as of each of the shades in the circle. He conceives that each of the 1440 colors may yield nine distinct shades by mixing with black, and these are called the broken shades. The first broken shade of any of the 72 colors contains $\frac{1}{10}$ of black to $\frac{9}{10}$ color, and the last $\frac{9}{10}$ of black to $\frac{1}{10}$ color; the first yields colors only a little darker than the normals, the latter yields shades hardly distinguishable from black. Multiplying the 1440 by 9, we have 12,960 broken shades, to which adding the 1440 pure shades we have a total of 14,400 colors all named and defined, besides twenty shades of gray. The blackening of a color is represented, in M. Chevreul's nomenclature, by a fraction following the name and preceding the tint, as, for example, *slate color* is 1 blue $\frac{9}{10}$, 10 tint, which reads, number one blue, darkened with nine-tenths black, the mixture being of the tenth tint.

It does not appear that M. Chevreul has constructed, or that there exists anywhere a complete series of these 14,400 shades. In the plates to his memoir there are circles printed in color, which represent the 72 pure shades of a medium tint, and these broken down by mixture with black in nine different circles, making in all 720 shades; the twenty different tints of each shade are not represented in the plates; probably the present state of chromo-lithography does not permit of it being done with sufficient exactness. The learned author of this system has projected the idea of a chromatic hemisphere, to be made of pieces of porcelain tinted by firing to serve as an unalterable standard of color; it is to be feared that the realization of this idea is very far distant, and as all dyed colors change by age the circles of M. Chevreul, constructed with so much labor and skill, will perish, and his labors remain a splendid but nearly useless monument of ingenuity.

In order to illustrate the applicability of this system to actual colors, I give a list of several colors and colored bodies which are pretty well defined in common language with the names of the colors in Chevreul's system—

Amber in mass=2 orange 12 tone or tint.
Amber in a thin slice=2 yellow-orange 11 tone.
Amaranth=red-violet 12 tone.
Amethyst=5 blue-violet from 3 to 16 tone.
Apricot=orange 6 tone.
Aventurine=1 orange 14 tone.
Blood, ox=1 red 13 and 14 tones.
Blue, indigo on wool=3 blue 14 tone.
Blue, pigeon=3 violet 10 tone.
Blue, royal=3 blue 12 and 13 tones.
Butter=yellow-orange 2 to 3 tones.
Brick color=3 red orange $\frac{5}{10}$ 12 tone.
Bronze=3 yellow 20 tone.
Brown=any one of the 72 colors when at 18, 19, and 20 tones are browns.
Coffee roasted=3 orange 18 and 19 tones.
Cinnamon=3 orange 14 tone, and 2 orange $\frac{5}{10}$ 9 to 12 tones.
Capucin=3 red-orange, 10, 11, and 12 tones.
Carmelite=3 orange 15 tone.
Carrot=orange 7 tone.
Cherry=red 9 and 10 tones.
Chamois=2 orange 2 to 6 tones.
Chestnut (the fruit)=2 orange 16, 17, and 18 tones.
Chocolate in cake=5 orange 18 tone.
Cigar color=2 orange $\frac{5}{10}$ 11 tone.
Citron=4 yellow orange 6 tone.
Crimson=3 red-violet 10 tone.
Emerald=2 green 11 tone.
Gold lace=4 yellow-orange 9 tone.
Gray, silver=3 orange $\frac{9}{10}$.
Gray, blue=5 blue $\frac{9}{10}$ 10 tone.
Gray, brown=normal gray 12 to 15 tones.
Gray, flesh=1 red-orange $\frac{7}{10}$.

Gray, iron $= 3$ blue $\frac{9}{10}$ 10 tone.

Gray, lavender $= 2$ blue-violet $\frac{3}{10}$ 6 tone.

Gray, pearl $= 2$ blue-violet $\frac{7}{10}$ 2 and 3 tones.

Gray, tan $= 5$ yellow-orange $\frac{9}{10}$ 10 tone.

Green, cabbage $= 3$ yellow-green 6 tone.

Green, turf $= 1$ yellow-green $\frac{4}{10}$ 10 tone.

Green, myrtle $= 3$ yellow-green 12 tone.

Green, apple $= 4$ yellow-green 8 tone.

Green, grass $= 5$ yellow-green 9 tone.

Isabelle $= 1$ yellow-orange.

Lavender flowers $= 3$ blue-violet 7, 8, and 9 tones.

Leather $= 1$ orange $\frac{4}{10}$ to $\frac{8}{10}$ 7 tone.

Lilac flowers $= 1$ blue-violet 1 and 2 tones.

Lilac on stuffs $= 1$ violet 7 tone.

Maize $= 1$ yellow-orange 7 tone, and 3 yellow orange 6 tone.

Malachite $= 3$ green 6 to 8 tones.

Mauve $= 3$ violet 8 tone.

Myrtle $= 3$ yellow-green 11 and 12 tones.

Nankeen $= 1$ orange $\frac{8}{10}$ 3 tone.

Nut color $=$ yellow-orange 16 tone.

Olive $= 3$ yellow $\frac{6}{10}$ 10 and 11 tones.

Pink $=$ Rose.

Puce $= 4$ blue-violet 13 tone.

Red lead $=$ yellow-orange 20 tone.

Rose $= 5$ violet, red-violet, and 1 red-violet 3 to 7 tones.

Ruby $=$ red 11 tone.

Sapphire $= 5$ blue 11 tone.

Scarlet, common $= 1$ red 12 to 14 tones.

Scarlet, flame colored $= 3$ red 10 tone.

Slate $= 1$ blue $\frac{9}{10}$ 10 tone.

Straw $= 2$ and 3 yellow-orange 3 to 5 tones.

Tobacco $= 3$ orange $\frac{6}{10}$ 15 tone.

Vermilion $= 3$ red 15 tone.

Wood color $= 1$ yellow-orange 15 tone.

Yellow, canary $= 1$ yellow 6 tone.

M. Chevreul has compared the colors of many thousands of natural objects with his scale, and defined them in his memoir. If this system has a radical defect which will prevent its becoming generally used, I am convinced it resides in the fact that the author has only combined the six colors in two and two to produce his *gamme* or scale of 72 colors; hence the necessity of complicating the system by the broken or *rabattue* colors obtained by mixing black with the shades. M. Chevreul must know that in practical dyeing and printing, as well as by the indications of theory, blue serves to darken colors; and I believe there was no need to introduce the black element as a separate one, but that green, blue, and violet would have done all that was required. But in adopting a single circle it was impossible to make the colors combine three and three, which, if it could have been done, would have doubtless, produced every imaginable shade.

Influence of Colors upon one another.—It is very well known that certain colors agree with one another, producing a pleasant effect in combination, and that there are also colors which in the common phrase do not go well together. When a person has a good eye for colors (and this is quite as rare as a good ear for music), he will naturally make harmonious combinations in his dress or furniture, which combinations will be re-cognized as good and agreeable by all cultivated or naturally good eyes. These colors can only be varied in quality or in depth to a small extent without departing from the limits of excellence; so that it would appear that this agreeable or dis-agreeable impression of certain associated shades upon the eye is in accordance with some law of nature.

The research of many scientific observers has succeeded in discovering and defining this law with a considerable degree of exactness, and there can be no doubt that a study of these researches will be of benefit to all who are connected with the arrangement of colored objects; they are also of high interest to the color mixer, who, though he has not to judge of the har-mony of colors, is expected to produce colors which shall be harmonious in the design, and which, being once produced, are not generally alterable. For example, suppose we have a delaine pattern with a blotch ground, and a narrow trail or stripe of another color running up the piece, and that there are fifty pieces to be printed all with a chocolate blotch, but the trail or stripe is to be changed every ten pieces—say ten pieces with scarlet, ten with orange, ten with blue, ten with green, and ten with lilac or purple. If exactly the same chocolate color be used for the whole fifty pieces, every experienced color mixer knows that when finished it will appear, upon inspection of the pieces, as if they were five different chocolate blotches; and, in consequence, such modifications are made in the color mixing as experience has pointed out to be necessary. I think it likely that attention to a few points below will be of service in understanding the reason for this change, and give some-thing like a scientific accuracy to the modes of effecting it.

Accidental Colors.—If a red wafer be loosely placed on a white spot of the same size in the centre of a sheet of black paper so as to cover it, and then intently looked at for a few moments with one eye, the other being closed, and then the wafer suddenly removed, the eye, gazing still on the same spot,

will perceive not a white but a decided greenish circle. This
impression will last for a few moments. The experiment may
be varied by taking a smaller red wafer and placing it on the
white spot so as to leave a circle of white all around the wafer.
and then again looking closely with one eye at the spot; in the
course of a few moments the white circle will appear green.
The color so perceived is called an accidental color, to distinguish it from the real color.

By using wafers of different colors a variety of accidental
colors are obtained, always the same for the same real or ex-
citing color. The experiments of Sir D. Brewster, confirmed
by other observers, give the following as the chief cases of
color so produced :—

Color of Wafer.	Accidental Color.
Red	Bluish-green.
Orange	Blue.
Yellow	Indigo.
Green	Red-violet.
Blue	Red-orange.
Indigo	Yellow-orange.
Violet	Yellow green.
Black	White.
White	Black.

There are a great many ways in which these colors can be
experimentally produced besides the one given above; one of
the best and simplest is the following : Take two white and
thin cards and, placing them together, punch out squares so as
to leave a kind of lattice work; then take colored tissue paper
and put it between the cards and hold up to a strong light, the
real color will be seen in the squares, while the bars of the
lattice will appear of the accidental color.

Sir D. Brewster explains this phenomenon as follows (his
language being purposely modified) : The eye being strongly
excited by gazing on a colored body, as a red wafer, becomes
partially paralyzed and insensible to the rays of light carrying
that tint, and when the color is suddenly removed, the white
spot reflects white light into the eye; but the red constituents
of the white light fail to produce their usual effect on the
retina, fall dead so to speak, and the other rays being active
produce the colored appearance. From this we deduce the important conclusion that *the accidental color created by a real color
is that color which would be produced by compounding together all
the elementary colors excepting the real color.* Or, taking for
granted the statement made page 164, that the whole of the
elementary colors when combined produce white, the law may

be expressed as follows: *The accidental color of a real color is that which being added to the real color would produce white.* From this fact these colors are frequently called *complementary colors*, since they each make up the complement of the other color.

Now, to apply these facts in explanation of effects observed in printing, it is sufficient to know that in all cases real colors always produce their accidental colors, and throw the hue of them more or less over the surrounding colors. Thus a bold scarlet stripe in a chocolate blotch throws on each side a bluish green accidental color; and if the balance of colors in the chocolate is pretty even, it will give a greenish shade to it. If the stripe be orange, the accidental color is blue, which throws a bluish hue over it; and so on with other colors. Now, in order to preserve the chocolate at the same shade in each lot of colorings, it is necessary that it should be able to absorb and neutralize the accidental color. Thus, suppose a standard chocolate uninfluenced by any contiguous color as a sample ; to work it with a scarlet stripe, and to obtain the same shade, it must be made a little redder; this red will absorb the green accidental color and neutralize it, while at the same time it will enrich it, giving depth and lustre. To work with an orange trail, the chocolate must be made a little yellower or browner ; to work with blue, it must be a little more purplish; to work with a lilac, it must be made a little bluer, and so on.

The fact that some colors do not agree together so well as others is accounted for on the same general principles of accidental or complementary colors being created, and interfering with the shades, besides which a great deal lies in the depth of tone of the contiguous colors. Many colors which have a bad effect together, when of equal depth of shade, agree very well when a difference is made in this respect; so that when it is stated, for example, that blue and violet are colors that go badly together, it is not meant to intimate that they will not form an agreeable contrast under any circumstance, but that usually they do not agree. The art of the designer and colorist consists in so adapting shades as to produce new and agreeable effects ; and the contrivances for accomplishing this end are so numerous that it is impossible to describe them in detail. The general maxim is this : *Colors look best when near their complementary colors ;* or, in other words, complementary colors are those which agree best together, as (for example) red with green, blue with orange, etc. The reason for this appears to be, that the accidental color created by each of the real colors is the same as the contiguous color upon which the accidental falls, thus adding to its depth. But this very limited explanation will not suffice for all the agreeable contrasts which are obtainable in the complex shades : it, however, throws a general light on the principles governing the effects of colors in juxtaposition, and may serve as a starting point for the colorist.

The limits of this book forbid going into further details upon a subject which, though possessing the highest interest for designers and "getters-up" of styles, touches but remotely the really practical dyer or color mixer.

The Stability or Fastness of Colors.—The term "fast color" has a very wide and somewhat indefinite signification. Under the old *regime* the French government attempted to define what was a fast and what a loose color, enacted laws bearing upon the matter, and prescribed tests to distinguish between one and the other. They prescribed a boiling in alum and water, then a treatment with weak vinegar and an exposure to the air, along with some other tests. If the dyed color did not satisfactorily resist these tests it was condemned. Dyers were compelled to confine themselves to either fast or loose colors ; the Government determined for the trade what drugs gave fast and what gave loose colors, and prohibited, under a penalty, the dyer of the "*bon teint*" to have in his premises the drugs which were used in the "*petit teint*," and *vice versa*. This was a very fallacious system of testing the colors, most ineffective in its results, and speedily fell into disuse. Soap is considered now as a test for fastness in colors ; but that is only a degree less absurd than alum and vinegar, though it gives a somewhat better idea of the nature of the colors. *A fast color may be defined as one which will resist the destructive agencies of the position in which it is intended to be placed.* What is a loose color on a print for a dress, may be fast for hangings or curtains ; what would be inadmissible on calico which has to be washed frequently, would be a fast and good color on silk velvet, which is not expected to be washed at all ; a color which would be loose for a silken banner, exposed to sun, and air, and rain, would do excellently as an article of furniture, or even as a pocket handkerchief ; and a color which is well adapted to resist all the silent influences of air, damp, light, and heat, may totally disappear in a wash tub. If a color is to be used upon an article of dress, its fastness should be tested by the way in which it will resist the agents to which it will be inevitably exposed. If a cheap calico dress pattern, it will have to resist much friction, and the detergent action of soda and soap ; if a superior kind of calico or muslin, it will not be always necessary for it to resist the action of soap in a perfect manner, but it is expected that it will not fade upon exposure to good bright daylight, that it will not spot or stain with pure water, and that the color will not be detached by simple friction either dry or in water. There would soon be an end to calico print-ing and dyeing if colored muslins and calicoes were to be washed as sheets and blankets are washed, or expected to go through the same treatment uninjured. It would doubtless be desirable that colors should be produced of that degree of fast-ness; but there is no hope of such a result at present. All printed and dyed goods which are intended as wearing appa-rel should, as a primary point, be required to withstand the action of air and water; but that is not required in an equal degree for all kinds of dresses. Promenade robes must be of more stable colors than is necessary for evening or home dress; many colors which will be faded in two or three days' expo-sure to air will be good for weeks in the diffused light of a house, or by gas or candlelight; and for this kind of wear it is possible to produce rich and delicate shades which would be inapplicable for other purposes. There is no abstract fast or loose color; they are all comparative, and all have their proper place. Turkey red and indigo blue are the fastest of all dyed colors—that is, when exposed along with other colors to the action of water, friction, air, and light, they will remain unin-jured, or but little injured, when the other colors are faded or completely discharged. The fastness of a color does not depend altogether upon the coloring principle, for it is found that the same color has very different hold upon different materials: a cochineal color is faster upon wool than upon silk, and faster upon silk than upon cotton; and this is generally the arrange-ment of the relative durability of the same vegetable or animal coloring matters, viz.: more stable on wool than on silk, more stable on silk than on cotton. There are exceptions to this rule depending upon the action of the animal matters in silk and wool upon certain colors: for example, Chevreul found that indigo upon cotton was capable of resisting higher tempera-tures than upon either wool or silk, and the same with safflower. But it requires no scientific test to know that some of the coloring matters will not fix upon some fabrics, and that if fixed upon others, they are, very loose. *The archil colors enjoy a fair amount of stability upon woollen, on silk less, upon cotton none at all. Silks dyed shades of violet blue with orchil and cudbear are bleached by exposure to strong sun-shine; but the color is not always destroyed—it can be revived again, or it revives spontaneously, if kept in a dark place for some days. The silk, under the influence of heat and light, seems to exert some chemical action upon the coloring matter, similar to the lime and copperas action upon indigo; and pro-

bably it is a real deoxidation which takes place.   Upon cotton,
colors seem to be affected in an opposite direction, and appear
to be destroyed by oxidation.   It is the nature of silk and
woollen, by virtue of their higher organization, to exert under
the influence of chemical agents an attraction for oxygen ; this
property, which is common to all organic substances, is possessed in only a feeble degree by cotton, and its organic nature
does not, under general circumstances, present an obstruction
to the oxidation which is always going on in nature.   Mineral
colors, for the most part, are oxidized bodies, and owe their
color to the particular balance of the metal and oxygen in them ;
upon cotton fabrics this remains undisturbed, as far as the
cotton is concerned, and it is only external influences which
can change them ; but, on animal tissues, the fabric itself is a
quasi chemical agent, which, having an affinity for oxygen,
generally takes it from any oxidized mineral substance, and
thus altering the relative proportions of the elements destroys
the color.   It is for this reason that the iron buff, which is an
oxide of iron ; the manganese brown, which is a peroxide of
manganese ; and the lead puce, also a peroxide, cannot be
applied on silk and wool with any good effect.   It is the same
with other mineral colors, which are not so simple in their
composition as the chrome yellow and orange.   Other mineral
colors are inapplicable on wool, because of the sulphur it contains acting upon them and injuring them.   It seems probable
that these properties of the different fibres may explain various
behaviors of the different coloring matters, both as affecting
the union of them with the fibre and their degree of permanency
when there.

   There is no general principle to guide the inquirer as to how
a fast color is to be produced.   We know too little of the nature of the union between the mordant and the coloring matter,
and between either, or both of these, and the fibre, to enter into
anything better than metaphysical or hypothetical explanations
upon this part of the matter.   No information has been obtained in that direction, except what practice reveals, viz., that
some compounds are less stable than others, but without any
information as to why.   There is a reason for this without
doubt, and, when it shall have been discovered, it is not unlikely it may lead to some improvements in the application of
colors which shall make all fast alike, or at least add to the
stability of those which are now deficient in this respect.

   Copper.—This metal is the best of all the common metals
for small vessels employed in holding or measuring chemical
substances ; it is less easily acted upon by acids than iron ; it
is harder and stronger than tin, and more economical, because

requiring less weight and enduring longer. Pure clean copper is not acted upon in the cold by muriatic or sulphuric acids; if the copper is tarnished, either of these act upon it in so far as to dissolve the oxide which tarnishes the metal, but they do not touch the metal itself. If exposed to the action of the air and acids simultaneously, the copper is rapidly corroded. Copper pans or boilers which are accustomed to hold acid liquors are more strongly corroded above the water mark than below it, and, if the level of the contained liquor is pretty constant, the corrosion is most active at or slightly above the water mark. This is owing to a natural chemical law, that a metal must be oxidized before it can dissolve in an acid; some metals take oxygen from the acids themselves, or from the water in them, but this is not the case with copper and the two acids mentioned. It must obtain the oxygen from some other source, and that source is the air. Copper is not hurtful to most colors, as iron is; many colors are improved by it, and but few really injured. Notwithstanding this, it is not wise to leave colors in copper vessels longer than necessary, especially steam and spirit colors, which contain acid salts; the copper dissolved, though little, alters delicate shades, and frequently puzzles the color mixer. Nitric acid, or aquafortis, acts rapidly on copper, hot or cold, itself supplying the oxygen necessary to the solution of the metal. Liquid ammonia in contact with the air acts on copper, dissolving it and becoming blue; it should not, therefore, be kept, or even measured in copper vessels. Colors which are alkaline with ammonia should not be allowed to remain in copper vessels, nor be worked in the printing machine out of copper boxes. Liquids which contain sulphur in alkaline solution also act upon copper, turning it quite black, and corroding it, although not to a very perceptible extent; the black body produced is a sulphide of copper, and if kept exposed to the air and moisture, it eventually changes into sulphate of copper, or bluestone. But, in dry situations, the sulphide remains unaltered for a long time: for example, copper rollers blackened in the machine by using pencil blue or impure alkaline pink.

Copper combines with oxygen in two proportions, forming the red or sub-oxide, that is, an oxide with less than an equal atom of oxygen, and the black oxide, which contains an atom of oxygen for an atom of copper. Neither oxides are ordinary commercial articles, nor used in dyeing or printing; but the sub-oxide yields a color which may be produced by printing acetate of copper, raising in lime or soda, and then boiling in a strong caustic ash, mixed with sugar, treacle, or calcined farina, until it has changed into a yellowish orange. The sub-oxide

in this state can act as a mordant, but it does not give bright
or fast colors with any of the dyewoods I tried. The production of the sub oxide above from the black oxide is owing to
the reducing or deoxidizing power of the alkali and organic
matter. The black oxide is used to make the salts of copper
from. The salts of this oxide have all a green or blue color
in the hydrated state, but when deprived of water are often
colorless. They have a feeble oxidizing power, which acts
under favorable circumstances upon many organic substances.
It would appear that a given portion of copper salt is able to
oxidize an infinite amount of some organic substances under
favorable conditions; it is supposed to act first by giving up
the half of the oxygen in its oxide, and the sub-oxide formed
absorbs oxygen from the air, transferring it to the oxidizable
substance *ad infinitum*. It is as oxidizing agents that copper
salts are used in most cases of dyeing and printing, as in catechu and steam colors.

Sulphate of Copper, Blue Vitriol, Blue Stone.—This is a compound of vitriol, the black oxide of copper, and water. It is
used to make the acetate and nitrate of copper from, by means
of the acetate and nitrate of lead; it is employed in a few cases
of color mixing and dyeing, but altogether only a small quantity is consumed in these arts, as the acetate and nitrate of
copper are mostly purchased from the manufacturers. The
greatest quantity is used by the indigo dyers as a resist, in
combination with other matters.

Chloride of Copper, Muriate of Copper.—This salt, although
much used on the continent in color mixing, is almost unknown
in this country. It is used in nearly all cases where English
receipts give some other copper salt, as the sulphate and nitrate in combination with sal-ammoniac; although I believe
sal ammoniac is necessary even with the muriate of copper, but
smaller quantities suffice. It is made by dissolving oxide of
copper in muriatic acid, and concentrating until it crystallizes;
liquid muriate of copper may be made from muriate of lime
and sulphate of copper.

Nitrate of Copper.—This is mostly sold as a concentrated
liquid; its principal use is in calico printing, where it is employed as an oxidizing agent, as in the case of catechu browns,
some steam colors containing logwood, chiefly browns and
chocolates, and in several spirit colors. Nitrate of copper
enters into the composition of a resist for China blue. The
value of nitrate of copper is usually judged of from the strength
indicated by the hydrometer: unless purposely falsified by
other substances, this test is satisfactory; if adulterated, only
regular analysis can give its value. It is sometimes sent out

too acid—not properly killed, as it is expressed. It is always acid, but there should not be any great excess present. A practical test of the comparative acidity of different samples consists in trying how much weak caustic must be put in a gill of nitrate of copper, before it begins to give a precipitate which does not dissolve on stirring : the more that is required, the more acid the liquor, and *vice versa.*

Acetate of Copper, Verdigris. (See ACETATE.)

Ammoniuret of Copper.—When liquid ammonia is mixed with a solution of a salt of copper it gives at first a pale blue precipitate, but a little more ammonia causes this to dissolve, producing a rich purple color. This solution, which may be called an ammoniuret of copper, has been employed to give a light shade of green, by simply padding in it, drying, and washing off. Ammoniuret of copper has the peculiar property of dissolving cotton, reducing it first into a pulp, and then clearly dissolving it. This fact, but recently discovered, may possibly be applied to some useful purpose hereafter.

Arsenite of Copper or Scheele's Green.—Arsenious acid can, by proper management, be made to form a salt with oxide of copper, which has a pleasing green color ; it is extensively used for printing paper-hangings, and as an oil color. It can be produced upon cloth, but not with the same brilliancy as the dry powder. The method of dyeing this color is to pad in a salt of copper, dry, and fix the copper by passing in a caustic bath, then pass into a beck containing the white arsenic dissolved, and keep in until the desired shade is obtained. Another method consists in padding in arseniate of soda, drying, then padding in nitrate of copper, washing off, and finishing with a short passage in weak acetic or nitric acid. This color is very little worked now on calico or other fabrics.

Fatty Compounds of Copper.—Green colors have been used made from the fatty compound of copper, which is produced when blue vitriol is mixed with a hot and strong solution of soap. A soap of copper is produced which is not soluble in water, but which can be dissolved in turpentine, and so printed or padded ; hung up in a hot stove the turpentine volatilizes, and leaves the green copper compound upon the cloth. The colors thus produced are not remarkable for either brilliancy or fastness, and are now hardly ever made. Some samples in my possession have gone much yellower by age, though they have stood better than most metallic colors in the same conditions.

Copper is particularly injurious in madder dyeing in several states, and in very small quantities. One per cent. of the hydrated oxide, or carbonate of copper, is sufficient to destroy

13

the tinctorial power of madder altogether, under ordinary circumstances. The soluble salts act quite as injuriously.

Copperas.—This is a generic name in common use for the metallic sulphates. When used without a qualifying adjective, it means the sulphate of iron, but the term *green copperas* is frequently employed to prevent misunderstanding.

Blue Copperas is sulphate of copper. (See COPPER.)

White Copperas is sulphate of zinc. (See ZINC.)

Cork-Tree Bark.—According to an expired patent, the bark of the cork tree may be used as a dyewood for dyeing nankeen shades on cotton, wool, and other articles. There is nothing special in the method of its application, excepting that the color is directed to be finished off in soap and warm water or hartshorn and warm water. The patent is dated Feb. 18, 1823. The material is not now found in the list of our dyeing drugs.

Cotton.—This valuable fibrous material consists of small filaments about $1\frac{1}{2}$ inch long for the long-staple cotton, and about $\frac{17}{30}$ of an inch for short-staple (East Indian) cotton. The diameter of the filaments of the finer cotton is very small, not exceeding $\frac{1}{2000}$ of an inch; the coarser kinds are thicker, Surats varying from $\frac{1}{800}$ to $\frac{1}{1000}$ of an inch. Under the microscope, cotton filaments look like flattened cylinders, or hollow tubes which have collapsed by drying, so that the sides adhere together. There is some reason to think—though it is a disputed question—that in the operations of dyeing and printing, the coloring matter finds its way between the walls or sides of the tube, filling up the vessels by which the cotton drew nourishment during its growth. Some qualities of cotton do not take coloring matters as well as others; for instance, it will sometimes be observed in looking over printed goods that here and there a single thread, or two or three threads together, have not taken color, though from their position they must have had every chance of doing so. This cotton, upon examination, appears different from the ordinary qualities: it has been called *dead cotton* from a belief that it is cotton which has not arrived at maturity in the pod, but through some cause or other has died while still in an undeveloped condition. The inner tube is defective, or partly filled with some substance, and so seems to prevent the admission of the dye.

As the civil war in America seems likely to cut off our usual supplies of cotton for an indefinite period, it is a matter of interest for the printer to know whether cotton from other quarters of the globe will absorb the dyes equally well with American cotton. A large quantity of cotton cloths made exclusively from East Indian cotton have been printed and dyed with tolerably satisfactory results; but the comparative merits of the various supplies do not appear to have been strictly tested yet. Amongst the numerous interesting objects in printing and dyeing contained in the International Exhibition, now open (1862), there is a case of prints, exhibited by the Cotton Supply Association, to show that Surat cotton is well adapted for printing on, but there is no American cotton to compare with. In the Swiss court there is exhibited a small but interesting series of trials in Turkey red dyeing. in which the American, East Indian, and Egyptian cottons are contrasted; from these experiments it appears that American cotton takes decidedly the best color, East Indian cotton is but little inferior, while the Egyptian appears considerably below either in its affinity for coloring matters. This is only a single experiment, and too much importance must not be attached to it. For the behavior of cotton to chemical agents, see FIBROUS MATTERS.

Cow-Dung.—Excrementitious matters have from time immemorial been employed in various operations in dyeing and printing. Their precise action is not generally understood; but it must be admitted that the operations in which they were employed were essential to the successful carrying out of the end aimed at. The progress of chemistry has enabled us to find substitutes for most of these substances, but not to dispense with the operations. Cow-dung is now but little used compared with former times; the introduction of substitutes acting quite as well, which are easily stored, always obtainable, and scarcely dearer, have tended to make the operation of *dunging* properly so called nearly obsolete. The explanations of the uses of cow-dung have been given in the article on CLEANSING.

Cream of Tartar.—This is the common name given to the acid combination of tartaric acid and potash, known in scientific works as bitartrate of potash. (See TARTARIC ACID.)

Crimson Color.—This is a red color with a tendency towards purple or blue, just as scarlet has the contrary tendency towards orange or yellow. The chief and most valuable crimson colors are obtained from cochineal, and have been treated under that head; it only remains here to add some of the crimson shades obtained from woods.

A common crimson may be dyed upon cotton by steeping in sumac for some hours, and then working in tin spirits at 2° Twaddle, in the manner common to so many shades; then washing out and working in a decoction of three parts of peach or Lima wood and one part of logwood, at a heat of about 120°, and then raising with tin spirits (oxymuriate of tin). Here the pure red, or rather scarlet, which would be produced by the red wood alone, is modified by the purple of the logwood, and made into crimson. By varying the proportions of the woods the shades may be modified at will.

Quantities.—100 lbs. cotton, 30 lbs. sumac, steep ten hours; tin solution at 2°, 45 minutes; 30 lbs. red woods, 10 lbs. logwood, 30 minutes; tin spirits to raise, 15 minutes.

When the coloring matter of safflower is deposited in considerable quantity it produces a crimson shade. (See SAFFLOWER.)

There are no wood reds for calico printing which can properly be called crimson. The whole of this class of reds are given under RED.

A crimson on delaine, or mixed cotton and wool, is obtained by passing the goods for two hours in the cold in a bath of oxymuriate of tin, at 6° Tw., and leaving them a reasonable time to allow the tin to fix. They are then dyed in one of the red woods, chiefly peachwood, at a temperature of 150°, for 45 minutes. If the wool is dyeing faster than the cotton, the temperature must be lowered; if the cotton dyes faster than the wool, the temperature must be raised.

The crimson shades obtainable on silk by means of the woods, have but little beauty, and so fugitive as to be rarely dyed.

Crocine.—Name given by Rochleder to the coloring matter from *gandenia grandiflora*, being identical with the coloring matter existing in the crocus. By the action of strong acids, *crocetine* is formed, which dyes a dark and bright yellow upon silk with salts of tin. Said to be used by the Chinese for dyeing the yellow mandarin robes.

Crystals of Soda.—This is the common name for crystallized carbonate of soda, as crystals of tin is the common name for crystallized protochloride of tin. *Pink crystals* are the double chloride of tin and ammonium, so called because used in preparing pink colors; they are hardly at all used in England.

Cudbear.—This tinctorial substance is a modified extract of colorable lichens, similar in its behavior to archil. It is sold in a state of fine powder, of a reddish color, and peculiar odor, easily remembered. Its use is confined to a few cases of silk dyeing, where it is employed to yield shades of ruby and maroon; upon wool it gives deep red shades. The colors produced by it are very fugitive.

Curcumine.—The name of the pure coloring matter of turmeric, from the designation of the plant *curcuma longa*, from which it is obtained.

Cyanogen.—This is a chemical compound of carbon and nitrogen, which enters into combinations with metals, forming salts called cyanides. Neither cyanogen nor the simple cyanides are yet employed in the tinctorial arts, but the double cyanides, under the names of red and yellow prussiates, are extensively used.

D.

Dahlia Color.—The shade of color which in common language is called dahlia is a reddish lilac rather low toned : it is produced by combining a blue or purple color with red when a compound color is used ; upon woollen and silk it can be obtained directly by means of an archil or cudbear either alone or blued by a small quantity of sulphate of indigo; upon cotton goods indifferent shades of dahlia are obtained by macerating in sumac liquor, working in tin solution, and dyeing in logwood mixed with some red wood. In printing, the shades are obtained by simply mixing red and purple colors previously prepared, as in the following examples upon cotton and delaine :—

Dark Dahlia for Calico.

6 quarts pink standard,
1 quart pale purple standard,
5 lbs. of gum substitute; boil well.

Light Dublin for Calico.

6 quarts pink standard,
2 quarts light purple standard,
6 quarts gum water.

The pink and purple standards referred to are made as follows :—

Pink Standard.

1 gallon cochineal liquor at 6° ; heat to
  170°F., and add
6 oz. alum,
3 oz. cream of tartar,
½ oz. oxalic acid ; dissolve the salts.

Pale Purple Standard.

1 gallon logwood liquor at 6° ; heat to
  180° F., and add
4 lbs. gum senegal,
8 oz. red prussiate of potash,
12 oz. alum,
1 oz. oxalic acid,
2 oz. oxalate of potash.

The dahlia color for delaines does not differ greatly from the above, but still sufficiently so to make the receipts interesting.

Dark Dahlia for Delaine.

3 quarts dark lilac,
3 quarts pink standard,
2 lbs. gum substitute.

The colors referred to in this receipt are made as follows:—

Pink Standard for Dahlia.

5 lbs. cochineal,
10 quarts water; boil 30 minutes and strain;
    the clear liquor is the standard.

Dark Lilac.

6 lbs. ground logwood,
1 gallon red liquor at 18°,
1 oz. oxalic acid,
1 oz. binoxalate of potash.

Steep all night; strain off, and make up to 3 quarts with water; thicken with gum substitute, at 2 or 3 lbs. per gallon.

Medium Dahlia for Delaine.

1 gallon dark purple,
1 gallon crimson standard,
3 lbs. gum senegal.

The colors which compose this mixture are prepared as follows:—

Crimson Standard for Dahlia.

10 lbs. cochineal,
1¼ lb. cream of tartar,
4½ gallons water,
7 quarts strong ammonia.

Keep hot for 40 minutes, and strain; use the clear liquor.

Dark Purple for Dahlia.

1 gallon logwood liquor at 18°,
1 gallon red liquor at 18°,
8 oz. red prussiate of potash,
4 oz. soda crystals,
12 oz. oxalic acid,
6 lbs. gum senegal; mix well.

The use of soda crystals here is only an indirect method of adding the binoxalate of soda to the color; experience shows that the binoxalate in small quantity is useful in logwood and alumina colors. Instead of this dark purple the dark lilac given above would answer, or some of the other purples given under that head.

Delaine : *Muslin de Laine, Mousseline de Laine.*—Although these words in French distinctly indicate a fabric made of wool, they are employed in English to indicate a material of which the weft only is wool, the warp being cotton. In French, this fabric is called *chaine cotton*, that is "cotton warp," and sometimes *mi-laine*, that is "half wool." It has been extensively used as a fabric for printing upon, and also for dyeing. Since the two materials, cotton and wool, present a very different affinity for coloring matters, there is some little difficulty in obtaining the two threads of an exactly similar shade; this difficulty, however, has been overcome by a multitude of ingenious devices, the chief of which depend upon the fact that cotton is best mordanted and best dyed at a moderately low temperature, and in neutral liquids; and wool is best dyed at a temperature close upon the boiling point, and in acidulated liquids. In delaine printing the affinities of the different fibres have been tolerably well equalized by depositing in their pores a considerable quantity of oxide of tin, and compensation is made in mixing colors for the different properties of the wool and cotton; one ingredient is added because it goes to the wool, another because it goes to the cotton; as in colors where blue enters as an ingredient, extract of indigo fixes upon the wool but not at all upon the cotton, therefore the elements of Prussian blue are added which give the blue part to the cotton. When delaine colors are unskilfully mixed, the tint of the cotton and wool is very different, producing the effect called "threadiness." The colors suitable for delaines are generally given distinctly in this book; but when not given it may be understood that the colors given for wool or cotton are equally applicable to delaine.

Delph, or *Delft color; Crockeryware Blue; Bleu de Faience.*—The shade of blue which is common upon the ordinary kind of pottery ware, more generally known in England as China blue. (See INDIGO.)

Deliquescent.—Having the property of attracting moisture from the air and becoming damp or running to a liquid. The most familiar example of a deliquescent body in a print or dye works is American potash. Deliquescent bodies of a neutral nature are very numerous, and sometimes employed in color mixing in order to make the color keep soft or attract moisture when it is on the fabric. The most common deliquescent salts are muriate of lime, nitrate of lime, chloride of zinc, nitrate of zinc, acetate of potash, and common salt.

Dextrine.—The name given to a species of gum substitute made by the action of diastase or an acid upon starch or farina. (See GUM.)

Diastase.—A peculiar substance contained in malted grain which has the property of converting starch into gum; it is used for this purpose in some printing establishments. (See GUM SUBSTITUTES.)

Dip Blue.—A style of calico printing obtained by dipping pieces into a solution of indigo properly prepared. (See INDIGO.)

Divi-Divi.—This is an astringent substance possessing some of the properties of sumac; attempts were made to extend its use in dyeing, but not with much success; it is used by the tanners, but scarcely, if at all, by the dyers or printers.

Discharge.—The name is given to a composition which has the power of bleaching or discharging the color already communicated to a fabric, as for example, it may be desired to produce a design consisting of white figures upon a colored ground. In many cases the best way of attaining this object is to dye the whole fabric of an uniform color, and then print the design of the white spots with a discharging substance which, removing the color at the parts where it was applied, lays bare the white of the fabric and makes the design. The composition of the discharge, and the processes which follow its application, must evidently depend upon the nature of the color which has to be discharged. I give a number of examples with explanations of the reactions which ensue.

Discharge upon Indigo Blue.—The process generally in use was invented by Mr. Thompson, of Clithero. It consists in padding the indigo pieces in solution of bichromate of potash, drying, and then printing on a highly acid paste, which liberates chromic acid from the potash. This is an acid of great energy, and highly destructive of coloring matters. As soon as liberated it acts upon the indigo blue, and destroys it by oxidizing it. The bichromate of potash without acid does not destroy indigo colors, and consequently, the parts not touched by the acid composition retain their color. But it is found that the bichromate is capable of injuring the blue if the piece is exposed to a strong light, as sunlight; it is necessary, therefore, to provide some place for hanging the pieces, where the light is not too direct or too brilliant. After the contact of the acid and cloth is judged to have been sufficiently prolonged to destroy all the color, the pieces are washed off in warm chalk and water, the chalk being necessary to neutralize the still acid character of the discharge. The following receipts

will sufficiently illustrate the methods of preparing the colors
or resists.

Pad the pieces to be discharged in a solution of bichromate,
made by dissolving half a pound of the salt in each gallon of
water, the solution standing between 5° and 6° Tw.; dry
quickly.

White Discharge on Indigo Blue.

1 gallon boiling water,
2 lbs. oxalic acid,
10 lbs. sulphate of lead pulp,
1¼ lb. oil of vitriol,
8 lbs. calcined farina.

The oxalic acid in this proportion will crystallize out if the
color be allowed to get too cold in the working. The practical difficulty in this style is to obtain an acid sufficiently energetic to accomplish the object without endangering the cloth;
oxalic would be the best acid, but it is only little soluble in
cold water, and it has, therefore, to be combined with other
acids. The sulphate of lead is added to give more body to the
color, and to allow a greater quantity to be deposited by the
machine; for the same purpose the rollers for printing this
style must be deeply engraved.

A process lately in use in Manchester was as follows: Two
parts of bichromate and one part of pearlash were dissolved in
water to mark about 6°; the cloth padded in this solution,
dried, and printed with a color made with one gallon of starch
thickening and 2 lbs. oxalic acid, worked warm, hung up all
night, and passed through warm and very weak caustic soda,
and then finished in weak sours. This would not give good
whites on the darkest styles, being deficient in power.

Another Discharge for Indigo Blue.

1 gallon of water,
2 lbs. oxalic acid,
1 lb. tartaric acid,
3 lbs. gum,
6 lbs. pipeclay,
8 oz. muriatic acid.

This prescription differs from the first by the addition of tartaric acid, and the substitution of pipeclay for sulphate of lead,
and muriatic for sulphuric acid; it is more suitable for blocking than for machine. By using weaker acid solutions the
discharge would be imperfect, only reducing the color; this is
sometimes aimed at, as in the following receipt:—

Half Discharge for Indigo Blue.

1 gallon lime juice at 80°,
10 lbs. bisulphate of potash,
10 lbs. calcined farina.

Mr. Mercer discovered that a mixture of red prussiate of potash and caustic potash would discharge indigo blue in a very perfect manner; but on account of difficulties in the application and the greater expense of this process over the chrome discharge it has not been much applied. If the materials be mixed and printed on a blue, the color is at once discharged and only requires washing off; but the mixture has a very energetic action upon all thickening matters, and is very irregular in its results.

The discharge upon indigo blue can be made to act as mordants or colors to produce a colored design instead of a white one; for example, if acetate of alumina be mixed with the discharge, and the alumina fixed in the washing off, it may be dyed up in madder, garancine, etc., to produce a colored instead of a white discharge.

Discharge upon Madder Colors.—The chief case in which these discharges are employed is upon Turkey red, but they are also applicable, if required, upon lilac and other madder colors. I omit the discharges to be printed on mordants before dyeing, since that style is hardly ever practised, not giving so good results as resisting, and being more difficult in its application. The bichromate process of discharging blue is applicable to madder colors, but, not being nearly so good as the chlorine process, is never used.

The process generally known as Monteith's, consists in the direct application of solution of chlorine to the dyed fabric. The Turkey red cloth, in folds properly smoothed and arranged, is placed between two thick cast lead plates, which are both perforated with the design to be produced, the perforations exactly corresponding; the cloth is then pressed between these plates with considerable force, and the plates secured. This is a mechanical contrivance which effectually prevents the wetting of the parts of the cloth tightly nipped between the unperforated parts, while any liquid poured on the upper plate will gradually filter through the cloth from the perforation on one plate to the corresponding one on the other. The passage of the discharging liquor, chloride of lime with an excess of free acid, is assisted by currents of air. When the color is discharged clear water is passed through, and the pieces then washed and finished. As might be expected from the nature of the process, the edges of the design are not sharp and clear;

notwithstanding any practical amount of pressure the capillary attraction will assert itself to a greater or less degree, and the designs have, consequently, ragged edges. The imitation bandanna handkerchiefs are produced by this system, which admits of the discharging of large masses of white.

The discovery of the other process is claimed for both M. D. Koechlin and Mr. Thompson, and is in frequent use. It consists of printing a highly acid color upon the cloth to be discharged, and then plunging it into a mixture of bleaching powder and water. The acid acts upon the bleaching powder causing a disengagement of chlorine, which destroys the color upon the spot where the acid is printed. The difficulty in this process consists in confining the discharging to the points printed on, for it is not possible to hit on the quantity of chlorine just necessary to discharge the color, and any excess generated spends itself upon the color in the neighborhood, unless some precautions are taken. By having a considerable excess of lime in the bleaching powder, and by not leaving the pieces in any longer than is absolutely required, and seeing that they are in a proper state of dryness when entered, very regular results can be obtained. Besides obtaining a white object, there are discharges which, at the same time that they destroy the red color, leave another in its place, or the basis for producing another, as explained in the following receipts:—

White Discharge on Turkey Red.

1 gallon of water,
10 lbs. tartaric acid,
7½ lbs. pipeclay or China clay,
1 pint gum water,
1½ lb. bichloride of tin.

The pipeclay is the chief thickening matter in this receipt, and it will be found difficult to work by machine. It is highly important for these discharge colors that the thickening should be of a porous and easily penetrable nature; if, on the contrary, it is dense and compact, it resists, for a considerable period, the entrance of the bleaching liquor, and subjects the body of the cloth to injury from a too prolonged immersion in the decolorizing vat.

Another White Discharge on Turkey Red.

1 gallon hot water,
9 lbs. tartaric acid,
10 lbs. pipeclay.
3 quarts gum senegal water.

These colors can only be well applied by block. The following discharge for working by machine may do for light work :—

White Discharge on Turkey Red.

1 gallon water,
6 lbs. tartaric acid,
1½ lb. starch.

The following are compositions for discharging the red and producing some other color on the discharged spot :—

Blue Discharge for Turkey Red.

2 gallons muriate of tin,
1 gallon Prussian blue pulp,
1 gallon water,
5 lbs. tartaric acid; dissolve
2 gallons thick tragacanth gum water.

Another.

9 lbs. tartaric acid,
1 lb. oxalic acid,
4 lbs. yellow prussiate of potash,
4 oz. green copperas,
6 quarts water,
2 lbs. starch,
4 oz. gum tragacanth.

The excess of acid in both these receipts causes the discharging of the red color, while the blue part, not being acted upon by the chlorine, remains. Green discharges are composed the same as the blue, with addition of a salt of lead. After going through the dipping in the decolorizing vat, and washing, the pieces are worked in chrome, which converts the lead into the yellow chromate, which, combined with the blue, forms the green. The yellow discharge consists of the acid necessary to produce the white with nitrate of lead, to deposit the lead basis, which is afterwards converted into yellow by chroming.

Yellow Discharge for Turkey Red.

1 gallon water,
15 lbs. tartaric acid,
14 lbs. pipeclay,
1 gallon water, in which dissolve
7 lbs. nitrate of lead, and add
1 gallon thick gum water.

Another Yellow Discharge.

1 gallon lime juice 18°,
4 lbs. nitrate of lead,
2½ lbs. tartaric acid,
3 lbs. gum substitute.

This last receipt is workable by machine as is the following also:—

Yellow Discharge for Turkey Red.

1 gallon water,
1¼ lb. starch; boil, and add
2½ lbs. nitrate of lead,
7½ lbs. tartaric acid.

The green discharge is obtained by mixing suitable quantities of the blue and yellow discharges.

The blacks which are called discharge blacks are in reality only topical blacks without any discharging quality, for the red does not interfere with the shade of the black except to make it a little more intense. In order to complete the series, however, I give here receipts for the blacks which are used in conjunction with the above discharges, and commonly called discharge blacks.

Discharge Black for Turkey Red.

1 gallon logwood liquor at 4°,
2 lbs. yellow prussiate of potash,
1 quart thick gum tragacanth,
2 lbs. flour; boil and add
2 quarts iron liquor at 30°, and when cool
½ pint nitrate of iron at 80°.

The receipt for soda black or spermaceti black, on page 88, is suitable for this style of work. The yellow prussiate is added to give a blue shade to the black, as well as to correct any tendency to brownness which the red might have on the logwood color. The following black for the same purpose is composed partly of galls and partly of logwood, and though somewhat more expensive, is also more durable:—

Blue-black Discharge for Turkey Red.

3 gallons logwood liquor at 8°,
7 gallons gall liquor at 9°,
20 lbs. starch; boil, and while hot add
2½ lbs. yellow prussiate of potash, and when cool
3 quarts muriate of iron at 60°,
3 quarts nitrate of iron at 80°.

It has been proposed to substitute arsenic acid and sulphate of zinc for the more expensive tartaric acid in the above receipts; but the substitution has not been general.

Discharge upon Manganese Bronze.—The proper discharge for this color is muriate or crystals of tin, either with or without addition of acid: for the darkest shades the discharge must be rather strong, containing from six to eight pounds of crystals per gallon, and may be mixed with either muriatic acid or oil of vitriol. After printing, the pieces may be hung up for a few hours and then washed off in chalk and water.

Discharge upon Iron Buff.—The same discharge as for manganese brown, the strength to be proportioned to the darkness of the color; if only a light shade of buff, muriate of tin is not necessary, a mixture of oxalic, tartaric, and muriatic acids will suffice.

The difficulties attending the successful discharging of these two colors reside in having the discharge strong enough, and not too strong; if deficient in strength, the whites are of course bad; if too strong, the discharge will run and give a bad impression. The iron buff does not give way very readily, and the pieces must be hung up in a cool place for some hours; if the room is too dry the action of the discharge is suspended, if too moist the color deliquesces and marks off.

Discharge upon Prussian Blue.—This discharge is simply caustic potash thickened with gum substitute; the potash must be about 20° Tw. After printing, the pieces are washed in clear water, and though the blue is removed there is a buff left of oxide of iron which is cleared by a passage in warm sours.

Discharged whites are usually inferior to resisted whites, excepting in the case of indigo blue where the contrary is the case; consequently, most whites are now obtained by resisting, unless the ground color is one which can be better obtained by dyeing than by printing.

Discharging or Bleaching Printed Pieces.—This is frequently done upon a print works with spoiled pieces, etc., before dyeing. Iron and alumina mordants are easily discharged by a passage in warm sulphuric acid at about 5° Tw. Reds which have tin in are extremely difficult to discharge, especially if three or four days old; they are best soured and then bowked as in bleaching. Catechu brown requires a very strong solution of bleaching powder after souring and sometimes a prolonged steeping in spirits of salts. Chrome colors are discharged by passing in warm muriatic sours, washing, and then steeping in dilute caustic soda. Indigo blues are discharged by bowking under pressure with soap and soda, then souring and chemicking. Iron buffs and manganese browns are discharged by warm muriatic sours. Dyed madder colors can be discharged by a good warm souring and then bowking. In all cases the goods must receive a slight chemicking or bleaching steep, to take off a yellowness which they unavoidably acquire in the treatment.

Drab Color.—Drab is a kind of gray; it is generally said to be the color of fullers' earth. I give some receipts for producing shades known as drabs. Further information may be found under GRAY.

Drab Colors on Cotton:—

Catechu Drab for Madder or Garancine.

3 gallons water,
10 lbs. muriate of ammonia,
10 lbs. catechu ; boil, dissolve, and strain ;
1 gallon acetic acid,
½ gallon acetate of copper, p. 45.

The above would only yield a brown; but by the addition of muriate of iron it is turned to a drab, the shade of which varies according to the proportion of catechu and iron salt. The following proportion gives a light drab:—

2 gallons gum water,
5 quarts of above standard,
1 quart muriate of iron, at 10° Tw.

The muriate of iron being increased the drab becomes darker ; there may be six or eight well-defined shades between the above and the darkest shade, besides two or three yet lighter. The color is treated exactly as a madder or garancine color, and is worked with red, chocolate, etc.

Dark Drab for Calico. Steam.

1 gallon berry liquor at 12°,
7 lbs. gum substitute ; boil and cool ;
1½ lb. alum,
1 lb. green copperas, .
1 quart logwood liquor at 2°,
1 quart cochineal liquor at 3°.

In this receipt all the elementary colors are present, with the iron salt for a darkening agent; the coalescence of the whole destroys the particular shade of each, making a gray brown. This color being reduced by gum water yields the lighter shades of drab.

Another Drab for Calico. Steam.

3 quarts hot water,
1½ pint lilac standard,
¾ pint bark liquor at 30°,
1 oz. yellow prussiate of potash, dissolved in
1 pint hot water; thickened with gum.

Lilac Standard for above.

2 gallons logwood liquor at 12°,
2 gallons red liquor at 18°,
1 lb. oxalic acid,
1¼ lb. muriate of ammonia,
10 oz. acetate of copper crystals.

In this receipt, which is of French origin, the blue part is partly owing to Prussian blue, and partly to the violet of the logwood, the bark liquor giving the yellow part, the red part of course exists in the violet.

Dark Drab for Delaine.

3 quarts purple standard made from
4 lbs. ground logwood,
1 gallon red liquor, steeped all night,
2 quarts blue standard,
3 pints berry liquor at 6°,
5 lbs. gum substitute.

The blue standard, named above, is composed as follows:—

7 lbs. yellow prussiate of potash,
2 lbs. alum,
12 oz. oxalic acid,
2 gallons hot water,
½ pint muriate of tin.

To be well stirred before mixing.

Another Dark Drab for Delaine.

1 gallon gall liquor at 12°,
1 quart berry liquor at 10°,
1 pint cochineal liquor at 6°,
1 pint iron liquor at 24°,
5 quarts thick gum water.

Drab Colors upon Cotton by Dyeing.—Drab dyed upon cotton is obtained in a variety of ways, some of which are here given.

Iron and Catechu Drab.—Pad in buff liquor and raise, then

dye in beck taking four pieces and using one quart of solution of catechu, made by dissolving 1½ lb. catechu in one quart of acetic acid and one quart of water. Keep the pieces well open or they will be irregular.

Another Drab.—Steep in sumac, add some copperas, and work for fifteen minutes; wash out, and then work in a mixture of fustic, logwood, and peachwood, and raise with alum. The shade can be varied at will by varying the proportions of the woods.

Catechu Drab.—Work the cotton goods in weak catechu liquor for some time, then add a little copperas and work until the desired shade is obtained. By afterwards working in the woods and raising with alum, an endless number of shades may be obtained.

Drab Colors upon Wool by Dyeing.—By boiling the wool in bichromate of potash and then adding a small quantity of logwood, an agreeable soft shade of gray drab can be obtained It is modified to a slate-colored drab by addition of sulphate of indigo.

Another drab, nearly similar in shade, is obtained by taking the following, for about 12 lbs. of wool: Mordant first in 2 lbs. alum and 1 lb. tartar, then add 2 lbs. fustic, 6 oz. extract of indigo, and 8 oz. archil liquor, or in proportions thereabouts, and work at the boil for an hour.

Another drab is as follows, for 10 lbs. of wool: 2 oz. ground madder, 1 oz. peachwood, 6 oz. fustic ; work in hot for half an hour, then lift and add 3 oz. copperas dissolved in water; work for another half hour.

It is evident that by varying the proportions of woods any number of shades of drab can be readily produced by an experienced hand.

Drab Colors upon Silk by Dyeing.—The basis of drab upon silk is sumac and fustic fixed by copperas; by using other coloring matters in conjunction the shade may be varied to any degree. For 10 lbs. of silk the following proportions may be taken: 1 lb. sumac, 1 lb. fustic, ½ lb. copperas, afterwards 1 pint archil. If sulphate of indigo be used the drab will become greenish, especially if a greater quantity of fustic be employed.

Drying.—Great care is necessary in drying colored fabrics so that they may retain their highest bloom. The consideration of the machines used for drying does not enter into the plan of this part of the book ; but a few general observations present themselves worthy of attention. In drying mordants before dyeing, a distinction should be made as to whether thickened or not ; thickened mordants will bear harder drying than mordants not thickened ; if the mordant be but slightly

14

thickened, as for padding, it stands in a medium position. As a general rule, mordanted goods should be dried soft, in a stove or by passing before, and not in contact with, moderately heated iron plates. It is found that if mordanted goods, or steam colors, be dried hard or be allowed to come in direct contact with heated metallic surfaces, until all the moisture is expelled, the resulting colors are imperfect and irregular. If the steam color be a dark one and strongly thickened with flour or starch, it will stand a much harder drying than if a light color and lightly thickened. The very few colors or mordants which do not contain a volatilizable element are those least affected by the drying. For example, acetate of alumina mordant if hard dried gives bad colors in dyeing; the aluminate of potash employed for exactly the same colors is little influenced by any kind of drying; the former contains acetic acid which is volatile and becomes expelled suddenly and too quickly by hard drying, the latter contains no volatile element and suffers no decomposition by the heat of drying. The same in the case of lead mordants for yellow and orange. The reasons of this are very evident to the chemist, and the facts are quite familiar to the practical man. In drying finished colors much care is also necessary : in the case of fugitive shades the hot vapor may act chemically, injuring the color very considerably, especially in drying over steam cylinders when the machine is stopped with the goods on. Even in drying fast colors, such as Turkey red and indigo blue, the tone and brightness may be very much changed by too hasty and careless drying. The effect is partly chemical, and also probably partly mechanical; it is likely that when the water contained in the threads is suddenly converted into steam by running on a hot cylinder, a bursting or blowing-up of the fibres takes place, by which the uncolored or but slightly colored filaments of the interior are brought to the surface, and so mix with and injure the true surface color.

It is well understood that all colors look best and brightest when moist, some dyed goods (navy blues) are actually sent wet into the market; Turkey reds are also loaded with oil partly to increase their lustre and partly to increase their weight. All colored goods should have an opportunity of absorbing their proper hygrometric moisture, which may amount to $7\frac{1}{2}$ or 10 per cent. of the weight of the dry cloth; for this purpose goods are hung up in cool places after drying, or taken damp from the drying machine and hung up to part with superfluous moisture, or else damped by mechanical contrivances in connection with the finishing machine. Although this damped state is favorable to the hue of the colors, it is by no means favorable to their preservation if at all loose or fugitive ; therefore, fancy styles, spirit colors, and dyed goods for shipping should be dry, and especially if to be packed in air tight tin packages.

Dung.—There is hardly any animal whose dung has not been used in some way or other in dyeing. Cow dung, sheep dung, and dog dung are at present used in various ways; serpents' dung was the first and best source of the murexide colors, but gave way to fowl dung (guano) which was more plentiful and accessible. No satisfactory explanation has ever been given with regard to most of the uses of excrement; and there seems no doubt that a little enlightened inquiry would show that they might be substituted by cheaper and less objectionable chemical preparations.

Dung Substitutes.—This name refers to substances introduced as substitutes for cow dung, the chief of them and their appliances will be found under CLEANSING.

Dunging.—The operation of subjecting mordanted goods to the action of a mixture of cow dung and hot water. (See CLEANSING)

Dyeing.—The verb "to dye" appears to mean strictly the tinging or coloring of absorbant substances by impregnating them with solutions of coloring matters. It is thus opposed to painting, which consists in laying a color upon the surface colored; and is distinct from inlaying, embroidery, etc., which consist in working colored substances mechanically into another substance; and from staining of glass and porcelain, which is the effect of dry heat upon the coloring particles. Nearly all vegetable and animal substances are porous and absorbent, and can be dyed; also some minerals, as marble and stone. Metals are not perceptibly porous, nor glass, and such substances cannot be dyed.

The practical operations of dyeing are given in detail in these pages, and the general principles upon which the operations depend are specified for each color or dyewood. The art of dyeing, though one of the most ancient in the world, has not been successfully studied by scientific men, and its principles are not reduced to anything like scientific accuracy.

The nature of the combination which takes place between coloring matters, mordants, and fibrous substances, has been the subject of considerable difference of opinion among writers upon this subject. In the last century, Hellot and Le Pileur d'Alpigny enunciated an opinion which may be called the mechanical theory; they were opposed by Bergman and Macquer, who sustained a chemical theory. The ideas of the first two authors, as translated into current chemical language, are of such a nature as to command every respect. They do not admit of any chemical affinity on the part of the fibre, but rest their arguments upon an assumed porousness of all fibres, and, further, that the pores of one fibre are of a different size to the pores of another of a different origin, and that the number of pores is greater in some cases than others. Wool was held to have pores of greatest diameter, and in a given space the greatest number of them, and silk those of least diameter. These pores, which they supposed to pierce the fibres laterally, were the recipients of the colored particles; they were expand- ed by heat, and various chemical agents, in a such a way as to admit the particles of coloring matter, and then, being closed by cooling or other chemical agents, were enabled to retain them securely. They of course allowed the influence of mor- dants as changing the shades of color, and as forming insoluble lakes with the coloring matter. To explain why dyes would not enter into combination with all fibres equally, the theory of different sized pores was invented, and then different sizes attributed to the particles of coloring matters.

These assumptions serve to explain most phenomena in dyeing, and after the following manner: Wool has the greatest number of pores and of the largest size, therefore it stands at the head of all fibrous substance in its affinity for colors; silk has pores much smaller, and is more difficult to dye, and never attracts so much coloring matter as to yield the dark shades obtainable on wool; cotton holds an intermediate place. But there are coloring matters which do not dye wool so well as they dye silk and cotton fibre. This is explained by saying that those coloring matters do truly enter into the pores of the wool, but they are of too fine a nature to be retained there, they swim out again upon washing; they are deficient also in some astringent principle, which in other colors contracts or shuts up the pores; when they are applied to silk or cotton they meet pores of a narrowness suitable for them, and are retained. The difference between fast and loose colors is in connection with the above phenomena; a fast color is one in which the size of the colored particles is adapted to the size of the pores of the material on which it is applied, and in which the coloring particles either contain in themselves, or in the mordant, some astringent, or binding substance, which shall cause the con- traction of the pores; a loose color arises from an unfitness between the size of the pores and the coloring particles, and an absence of the astringent principle; the particles are either too small, and pass too readily in and out of the pores, or they are too large, and only partly enter them, and the astringent prin- ciple is either absent or too weak to close up the pores in a firm and enduring manner. Hellot insisted a good deal upon the astringent principle, and considered that the essential difference between dyewoods lay in their astringent power, saving that if this principle could be communicated to those which were deficient in it all colors would be alike fast colors.

The theory of Hellot rests entirely upon the supposition of the actual existence of pores in fibrous substances. It has been objected that this is gratuitous assumption, for the most perfect microscopes of the present day show no lateral pores, but only a central tubular structure in either vegetable or animal fibres. But, apart from microscopic evidence, or rather the want of it, there is every reason to deduce from analogy that the walls of the fibres are permeable to fluids, and, therefore, contain some kind of passages or pores. No real ground of objection can be taken on the score of want of microscopic confirmation; the microscope is a progressive instrument, and can only be of value so far as it approaches to perfection in its powers, and it is far short of that. From mathematical and physical grounds, if not from our knowledge of organized structures in general, there is reason to predicate the porosity of fibrous matters. The further development of the theory, which asserts the size and number of these pores to be different in different fibres, is in like manner supported by analogical reasoning. The naked eye discovers in the fibre of wood, sections of vegetables and fruits; and in some minerals a regular difference of structure —a coarseness or a fineness, which is always the same for the same substance. In amylaceous matters the microscope shows a regular difference in the relative sizes of the largest granules of any two qualities of different origin : the globules of potato starch are invariably larger than those of wheaten starch, and these are larger than the globules of other kinds of starch. It is not, therefore, exceeding the bounds of fair analogy to suppose that the pores of fibres of a different origin may be different in size. The further assumption that the pores of wool are greater in size than those of silk or any other fibrous matter, does not rest upon such general principles; it is deduced from the behavior of fibrous substances towards coloring matters, which behavior might be explained in other ways.

The chemical theory, supported by Berthollet and chemists generally, considers the fibre as having an attraction of a chemical nature for the coloring manner, or for the mordant; that a real intimate union of atom to atom takes place; it throws aside the question of porosity as not necessary to the explanation; and looks upon the dyeing as a union upon the surface of the fibre between it and the coloring principle with or without a mordant. The various conditions of the fixing of a coloring matter, the different mordants and treatments necessary, are

looked upon as arising from the different chemical properties
of the respective fibrous and coloring matters. This theory has
been the one accepted by the majority of writers upon dyeing;
but, as these writers are for the most part chemists, it is not,
perhaps, surprising that they should have adopted a chemical
explanation, and made a free use of such terms as affinity,
attraction, and combination—terms which are, for the most
part, very loosely used, and include too much to mean anything
precise.

Mr. Crum published a paper a few years ago upon this question, and supported, with great ability, the non-chemical theory
as far as the fibre was concerned, admitting of course a chemical action between the chemical materials used ; but, looking
upon the fibre as a neutral ground upon which the coloring
matter was formed, but retained there by purely physical
means. He rejects the idea of any deposition of any permanent coloring matter upon the surface of the fibre. To be able
to resist washing, the color must be, he says, actually inclosed
by the fibre as by a bag or fine network, the coloring particles must be in the cells or pores of the tissue. In madder
dyed goods there is a chemical compound of an oxide with the
coloring matter, but there is no chemical union of either of
them with the cotton fibre. There can be no chemical combination, he considers, with the fibre, for, "if we only consider
that chemical attraction necessarily involves combination, atom
to atom, and consequently disorganization of all vegetable
structure; that cotton wool may be dyed without injury to its
fibre, and that that fibre remains entire, when, by chemical
means, its color has again been removed, we shall find that
the union of cotton with its coloring matter must be accounted
for otherwise than by chemical affinity." He admits a power
of attraction on the part of cotton fibre, and a capability of its
withdrawing chemical or coloring substances from solutions of
them. But in this force he only recognizes that same action,
whatever it may be, which enables charcoal to absorb gases,
and remove coloring matters and some salts and metallic oxides
from solutions. The blue vat, the plombate of lime for chrome
oranges, the nitrate of iron and muriate of tin—passages so
familiar to dyers—are examples adduced ; but they are not
examples of chemical attraction, but of catalytic force.

Mr. Crum's theory has been controverted by Mr. Persoz at
some length in the second volume of his work upon calico
printing; but in my opinion he does not meet Mr. Crum's
arguments upon the exact ground of the experiments and de-
ductions of the latter ; and most of his illustrations from calico
printing and dyeing are directed against ideas which do not

appear to be contained in Mr. Crum's paper, and which are not fairly deducible from what it does contain. Mr. Persoz considers that cotton fibre, and other fibres, have an actual influence of a chemical nature upon salts; he states that cubical alum left in contact with cotton fibre loses alumina and becomes converted into ordinary alum, and that acetate of alumina is much more completely decomposed upon cotton cloth than when it is exposed spread over a similar surface of glass, mica, or platinum, which are chemically inactive. He rejects the idea of coloring matters being held by cells, pores, or bags, chiefly on the ground that colors when deposited on cloth are as easily acted upon as in any other condition; that if there were such cavities they would be filled up in many cases by the materials used preparatory to the application of the actual color, as by the oil and astringents in Turkey red, by the oxide of tin in steam colors, and by the oxide of manganese in indigo dyeing. He states that upon a careful examination of dyed or printed colors, it may be seen that the color is upon the surface and in relief; if it were otherwise, he demands, how could thickened discharges perform their office, and how would the slightest washings suffice to remove colors so made soluble? He considers that there is a chemical combination between the coloring matter and the cloth, even in such a case as indigo dyeing, and evidently considers the fact of the fibre being unchanged or uninjured as forming no argument against the supposition of chemical and atomic combination. He proceeds at great length to prove that the adhesion of colors to fibres is simply due to attraction; but it is not very clear what Mr. Persoz means by attraction in this particular case. Upon intelligible definitions of those two words "*catalysis*" and "*attraction*," used respectively by Crum and Persoz, the whole discussion rests.

The difference between the ideas of Persoz and Crum are very distinct upon the matter of pores; one thinks them all-essential, the other either considers them not to exist, or, if existing, to be of no essential use. Hellot uses the expression that colors are held by the pores of the fibre as a diamond is by its setting. In Persoz's theory the diamond is held without any setting, it adheres where it is applied, as two true surfaces to each other, or two edges of freshly cut caoutchouc; for he does not assume that it is by virtue of those forces commonly known as chemical, but, by some congruity of physical form, that the mordant or coloring matter, attaches itself to the external wall of the fibre. Both Crum and Persoz admit the existence of a positive attraction on the part of fibre for color-

ing matters and mordants; but neither of them considers this
attraction to be that of simple chemical combination, such as
takes place between an oxide and an acid.

Upon a full consideration of all the circumstances of absorption of color by fibre known to me, I conclude that the idea
of the fibre taking a chemical part is the one which agrees the
best with all the phenomena. The great difficulty appears to
be in that objection of Mr. Crum's, that if there was chemical
combination there should be disorganization of the fibrous
structure when this compound is destroyed by the removal of
one of its components, the oxide or the coloring matter;
whereas, on the contrary, it is not found that the fibre is altered
or injured, in a conspicuous manner, by several such successive
combinations and decompositions. I venture to think this
objection is not of so great a weight as to decide the question.
Admitting that chemical combination alters the physical characters of a body, I think it may be supported that there are
several cases known to chemists in which the alteration is of
such a nature as not to be observable without special precautions. I may cite gun-cotton for instance, which in the hands
of most persons would pass for simple cotton; and, if chemical
combination be anything but a word to play with, it surely
exists here. In the case of the fatty bodies, after saponification
and decomposition by acids, we are certain that the fatty acids
are not the same as the original fat, but our senses would
scarcely have given us the information—it required the genius
of Scheele and Chevreul to show what the difference was. In
organic chemistry we have numerous instances of feebly acid
bodies forming combinations not greatly differing in physical
aspect from the acid itself. Mr. Crum seems to assume that
the whole matter of the fibre must enter into combination with
the coloring matter or mordant, just as Persoz assumes that
the imagined pores are entirely filled in every case of dyeing;
but this never is the case in calico printing or dyeing, only a
small portion of the cotton or other fibre is supposed to be in
actual combination, and if all this part was to be detached in
dust upon the removal of the mordant or coloring matter
with which it is combined, the fabric would probably be only
weakened, not destroyed. Fibre is insoluble in all the men-
strua used in dyeing; such chemical affinities as it possesses
are very weak, and it is next to impossible that an atom to
atom combination of the whole fibre ever should be accomplished. When sulphuric acid acts upon carbonate of bayrta,
or lime, in mass, the action is soon suspended, though sulphuric
acid is one of the strongest and most soluble of acids; how

much sooner should the chemical action come to an end between bodies one of which is always insoluble, and of a texture
peculiarly opposed to the continuation of chemical action, and
other bodies, which are at the best indifferent to it in their ordinary combinations?

Mr. Crum merely says that dyed cotton fibre remains entire
when its supposed chemical combination is destroyed; that is
so without doubt, but I am strongly convinced that its integrity is injured; in numerous instances of discharging dyed
goods I believe to have found a diminished strength in the
cloth—it was thinner and softer to the feel, seeming to have
lost body—and this was not owing to the discharging process
as I have tested by sending white cloth through the same process. It is well known that it is dangerous to discharge heavy
printed goods twice over, the greatest care will hardly succeed
in turning them out fit to be printed a third time as sound
goods. But if all the phenomena of dyeing can be included
or explained under a single theory, must it not be the theory
of Hellot and d'Alpigny, with their various sized pores, their
plasters and cements, and their astringent substances tying up
the pores? This theory seems grossly mechanical, but if
granted it will include all cases that I know, and certainly no
other theory will. Picric acid dyes wool and silk, but not
cotton, sulphate of indigo the same; they are probably both
contained in the undecomposed state in those fibres—the picric
acid can be tasted, and the indigo compound easily extracted.
Do they combine chemically? That seems very unlikely; and
if so, why not with cotton? If it is a catalytic attraction of
surface, why does not cotton exert it here as elsewhere? If a
different chemical composition of these fibres is adduced as the
explanation, the pure chemical theory is at once granted; if a
distinct physical structure is suggested as the explanation,
Hellot's theory is adopted. The coloring matter of safflower is
thrown into the insoluble state in order to dye with, and the
operation is performed cold. What kind of chemical combination can take place under these circumstances, and how are the
colored particles to get into Mr. Crum's sacks and be secured
there? Hellot's pores are always gaping for them like an open
trap, but he admits not always shutting up when the object
enters. Freshly precipitated chromate and sulphate of lead,
when thickened and printed, contract at once an intimate connection with the fibre and are not to be removed from it by a
simple washing; yet these are insoluble salts and not known
to undergo any decomposition or form any compound with the
fibre. Turmeric and anotta are also difficulties hard to recon-

cile to any theory, but suggesting an active affinity on the part
of the fibre.

Besides the theorists already named, Messrs. Dumas and
Chevreul have also emitted theories upon dyeing, which may
be called physico-chemical theories—endeavoring to show that
the affinities exerted between tissues and mordants, though not
precisely the same as those between the chemical elements, are
not greatly different from them. In fact, we must confess, that
not only is there no theory of dyeing which has a respectable
claim to consideration, but there is also an entire absence of exact
data upon which to build a rational theory. Before using chemical terms for dyeing phenomena, we should be assured that we
are describing chemical operations; but no one can point to
anything like a satisfactory quantitative experiment upon this
subject. There is an absence of definite forms and crystallization about the combinations of fibres, mordants, and coloring
principles, which discourages analysis, so that we have no information as to whether there is any atomic combination
between the mordanting oxides and cotton, wool, or silk; or,
on the other hand, whether the pure coloring matters, most
of which have been carefully analyzed, combine with these
mordants in equivalents or not; that is, whether there is any
real chemical combination, or whether the union is simply
mechanical—a species of absorption or adherence by cohesion.
Convinced that all attempts at general theorizing, without some
exact data, would only result in unsatisfactory hypothesis, I
commenced a series of experiments to determine the quantity
of mordant coloring matter and fibre which went to build up
dyed materials. After numerous and unsatisfactory experiments, I was compelled to abandon for the present at least, the
hope of obtaining exact quantitative data; the difficulties are
very considerable, and it would require the devotion of uninterrupted attention for a considerable period of time to obtain
the results I aimed at.

Dyers' Spirits.—Solutions of tin in muriatic acid and nitric
acid, or in nitric acid and some muriate or chloride, as sal ammoniac or common salt; or, sometimes, but rarely, in muriatic
acid and sulphuric acid. (See TIN.) The name, rather curious
in appearance, is derived from the solvents used; muriatic acid
is yet commonly known as *spirits of salts*, and nitric acid sometimes as *spirits of nitre*. These *spirits* were said to be killed
by dissolving tin in them; and as they were used by the dyers,
and made either by or for them exclusively, they were called
dyers' spirits, a name they still retain.

Dyers' Weed.—Probably more than one substance used

in dyeing receives this name, but it is generally applied to WELD, the source of a yellow coloring matter, which see.

E.

Ebony Wood, *Green Ebony Wood.*—A wood, known in England as green ebony wood, is used by the dyers as a yellow coloring matter, principally in dyeing greens and other compound shades.

Efflorescence.—The spontaneous drying of crystals when exposed to the air and falling into white powder. Crystals of soda in dry air, give a good example of an efflorescent salt.

Emeraldine.—This name was given to a green color obtained by the direct application of a salt of aniline to calico prepared by steeping in chlorate of potash; it was the subject of a patent granted to Calvert and others, but has not been successfully applied as yet upon the large scale.

Epsom Salts.—The common name for sulphate of magnesia. (See MAGNESIA SULPHATE.)

Equivalent Weights, *Atomic Weights.*—One of the fundamental principles of chemical science is that chemical substances combine with one another in certain weights fixed by nature; for example, oxalic acid combines with crystals of soda to form a compound in which the distinctive properties of both bodies are lost or emerged in the production of the compound called oxalate of soda. The proportion of oxalic acid to soda is fixed and unalterable; in like manner other acids—as sulphuric, muriatic, nitric, tartaric, and citric—form compounds with soda, and there is a fixed proportion of each acid to soda in the compound, but different for each one of these compounds; supposing the soda to remain constant, and taking its equivalent as 31, we should find sulphuric acid 40, muriatic acid $36\frac{1}{2}$, nitric acid 54, oxalic acid crystals 63, and so on. Now, these weights of the different acids are their *equivalent weights,* that is, they have an equal power when used in these weights. Of sulphuric acid, in the dry state, 40 lbs. are equal to accomplishing what 63 lbs. of oxalic acid crystals can do; and of muriatic acid, in the dry state, $36\frac{1}{2}$ lbs. are capable of performing what 54 lbs. of nitric acid, equally dry, would be required to do. Further than this brief explanation of what is meant by equivalent weights I deem it necessary to proceed; for, although the applications of a complete knowledge of these weights upon a print works would often lead to economy and regularity, it is a point in which more harm than good will be done unless the applier is perfectly master of both the theory and practice of what he is about. The equivalents given in chemical treatises are for perfectly pure substances, which are never found in trade, and are consequently, not applicable to them without a compensation made for the impurity present. This compensation can only be made after careful analysis, which presupposes chemical skill of a high order, and access to the standard works of chemistry.

Erica ; or *Heath.*—British heath is capable of dyeing a yellow color upon woollen, the same as dyers' broom ; but it is very poor in coloring matter, and never used when the richer tinctorial substances, as fustic, bark, etc., can be obtained.

Euxanthic Acid, *Purreic Acid.*—A yellow-colored acid which, in combination with alumina and magnesia, constitutes the pigment called Indian yellow. It is capable of dyeing yellow colors, but has not yet been applied for that purpose.

Extract of Indigo.—A common name for sulphate of indigo. (See INDIGO.)

F.

Farina, *Potato Flour.*—This term is usually applied in trade to the starch obtained from potatoes, often called potato starch. It has most of the properties of flour starch in a chemical point of view ; but on account of a want of tenacity in the paste, which it gives when boiled with water, it cannot be well used as a thickening material. It is used in combination with other substances for finishing or stiffening printed goods. Its tendency is to give a hard finish. When roasted, it produces the well-know thickening matter or gum substitute called calcined farina. (See GUM SUBSTITUTES.)

Fawn Color.—The color of the fawn ; a yellowish brown, rather light colored. The color upon dyed cotton goods, distinctively known as fawn, is dyed with catechu by very simple operations. The cotton goods are worked for twenty minutes or so in a hot solution of catechu containing a little nitrate or sulphate of copper, and then worked in water containing a small quantity of bichromate of potash, which fixes and raises the colors. For some light shades of fawn, bichromate is dispensed with ; and in some cases, to obtain modified shades, it is replaced by acetate of lead or sulphate of iron. A final rinse in soap is sometimes used to soften the color. Fawn colors can also be obtained in the same way as browns, from anotta and the dyewoods, keeping an excess of yellow.

The term fawn is not usually applied to any of the colors used in calico printing ; similar shades are tan, snuff, nut, etc.

Fenugreek.—The seeds of a plant; they are of a mucilaginous and bitter nature, and were formerly supposed to have a beneficial influence in madder dyeing. Not at all used in dyeing at present.

Fernambuc Wood.—One of the varieties of Brazil wood, so called because imported from Pernambuco or Fernambuca. According to continental authorities it is the richest variety of the red woods. (See BRAZIL WOOD.)

Ferridcyanide of Potassium.—This is the present scientific name for the red prussiate of potash. (See PRUSSIATE.)

Ferrocyanide of Potassium.—Scientific name for yellow prussiate of potash. (See PRUSSIATE.)

Fibrous Substances.—The only fibrous matters necessary to be noticed in connection with our subject are cotton, silk, and wool; the few other substances that come under the dyer's hands, are either so rare, or so similar to one of these three, as not to call for a distinct article. It is convenient here to bring together the most conspicuous or important relations of the fibrous matters with the chemical agents they are constantly brought into contact with in printing and dyeing. In the article upon DYEING we have seen that there is a great difference of opinion among chemists as to whether the fibrous matters act chemically at all upon the mordants or colors used to dye them; but that there is an action of some kind no one can deny, because the evidences are plain. Without, then, entering into the disputed question as to what the actions should be called, we will proceed to describe what the visibly are.

Chemists recognize under the name of Lignine a matter which is looked upon as the basis of all fibrous substance of a vegetable nature. It can be obtained from wood by treating it successively with ether, alcohol, water, acids, and alkalies, in order to remove all soluble substances. It finally remains as a white substance of regular composition, which is the pure fibre of wood, and stands as the type and representative of all other vegetable fibres. It is composed of carbon, oxygen, and hydrogen, in proportions which give little clue to its internal composition; and its behavior with chemical bodies is so feebly marked by the expression of affinity in any direction, that it can only be put down as an indifferent substance, forming very few combinations with other chemical bodies. Lignine is of no importance except in a scientific or theoretical point of view, as showing that in all vegetable matter there is a substance of a fibrous nature, and that it has everywhere nearly the same chemical composition and properties. Flax and cotton, along with a great number of other substances of less repute extracted from vegetables, are of the same chemical composition

as lignine, when, like it, they are purified from the resinous and
coloring matters which naturally adhere to them. In our remarks upon fibrous substances, cotton will be taken as representing all the vegetable fibres, being by far the most important of them, and all the others possessing nearly the same properties and characteristics.

The animal fibrous substances, of which the principal are
wool and silk, do not appear to have any common origin of
the nature of lignine. They are very complex in their chemical composition; and besides the carbon, hydrogen, and oxygen contained in vegetable substances, they have also nitrogen,
and some of them sulphur, apparently as an integral portion of
their structure.

Action of Acids upon Fibrous Substances.—Weak acids have
no special action upon cotton. If cotton be steeped in sulphuric acid, or any other acid weakened by water until it does
not act injuriously at once, it may remain in contact with it
for almost any length of time without being injured. If the
cotton cloth which has been steeped in a dilute mineral acid,
be placed in any situation where it will become dry, in proportion as the water leaves it the acid gets more concentrated,
until it becomes strong enough to act upon the cloth and rot
it. The same remarks are applicable to wool and silk. Heated
dilute acids act differently than when in the cold state; being,
as is usually the case with hot liquids, more powerful than
cold, and a strength is soon arrived at where acids do act destructively upon fibres. But it is not the same for each of the
fibrous substances treated of. Cotton is the soonest affected,
and wool the last, silk holding an intermediate state. An acid
liquor which would rot cotton in a short time will not injure
wool, nor act much upon silk; the animal fibres not being so
soon broken up by acids as the vegetable. This property is taken
advantage of in separating wool from cotton in worn mixed
fabrics. The whole fabric is steeped in dilute acid, wrung out,
and dried in a hot stove; after a certain length of time it is
beaten with rods, all the cotton flies off as dust, being corroded
and disorganized by the acid, while the woollen threads are
not considerably injured in strength. It does not appear that
dilute acids, though they may destroy the structure, alter the
chemical composition of vegetable fibres; the powder to which
the fibre is reduced being cotton just as much as it was before.
The action must be of a purely physical or mechanical nature.
It is as easy to suppose a single fibre of cotton as consisting of
a vast number of small particles, held together by some kind
of mechanical adhesion, as by fitting into one another, as it
were, or holding together by power of contact. The acids may

interfere in some unknown manner to loosen this adhesion, when the particles will fall asunder, still remaining the same in chemical nature, but wanting the bond which made them into a fibre ; as a machine may all fall to pieces by the loosening of a screw, and become a heap of iron—still iron, but no longer a machine.

Of the principal acids in use among bleachers and others the muriatic acid, or spirits of salts, seems to be the most energetic for its strength or degree of acidity in disorganizing fibres. In cases where it is employed special care should be taken that pieces are not exposed to chances of drying in places where they are laid before washing off.

Action of Strong Acids upon Fibre.—Cotton can be dissolved in strong oil of vitriol, if added in small portions at a time, being dried previously ; it forms a gummy solution, not going black. When water is mixed with it, the cotton falls out as a white powder, which appears to contain some of the sulphuric acid in combination with it. If heat be applied to the gummy solution, or if the liquor should become hot by mixing too much cotton at once, it will go black and give off fumes of sulphurous acid, changing the cotton into a black charcoal. If cotton be put in a mixture of equal parts of vitriol and water, it can be dissolved, and in this case it is changed into grape sugar. If cotton cloth be quickly passed through a cold mixture of oil of vitriol and water at about 140° Twaddle, it is not injured, but somewhat strengthened. This was one of the processes for Mercerizing cloth.

Muriatic acid destroys the fibre of cotton without blackening it. There are cases in which it does blacken it, but that is only when a high temperature has been employed, or when the acid is impure. This acid, it will be remembered, is only a mixture of an acid gas with water, and it cannot be made above a certain strength ; that is, in the strongest liquid acid, there is only one part of acid, by weight, to two parts of water. It has not, therefore, the strong action of sulphuric acid upon cotton ; and a drop of the strongest spirits of salts may be applied to calico, or calico may be steeped in spirits of salts for some time, without being destroyed, but it must not be exposed to the air.

The action of nitric acid is different in kind to that of the other two acids. The very strongest acid standing at about 100° Tw., converts cotton into an inflammable substance (gun cotton) without much injuring its strength as a fibre, and not at all affecting its appearance; it acts in the same manner if mixed with about an equal part of strong vitriol. Such a liquid acting upon paper converts it into a tough, horny substance, resembling parchment, which, in some respects at least, is very much stronger than the original paper. In the case of gun cotton, the cotton is no longer simply the vegetable substance which it was before the action of the nitric acid : when analyzed, it is found to contain nitric acid, or the elements of nitric acid, in some peculiar form of combination with the vegetable matter, the weight being increased 30 or 40 per cent. by the steeping in the acid. Mr. Kuhlmann, of Lille, in the course of some experiments he made, found that cotton treated with nitric and sulphuric acids (pyroxilized cotton) did not show as much affinity for coloring matters as ordinary cotton ; but that cotton so treated, when kept for some time, underwent a spontaneous decomposition, evolving nitrous gases, and that, when then tried, it showed a considerably increased affinity for coloring matters. But it is a question only interesting in a scientific or theoretical point of view; the dearness of the acids, and the injury to the cloth, both with regard to strength and being made more combustible by the treatment, put it entirely out of the possibility of practical application. When nitric acid acts both strong and hot upon vegetable or animal fibre it converts it into oxalic acid. Strong cold nitric acid tinges wool and silk of a durable yellow color. Upon silk it has been applied practically as a dye, but I believe it is little used now : the silk must not stop more than a few moments in contact with the acid or it will be destroyed. What is said of acids is true with respect to acid salts, in proportion as the salts contain free acid, or as they may give it off in a free state under the processes to which they are submitted upon, or in contact with the cloth. Such salt as bisulphate of potash, sulphate of copper, muriate of tin, nitrate of iron, nitrate of alumina and alum, are, under certain circumstances, likely to rot the fabrics they are applied to, because they either contain free acid, or are so constituted that the acid becomes easily separated from the base, and acts with destructive energy. In all these cases wool resists better than cotton ; and many a color, safe and good, applied upon wool, will destroy calico, if applied to it. Silk holds an intermediate position, nearer to cotton than to wool.

Chlorine acts upon fibres when in a strong state, but it is a question whether it acts as such, or by forming muriatic acid with the hydrogen of a portion of the material acted upon ; it colors silk and woollen yellow, which muriatic acid does not, but its action upon cotton fibre is very similar to the action of muriatic acid. It is not likely that under ordinary circumstances it will ever be required to use chlorine in that state of purity where its action becomes destructive. Chlorine as developed from chloride of lime and sours, acts beneficially upon wool for receiving colors; but whether it is upon the pure fibre of the wool, or only upon the tin prepare, is not clear. The action of chlorine upon cotton turns practically upon the action of bleaching powder upon it. This is known to be destructive in some cases, as where it is exposed to the air in contact with it, or kept hot in strong chemic solutions, etc.; it acts chemically, and according to the observations of some chemists, a kind of combustion goes on, the fibre is oxidized, and carbonic acid gas given off. Under all ordinary circumstances of bleaching, raising colors, clearing, etc., the chloride of lime has no action at all upon the fibre, but only upon foreign substances in or upon it, or if it has an action it is too obscure to be perceived.

Action of the Alkalies upon Fibrous Substances.—The alkaline substances with which fibrous matters are likely to be placed in contact in ordinary circumstances are soda, as soda ash or carbonate, and also in the caustic state; potash in the same conditions; ammonia or carbonate of ammonia as in putrid urine; lime as in bleaching, both in the state of clear lime water and as milk of lime containing solid lime diffused through water; also some kinds of soap, which are very alkaline.

Action of Potash and Soda upon Cotton Fibre.—Potash and soda, in their natural state of carbonates, before being made caustic with lime, have no particular action under ordinary circumstances upon cotton fibre, that is, whether weak or strong, hot or cold. American potashes often contain some portion of caustic potash, whose action is more energetic than the carbonate, and may produce more marked effects. Soda and potash, in the caustic state, do act upon cotton, and differently as they are hot or cold, strong or weak. Weak cold caustic, say as high as 10° Twaddle, will not act injuriously upon cotton steeped in it for a considerable length of time; if the cotton be repeatedly exposed to the air while it is wet with the caustic it is liable to be affected, but not strikingly so; if the same strength be made hot it will injure the fibre very much, and in no long time make it as tender as wet paper. What the chemical change is which takes place under these circumstances, if indeed any such change takes place, is not known. Some chemists suppose an oxidation, but that is not satisfactorily demonstrated; and some cases which have occurred under my observation where oxidation could not take place (because there were deoxidizing agents in the caustic, as grape sugar), make me doubt if the oxidation has anything to do with it. Some cotton fibres resist longer than others, the disintegrating action of caustic solutions producing results of a

15

very anomalous nature on various kinds of cloth; sometimes pieces are rotted in alkaline solutions which do not mark more than two degrees on the glass, while other pieces are not affected at four times that strength. Cloth which has been subjected to several treatments—as in dyeing, to the action of acids, soap, lime, chloride of lime, etc.—seems to be more easily acted upon than cloth which has not been through so many operations. Very strong caustic has a peculiar effect upon cotton fibre in causing it to contract or shrink up: if a few drops of caustic soda, at 50 degrees Twaddle, be dropped in the centre of a bit of calico it will be seen that it causes a puckering, owing to the fibres touched by the alkali contracting in length and pulling the others up. This effect remains upon washing, and is not removed if the calico be steeped in sours. This action was known long ago, but no notice was taken of it until Mercer, to whom calico printing and dyeing owes several happy discoveries, announced that the action of strong alkalies was to make cotton fibres stronger than they were before, and to endow them with a superior attraction for most colors. He patented the process, which seemed to promise great results, but unfortunately it has not turned out so valuable or useful as was expected. Upon passing the pieces in caustic soda, at from 40° to 50° Twaddle, they appear to become transparent and gelatinous, at the same time contracting both in width and length. They are not allowed to rest in the caustic more than a few minutes, and then washed off in water, and the last remains of the alkali removed by a weak sour. When dry, the cloth has a rather different aspect with regard to color than before treating, it does not reflect so much white light, but has a translucent appearance; the threads appear rounder, firmer, and closer together. The contraction in breadth, upon good printing cloth, would be about one-fifteenth, and the same in length which accounts for the closer texture. Analysis shows that some soda remains combined with the cotton however well washed, and thus points to a chemical combination between the alkali and the fibre. The souring takes this soda out. Cloth thus treated shows a superior affinity for some colors, especially the indigo blue; it takes as deep a shade of blue in one dip as common cloth takes in six, and generally speaking, colors look better on this than on untreated cloth. The strength of the fibre is improved, but not more, I believe, than could be explained by the contraction of it. The expense of the soda, a large quantity being required, mainly caused the abandonment of this plan of treatment; and, besides this, the advantages were not of that distinct nature which would justify anything beyond a very small expenditure in the treatment. The contraction of the fibre was also an objection, for though it made the cloth look finer and closer, that is an effect which can be more surely and economically produced by the loom.

Action of Lime upon Cotton.—Lime has no injurious action upon cotton in any moderate quantity, or applied under any ordinary circumstances. Injuries to cotton by lime will be found to exist generally where the cloth has been exposed to the action of the air and lime at the same time, in conjunction with heat. It is not known that the air acts chemically in assisting the lime to injure the cloth: I think it only dries up the water, and brings the lime more closely in contact with the fibre. Some degree of heat higher than the natural temperature seems to be necessary to the destructive action of the lime. As a general instruction, it may be said that it is injurious to expose pieces in contact with lime to the action of the air. It sometimes happens in bleaching calicoes that the kier is run off and the pieces left some time before running up with water on being taken out. This is a condition very likely to injure the cloth, especially those parts which are in contact with the hot sides of the kier, for there is lime on the pieces, the air penetrates as the liquor runs off, and the hot iron gives the heat; the vacant spaces caused by the folds of the pieces being for the most part filled with steam from the hot liquor still remaining, often saves them, when otherwise they would be tender.

Action of Potash and Soda upon Wool and Silk.—Potash and soda, whether carbonated or caustic, act destructively upon both silk and woollen, and must be employed in contact with them only in very weak states, and with much care. Tolerably strong ash dissolves woollen cloth up altogether; and when weak it is apt to injure the texture and properties of silk and woollen long before it arrives at the disintegrating point. Even lime is injurious in this way. The caustic alkalies make a kind of soap when boiled on wool. The addition of acids throws down a white pulp, and a disagreeable sulphurous smell comes from the mixture, owing to the sulphur naturally ex- isting in wool taking the form of gas. Soap of a slightly alkaline nature, and solutions of crystals of soda, rather weak, are the strongest alkaline substances which should be permitted long in contact with wool. Ammonia in its strong state has an injurious action upon wool: when mixed with water it is not injurious, if it is not kept too long in contact with it. Fermented or putrid urine, which is used in bleaching wool, is alkaline, owing to the carbonate of ammonia contained in it. This does not appear to exercise any injurious action upon the woollen fibre, while at the same time it acts upon the grease in it. Ammonia is rather hurtful to silk in the strong state; but moderately diluted, it may be used for any purpose for which it is adapted without inflicting injury upon it. As a general rule it should be understood that woollen is injured by alkalies, and not by acids; on the contrary, cotton fibre is not affected by weak alkalies, but injured by acids sooner than wool.

Action of Salts upon Fibrous Substances.—The phenomena of the decomposition of metallic salts by fibrous matter, or in presence of fibrous matters, form one of the most interesting and difficult branches of the chemistry of dyeing and calico printing. The chief cases as presented in practice are given here, and some further information may be found under MORDANTS.

A Salt may be Entirely Absorbed or Retained by the Fibre.—This case is only common when the salt is insoluble in water. To be considered as absorbed or combined with the fibre it is necessary that the fibre should retain the salt, or a portion of it, when washed in water. The cases in which a whole salt, in all its parts, is retained upon the fibre are few, not considering at present that the coloring matters form salts. The chrome yellow and orange, being respectively the chromate and subchromate of lead, are salts; Prussian blue is the ferrocyanide of iron, and a salt. These two bodies being insoluble salts, and produced from two soluble salts, are formed as it were in the fibre, and not being dissolvable in water, are retained by it. Other insoluble salts, which can be formed in the same way, have not the same hold upon the fibre, and can be removed by brisk washing and agitation. What the difference of adhesion may be owing to is not well known. Woollen fibre is able to retain some salts in it which are soluble in water, for example, alum; if wool be boiled with a weak alum water, it takes up some of it, which cannot be removed by washing, and is said by some chemists to be there in the state of alum, the same as common alum. Cotton can attract to itself some insoluble or slightly soluble salts, and retain them with considerable tenacity, such as the sulphate of lead, newly precipitated, and the chromate of lead in the same condition; when once these have been printed on calico, and left for a short time, they have become fast, and cannot be washed off. But altogether the cases in which complete salts, both acid and base, are retained by the fibre are few, and form only an unimportant series.

Several Salts are Decomposed under the Influence of Fibrous Substances alone.—This case frequently happens with all three kinds of fibre, and is much taken advantage of for mordanting them. Calico, passed through a clear solution of nitrate of iron, decomposes it by taking up some of the oxide to itself. It holds firmly to this, no washing in water will remove it, and it may even be passed in dilute acids without losing its iron; silk does the same. If calico be passed through hot alum liquor, it does not take up any of it that cannot be removed by simply washing. Wool, as just stated, does take up some of it, but not much; for, unless assisted by some other agent, it cannot remove either the alum from the water, or take the aluminous base from the strong sulphuric acid in any useful quantity. Substances are therefore added which will fill up the acidity of the salt caused by the abstraction of the base; when thus assisted, wool can withdraw a considerable amount of the aluminous base from the alum, but in what precise state it exists in the wool is not known. The salts of tin are easily decomposed, and portions of them retained in the fibre, by simply passing the fibrous substance through solutions of them. Many other metallic salts are acted upon in the same way, and the fibrous matters are able to take the metal of the salt easier and in greater quantity, as the combinations between the acids and oxides are feebler. Many nice chemical affinities and balancing of attractions are shown by the behavior of fibrous substances with solutions. The aim to be kept in view in depositing metallic oxides in this manner is generally to render the metallic salt itself unstable, to substitute weak acids for strong acids, and to induce, as far as possible, the formation of that class of oxides which have the smallest force of attraction for the acid. When a fibrous substance takes up metal from a solution, it may be looked upon as an acid body entering into competition with the acid already existing in the salt, for a portion of the metallic oxide combined with it. If its attraction was very superior, it might take all the metal from the acid, but this never occurs; so long as there is acid present, it will retain a portion, and the largest portion, of metal. If the excess of acid is kept naturalized by any means, it, of course, is continually putting the fibre into a condition to remove more and more of the oxide, and this is what is practically done when tartar, chalk, and carbonate of soda are added to mordants, and also when acetates are mixed with 'them. The acid of the acetates is a comparatively weak acid, besides being easily volatilized; this volatility only takes place when it is dry and in absence of water, or when only very little water is present; it is, therefore, not much used in mordanting in liquids, but it is of essential use in cases where the cloth is dried with the mordant in it, as in calico printing, and in some cases of dyeing.

Some Salts are Decomposed by the Fibre, with assistance of Chemical Additions.—This, as partly explained in the last sec-

tion, is the most ordinary method by which salts are sought to
be decomposed for practical purposes on cotton, wool, and silk.
The chemical additions are usually of such a nature as to weaken
the existing affinities between the acid and the base, and then the
fibre steps in and helps the decomposition, attracting to itself
the insoluble, or slightly soluble, substances produced.　For
example, if clear alum liquor, as strong as it can be made,
were thickened and printed upon calico, it might be aged any
length of time, but would all wash off in water, leaving no
mordant attached to the cloth.　But if a certain amount of
acetate of potash was to be dissolved in the same alum liquor,
and then thickened, printed, and aged, it would be found that
a good deal of mordant was attached to the cloth, and that it
would dye good colors.　The chemical effects of this acetate
may be stated as follows, assuming for the sake of facility of
illustration, that the cotton fibre has a tendency to attract the
aluminous base, and is continually soliciting it from the alum.
Alum is a sulphate of alumina, and every particle of alumina
being combined with one of the strongest acids known to
chemists, the solicitations of the cloth are powerless to overcome the resistance of the sulphuric acid, just as the solicitations of gravity on a weight are overcome by the string which
suspends it.　But the addition of the acetate of potash changes
the internal structure of the alum; the sulphuric acid goes to
the potash in part to form sulphate of potash.　It cannot do
this without leaving the alumina in part, and as soon as it does
so the cloth acts upon it and seizes it; the acetic acid, which
has been driven out of the acetate of potash by the more powerful sulphuric acid, escapes in vapor, having nothing to detain
it; the decomposition proceeds until eventually all the alumina
is fixed by the cloth, and there remains only sulphate of potash
upon it in a loose state, all the acetic having gone away in
vapor.　Other acetates act in the same manner, and other
metallic salts the same as alumina.

Decomposition of Salts, the Fibre being present.—A number of
important decompositions of metallic salts take place in a
more direct and less complicated manner than when the acetates
are used.　These are cases where the salt which is wished to
be decomposed is at once placed in contact with substances
which have a stronger affinity for its acid than the oxide it
contains can exert, when, of course, the acid leaves the one
oxide to go to the other, and the oxide which is left without
acid remains attached to the fibre.　Lime is the oxide most
frequently employed, because it is cheap, and its affinities are
very powerful; but potash and soda act more powerfully still,
and their greater lesser use for this purpose, as compares with

lime, depends not so much upon chemical as upon economical
considerations. Cases in point here are the chrome green
shades, iron buff, manganese. brown, and the puce from lead
salts. Suppose the chromium salt which has been printed on
the calico is the sulphate: when it is passed in lime there is
formed sulphate of lime, by all the sulphuric acid going to the
lime in preference to remaining with the oxide of chromium;
in consequence, the oxide of chromium is left to itself, insoluble
in water, and adhering to the tissue. It withstands washing,
which the sulphate would not. I know it is a common error
to consider that lime fastens this class of colors, by combining
with them in some manner; but it is not so. The lime performs all its office in the raising beck, and when the piece is
washed, there is none left on it at all. In the case of acetate of
manganese, or acetate of iron, it is acetate of lime which forms
in the raising, because it is the acetic acid which was combined
with the metallic oxide.

Heat has a great influence in changing the affinities and composition of salts, especially in mixtures of them. It is for the
purpose of effecting such a change that steaming is employed
upon topical colors, and hot stoves, or baths, on dyed colors.
The easy destructability of all animal and vegetable fibre by
heat prevents the development upon it of several mineral colors,
which can only be accomplished by high temperatures, such as
red lead, vermilion, ultramarine blue, etc.

Wool does not behave, with regard to metallic bodies, in the
same way as cotton. Generally speaking, metallic colors are
not good upon wool; this seems to be owing to its chemical
properties, which show themselves mostly by deoxidizing substances placed on it. It is not possible, for example, to get any
good shade of buff on wool from the oxide of iron, neither a good
brown from manganese, for either the nature of the wool never
permits a full oxidation of these metallic colors, or else it
speedily deoxidizes and deteriorates them. The same remarks
are true concerning silk. In mordanting wool, either with
aluminous or tin salts, the use of tartar appears necessary.
Tartar is the compound of potash with tartaric acid; and tartaric acid belongs to the same class of acid bodies as acetic
acid, and with regard to the strong mineral acids has about the
same relations, excepting that it is a fixed acid. When tartar
is mixed with alum, and in contact with woollen fibre, there
seems no good reason to doubt that it acts much in the same
manner as acetate of potash would do; that is, part of the sulphuric acid of the alum goes to the potash of the tartar, leaving
the alumina in a weaker state of combination, so that the wool
can exert its attractive force for this body with less resistance

and greater effect; for supposing a tartrate of alumia to be
formed by the exchange of acids between the alum and the
tartar, the tartaric acid, being a so much weaker acid than the
sulphuric, could not restrain the alumina from entering into
combination with the wool. But it is not probable that a tartrate of alumina is ever actually formed, the decomposition is
not pushed so far; only so much tartar is required as will be
equivalent to taking away a small portion of the sulphuric
acid from the alum; for when alum is robbed of a little of its
acid, it becomes easily acted upon by substances which have an
affinity for the alumina, a compound being produced called
cubical alum, or basic alum. The same explanation may be
accepted in the case of tin mordants; the acid being partly
killed, lets its tin go to the wool with so much the greater ease.

In mordanting cotton with tin, the process that succeeds
with wool would give only very indifferent results, principally
because its affinity is less powerful, and because as explained
before, it does not stand contact with acids so well as wool. It
has to be done, then, in another manner, and the alkaline salt,
composed of tin and soda, and called stannate of soda, is employed instead. This salt differs from ordinary tin salts, for *it*
has the tin dissolved by alkali, while *they* have it in solution
by reason of strong acids. A different mode of fixing is then
required, depending upon the difference of chemical constitution. The cloth is impregnated with a solution of the stannate
of soda, and then passed through dilute sulphuric acid. The
action is simple—the soda goes to the acid, and the tin, being
set free, combines with the fibre of the cloth. The cotton fibre
would have no power of itself to withdraw tin from the stannate,
especially in the very alkaline condition of the commercial
article.

The great majority of salts known to chemists, as well as
those which are employed in manufactures, have no action at
all upon fibrous substances; this is the case with nearly every
neutral saline compound. It is only in the cases of oxides of
aluminum, iron, and tint, that fibres appear to exert a decidedly
attractive influence; it is difficult to leave either vegetable or
animal fibre in contact with salts of these metals without their
taking up and fixing more or less of them. The most powerful
affinity is for tin; it is absorbed under circumstances where the
other oxides are not, and it retains its hold upon the fibre under
influences which readily remove the other metals. There are
cases where the combination of the tin with the fibre is so intimate that nothing short of the complete disorganization of it
will enable the tin to be completely separated from it. Further

observations connected with these matters will be found under the head of MORDANTS.

Solubility of Fibrous Matters.—The case of disintegration of fibre is common enough under the influence of acids and alkalies; but a case of real solution of it was not known until quite recently, and has occurred in connection with a chemical mixture where *a priori* this action would never have been suspected. The oxide of copper is soluble in ammonia, and the solution thus produced may be called the ammoniuret of copper. It can be best prepared by dissolving sulphate of copper, at the rate of one pound per gallon, in cold water, and adding weak caustic soda until all the copper is thrown down; the pale blue sediment should be thrown upon a filter, washed, and drained; it will be soluble in strong ammonia liquor, with production of a magnificent purplish-blue color. This solution should be kept in a bottle or covered vessel, because it is injured by the air. This liquor is the solvent for cotton : it acts also upon silk, but not so strongly ; and upon wool, but still less so. There are other solutions of metals in ammonia better adapted for dissolving silk and wool, as those of cobalt and nickel. If a piece of calico be placed in the ammoniuret of copper it is soon acted upon, becomes gelatinous, is brought into a pulpy state, and finally, if not in too large a quantity, is dissolved. If this solution is mixed with acids the cotton is thrown down as a white powder, and, according to the statements of the chemist who discovered this solvent, it remains of the same chemical composition as before, but all traces of fibre are entirely gone. Before the discovery of this curious fact, cotton was considered as insoluble in all liquids. What practical results may flow from the knowledge of this solvent, or what insight it may give us into the nature and chemical bearings of the cotton itself, cannot yet be said.

Affinity of Fibrous Substances for Vegetable Substances and Coloring Matters.—All the fibrous substances receive coloring matters under some conditions, which vary for the particular fibre and the special coloring matter. As a general rule it may be laid down that the animal fibres have a stronger attraction for coloring matters than the vegetable fibres ; they not only imbibe them more easily, and with less preparation, but they hold them more firmly and protect them better from the destructive influences of air and light. There are some colors which cannot be fixed at all upon the vegetable fibrous substances, but which give good shades upon wool and silk ; picric acid is a notable example. Others fixed upon cotton are weak and unstable, easily destroyed by natural agencies, while the same coloring matter upon wool and silk enjoys a fair degree of permanency—the case of archil is an illustration. There are, on the other hand, some colors which fix better on cotton than on wool, principally the metallic colors, and some which are not good on wool, because of its too great affinity for coloring matters, as in the case of madder; the wool does not permit the removal of the brown matters, which deteriorate the madder red. Some vegetable substances, not coloring matters, have a strong affinity in certain cases for fibrous matter, such, for example, as the tannin matter of gall nuts, sumac, &c. It may be observed that these substances are all what are called astringents, and it is not known what kind of combination they form with the fibre; but it is one of a very intimate nature, and its permanency resembles the combinations which are formed by mineral matters with it. The combination, or, more properly speaking, the adhesion of certain other substances, such as albumens, oils, caseine, etc., is not of a chemical nature. They adhere by virtue of becoming entangled by coagulation with the network of fibre, or they adhere by their closeness of contact, like paint to a wall.

Flavine.—A preparation from quercitron bark, which may be used as a substitute for bark in printing and dyeing. The agent for Sanford's American Flavine gives the following examples of the quantities in which it is applied in practice. I give the words of his circular so far as regards the process, which are, no doubt, applicable to other preparations of bark.

To Dissolve Flavine.—Put in a tin or stone dish, pour a little hot water on it, and stir it well with a stick, it will then get into a paste; keep adding a little water till it becomes thin; it is then ready to be mixed with the whole quantity of water in the kettle or pan, and boiled up with the mordants, using the same as for bark. Or another way of mixing it, instead of water, take a solution of tin, and stir it the same way.

For Printing Cotton and Delaines (as used when first introduced).—One pound of flavine, mixed to a paste, in a basin, by adding hot water and stirring with a stick; then add 3½ oz. (measure) of ammonia, sp. gr. 894 (more or less, according to strength); stir it well up, and mix with one gallon of boiling water, and boil for twenty minutes, it will then be in solution and quite clear.

A standard yellow was made as follows:—

> 1 gallon of 1 lb. flavine liquor,
> 4 lbs. of gum substitute,
> 2 oz. of tin crystals,
> ½ lb. of alum,
> 4 oz. of acetic acid;

steamed with six inches (pressure) of steam for one-half hour, then washed in water.

Another Method.—If used to make an orange, proceed as follows: Take 8 or 9 oz. flavine, and mix 24 oz. dry starch; measure 4 quarts of water, and add gradually to the mixture of flavine and starch, making a perfect paste; then add water according to judgment, and boil. When boiled, add to half a pint of boiling water 1 oz. of turpentine. When dissolved and thoroughly mixed, add half a pint of oxide of tin, and a quarter of a pint of acetic acid at 4° Tw.; mix them together, and add to the flavine. Mix well, and add 4 oz. of tin crystals, boil until it is of a rich golden orange color; strain it, and it is ready for printing. Use the best starch, the amount of starch is very important, as if not properly thickened it will stick in the engraving.

Flavine Green upon Woollen, No. 1.—For sixty pounds of woollen yarn:—

> 2 lbs. flavine,
> 6 lbs. alum,
> 1 lb. tin crystals.

Dissolve in boiling water. Then dissolve separately four ounces red prussiate of potash, and add it to the above. Then add 4½ lbs. extract of indigo, and 12 oz. oxalic acid. Dye in the usual way. This makes a full green. It is perhaps better to prepare the yarn by steeping it first for half an hour in boiling water, in which is dissolved one half of the above quantities of alum and oxalic acid, and putting the remaining half in the vat.

No. 2.—A very good green can also be made by using

> 3 lbs. flavine,
> 12 lbs. alum,
> 12 oz. oxalic acid,
> 4½ lbs. extract of indigo.

No. 3.—3 lbs. flavine,
> 12 lbs. alum,
> 1 lb. cream of tartar,
> 4½ lbs. extract of indigo.

Light greens may be dyed in the same vats after the darker ones without renewing them, as is customary in using fustic.

. For very dark greens a prussiate blue may be made first, and it may be dyed afterwards with flavine, using alum or tin crystals, or tin spirits for mordants.

Flesh Color.—A light pink with a little yellow constitutes the color known as flesh. The term flesh color is more properly rendered skin color, since it is evidently intended to indicate the color of healthy skin, or the color of muscle as seen through skin. In dyeing it is obtained from safflower, used·in small quantities, from a spent scarlet bath, in which cochineal and tin have been used ; and from anotta, reddened by soaping and a little alum. These are sufficient indications for a shade of color which varies very considerably in different dye-houses, and has usually to be dyed to pattern.

Flour.—Wheaten flour forms a very useful thickening in color mixing, and on account of its cheapness is extensively used. Flour, as a chemical body, may be looked upon as a mixture of starch and a peculiar substance called *gluten.* It is in proportion to the amount of gluten contained in flour that it is valuable as an article of food ; but for the color mixer, other things being equal, it is the amount of starch that renders it more or less valuable. The gluten has no sensible thickening power of itself, but it certainly influences the starch in its manner of thickening colors. A given quantity of flour thickens much better than the amount of starch which can be obtained from it: thus, the thickening power of a flour, though in relation to the starch it contains, is as a rule greater than that would account for, and leads to the belief of the existence in the flour of some form of starch decomposable by weak chemical or mechanical agency, and from which the substance called gluten is formed. The very best kinds of flour are weaker for thickening than common·kinds, requiring from half a pound to three-quarters more to the gallon of water, but they have an advantage in the fineness and smoothness of the paste they give, and their consequent better working in the machine. A part of this is doubtless owing to the greater quantity of gluten they contain, which I consider is the substance, giving toughness and tenacity to paste colors; something is owing to the absence of inactive matters of a cellulose insoluble nature present in inferior flour, and a good deal to the fine dressing which such first quality of flour receives. As a chemical matter, flour has no particular affinities or actions; it is not affected so easily as starch by acids or other chemical bodies.

Flour is occasionally troublesome through grit or sand being present, which it is impossible wholly to remove by straining. The coarser or worst part of the grit may be thrown out by a little careful management before boiling, as it settles down rather quickly in the thin mixture of flour and water, or flour and red liquor. If a quantity be mixed in a tub, of a churn shape, wider at bottom than top, and the mixture be kept quietly revolving, just so much as to prevent the mass of flour sinking, the greatest part of the grit may be brought towards the bottom. The top part can then be drawn off by a spigot, placed about six inches above the bottom of the vessel. This plan requires a little address to make it answer, but it will sometimes remove all real causes of complaint. The portion at the bottom can be used up for designs or patterns that are open, and do not show the action of the grit so much. Flour, starch, and gum always contain some grit, but the amount varies very much; if it comes to one per cent. it is very bad. At five grains to a thousand it is still bad; but at two grains to a thousand I think it cannot be avoided by any care on the part of the miller. It is impossible to remove grit by straining. Sand and bran will come out in a fine strainer, or a fine printing fent, but not that which is called grit. It can be separated by chemical means, and shown separately; it is a fine powder, that looks and feels in the fingers like dust; but under the microscope, and between the teeth, which are the best practical detectors of grit, it can be proved to be hard crystalline sand, scratching copper and steel, and taking the polish or the face off a roller like fine emery or glass paper. In the article upon GUMS some more particulars may be found upon the matter. Flour sometimes gives much trouble through some stringy, fibrous, tenacious substance, which forms in the paste. Only damaged flours do this, and I attribute it to some altered state of the gluten. The matter I speak of collects under the doctor, and drags out the color from the engraving. It is found, on taking off the doctor, as round flattened globules (the shape, I believe, being given by the revolution of the roller), about the size of a pellet, or larger. They have an India-rubber elasticity and feel; a straining takes them out; very often they form again on working. This substance appears under certain conditions in the cooled colors before going to the machine. If a hot paste color be strained, it seems free from them, and in twelve hours a pint or so may be obtained from three or four gallons by a fresh straining. The shape and marks of this substance lead to the belief that it existed in the hot color in some soft plastic state, and was forced through the strainer by considerable pressure; never forming an integral part of the mass of the paste, but upon cooling getting hard and tough, and able to resist pressure on the strainer, while the real paste goes through. There seems to be no cure for this but changing the supply of flour; the mischief is done in the flour, and cannot be undone.

The best and only test of any practical value for flour is to make a trial of it in thickening, observing how much it takes to give a good consistent paste. About twenty four ounces to a gallon of water should make colors thick enough for all usual purposes; it should remain firm when cold, and not show any tendency to break or become watery.

Flowers of Madder ; *Fleur de Garance.*—A preparation of madder, by processes which remove from it a considerable portion of useless matters, and concentrate the coloring matter. (See MADDER.)

Fuchsine.—This was the name at first given to the red coloring matter obtained from aniline, by means of bi-chloride of tin ; although it was a very beautiful color, the more superior color called magenta, obtained by Medlock's patent (arsenic acid), has completely driven it from the market.

Fustic ; *Yellow Wood. Old Fustic.*—The wood of a tree called by the botanists *Morus tinctoria*, and formerly known as *Dyers' Mulberry.* This wood is very extensively used in dyeing, appearing to be the most suitable yellow coloring matter for working with other colors, in compound shades, preferable in this respect to weld and quercitron bark. It is not in general use in printing. With alumina mordant it gives yellow of an orange shade ; with iron it gives drabs, grays, and olives ; as a yellow coloring matter it is considered to be only about one-fourth as strong as an equal weight of quercitron bark, and considerably inferior to it in purity of color ; but it has the valuable property of withstanding the action of acids and acid salts more perfectly than bark, and on that account is used in greens, blacks, and all mixed colors where the yellow part is required, as can be seen by referring to the receipts for these colors. The colors it gives are not very stable ; they stain easily, fade by exposure to air and light, and do not very well resist the action of soap. Like logwood, and other hard woods, it is thought to be greatly improved by ageing or mastering in a damped state for several weeks before it is used in dyeing, or before it is boiled with water to extract its coloring matter. The pure coloring principle of fustic is called *Morine.*

Young Fustic, French Fustet.—This is quite a different substance from the above, much more resembling sumac than the yellow wood in its origin. It is an European shrub, and obtained its prefix "young" on account of the smallness of its branches compared with that of the yellow wood, which was distinguished as old fustic. It contains a considerable proportion of yellow coloring matter, and may be used for the same purposes as the yellow wood, but its colors are usually considered more fugitive. It is not applied in calico printing, and very little if at all in cotton dyeing. The plant which yields it is known to botanists as *rhus cotinus,* and is sometimes called *Venice sumac.*

G.

Gall Nuts, *Galls.*—This valuable dyeing material is an excrescence from certain trees similar to the oak; it is caused originally by the puncture of a little insect on the leaves or small branches of the tree, in order to deposit its egg in the cavity formed. The juices of the tree collect round the egg, and hardening, form the gall nut. In favorable circumstances, the egg comes to maturity, and the insect born from it eats its way out of the nut in order to take its flight. The gall nuts, however, are as much as possible collected before the insect has arrived at maturity, for, though smaller, they are heavier and better. The galls, in which the insect has been developed, are hollow, and have a small round hole about $\frac{1}{12}$ of an inch in diameter, from the exterior to the centre. The galls, which have been collected before the escape of the insect, are frequently known as black galls, or blue galls, also as true galls; while the other are called white galls or false galls. The former, as their names would intimate, are deemed preferable to the latter, and command a higher price; but in the English market both kinds are frequently found mixed together in the same parcel. The galls called Aleppo are considered the best; those from Morea and Smyrna next, while the galls obtained in more northern climates are so much inferior as to be scarcely useable in dyeing. Galls contain from 50 to 70 per cent. of tannic acid, and it appears that the whole of their valuable properties are attributable to the tannic acid present. The quantity of this acid present in any sample of galls may be ascertained with considerable accuracy by chemical analysis; but for a practical test of the quality of this article, it is usual to make a decoction with a known weight, and compound from this solution some black or gray color, comparing the shades with those produced by a known quality.

Galls are but little used in calico printing, their application being confined to a few gray and black colors. In dyeing they are somewhat more employed, but not nearly to the extent they were before the introduction of sumac as an astringent material, or the general substitution of logwood in dyeing black colors. In the better class of blacks upon silks, galls are still much used; they give a very durable but somewhat grayish shade of color, and possess a property, very much esteemed in certain trades, of weighting, *i. e.*, accumulating on the fibre in such quantity as to add very materially to the weight of the silk.

The coloring principle of galls is dissolved largely by hot

water, and the solution may be concentrated without deposition
or crystallizing. A weak infusion of galls soon loses power
by undergoing a kind of fermentation, which facilitates the
conversion of the tannic acid into gallic acid; strong gall liquor
does not change so rapidly. In making extracts of galls which
have to be kept for any length of time, the French writers
advise to use as little water as possible in making the extract,
because then the tannic acid is dissolved, and the fermenting
matter left behind. M. Persoz states that if the gall liquor be
raised to the boil, and kept in well closed bottles, quite full of
liquor, it will not be injured in the course of three years. Galls
cannot, in strictness, be said to contain any coloring matter; it
is more properly a coloring principle which, in contact with
certain mordants, produces the color for which it is used. Thus,
the solution of galls has only a dull yellow color, and the tannic acid to which its active properties are due is nearly colorless. With alumina mordants it only gives faint shades of
gray; it is with the iron mordants that it gives fast and durable
blacks. The tannic acid appears to have a direct affinity of
itself for fibrous matters, and to be able to accumulate upon
them to a considerable extent, increasing the weight and altering the physical properties of the fibre. It appears also to
form combinations with coloring matters, and to fulfil the place
of a mordant. Where galls and similar substances can be used
with advantage in dyeing, their action seems to extend beyond
the share they have in contributing to the depth of color, and
to consist also in causing the fibre to take up more color, and
to form a more intimate combination with it, and one more
capable of resisting the action of soap and atmospheric influences. It is on account of this property that galls and sumac
are employed on Turkey red dyeing, and on other cases of
madder and garancine dyeing.

Gallic Acid.—An acid which exists in very small quantity
in good galls, but which is produced from the tannic acid of
galls when the latter are left exposed to air and moisture. It
is nearly worthless as a dyeing material, and care should be
taken that it is not formed in the case of decoctions of galls
and sumac kept for a long time in open vessels.

Gallipoli Oil.—An impure sort of olive oil in considerable
use in calico printing and Turkey red dyeing. (See OIL.)

Garancine.—This name is applied to a preparation of madder obtained by heating it with sulphuric acid and water; it is
derived from the French *garance* for madder. The discovery
of this article appears to be due to Messrs. Robiquet and Colin,
and dates from 1827, but it did not come into extensive use
for several years afterwards. These chemists found that strong

sulphuric acid converted ground madder into a black mass like charcoal, but that, curiously enough and quite unexpectedly, the coloring matter was not injured by the acid, and when the excess of acid was removed by washing, the charcoal-like powder was as capable of dyeing as the madder itself, and in some respects superior to it. The product, at first called *charbon sulfurique*, was prepared on the large scale and offered in the trade as a substitute for madder; whether owing to actual defects in the manufacture of the article, and inexperience in its application, or whether, as seems most probable, it claimed more than it could perform, it was for a time a failure. In a little time its powers and values were better known—a style of work was created for it—it became very extensively used, and its discovery ranks as one of the great eras in the history of dyeing and printing. The method of making a kind of garancine from spent madder was patented in England by Steiner, in 1843, but five or six years afterwards, in a trial in the law courts for an infringement of the patent, it was upset as not being novel, on the ground of a previous publication in a French work known in England. Since this trial it has been open for any one to convert the spent madder into garancine, and most calico printers who consume largely of this dyeing matter, and who dye garancine styles, do themselves make it again available for dyeing. The French writers call the material made from spent madder by the name of *garanceux*, to distinguish it from the product of the fresh madder; and, as the articles have a very different value in trade and different properties in dyeing, I adopt and recommend the name of *garanceux* in default of an English word of the same meaning.

Making of Garancine.—The madder is put into a wooden, stone, or leaden cistern, mixed with a sufficient quantity of water to give it the consistence of a thin paste, and the acid added; as much as 40 lbs. of brown vitriol may be used for every hundred weight of madder, and a considerably smaller quantity may also be used with safety; but as vitriol is cheap, and the larger quantity is not injurious, it is frequently employed in order to be on the safe side. In my experiments I have found that a very small amount of acid is sufficient to make madder into garancine, but there are advantages in working which make it desirable to use a much larger quantity than is actually necessary. The mixture being made, the cistern is covered, a jet of steam introduced, and the whole mass raised to the boiling point, and kept boiling for two or three hours or even longer. The mass is then run upon properly constructed filters, the great excess of acid drained off, and clear water frequently passed through the acid mixture, so

16

that the whole of the acid may be washed away. The washing may take six to eight waters, and occupy from two to six
days, according to the nature of the madder operated upon.
The wet garancine is then subjected to pressure to expel the
water, the drying completed in stoves, and then ground and
sieved. During the grinding it is usually mixed with carbonate of soda, to neutralize a remaining trace of acid, the presence
of which in a free state would be injurious to the dyeing.

The madder in being thus treated loses from one-half to two-
thirds of its weight, and as the same amount of coloring matter
is present in the smaller weight as existed in the larger, garancine is much stronger than madder, weight for weight. But
this loss of weight is not constant; some kinds of madder contain more of the solid, woody fibre, and less of the soft, soluble
principle than others do. In such cases the loss is less, and the
garancine being encumbered with useless woody matter is not
so strong. The kind of madder called *munjeet* is one that contains a very small portion of soluble matters; it scarcely loses
weight in making into garancine, and not containing any chlorogenine, does not change color in a notable degree. It produces a very weak garancine, and is employed to mix with
others to reduce their strength, or to increase their weight.
What the real and essential action of the acids is in producing
garancine is not known. It has been imagined that they acted
by breaking up the woody fibre, tearing it asunder, as it
were, and letting the water get access to the coloring particles
previously enclosed, and so transferring them to the dyeing
bath. But the weakness of the acids which can be successfully
employed, throws doubt upon this explanation. It was supposed again that the coloring matter was in a state of combination with some earthy base, as lime and magnesia; that it
could not dye while so combined; and that the acids liberated
it by themselves taking the lime and magnesia. But this is
not supported by any appeal to facts or experiments which
bear directly upon the question; it may be the explanation, but
it is not proved. It seems probable that some chemical affinities of a more refined and subtle nature are brought into play:
it is likely that the acid employed forms a combination with
the coloring matter, which is itself easily decomposed, even by
water. It would not be in accordance with analogy to suppose
any formation or creation of coloring matter, but simply to
suppose that a portion of the coloring matter was present in
some form or other in which it was not soluble in water, but
by some agency not known it becomes soluble and capable of
combining with mordants.

The blackening of madder, when made into garancine by the

ordinary methods of a mixture of acid and water, is due to the presence of a peculiar matter in it, which is called *chlorogenine;* a substance which can become green by the action of several reagents, and especially by the action of acids. The intensely dark, almost black, color which some samples of garancine have in the damp state is due to the dark green substance thus produced; the true shade can only be seen when it is dry, by transmitted light. If a clear solution of madder be placed in a deep glass, and strong sulphuric acid poured on it without mixing, this latter will sink to the bottom; but at the line of junction of the two liquids a green color will be perceptible, which grows in extent and depth, becoming nearly black. If the liquor which filters from steeped madder be boiled in a glass flask with acid, it changes color, and the dark green substance alluded to settles down as a powder: it is so dark, that it is only when it has been washed and dried that it is plainly seen to be green and not black. In the state in which it exists in the madder root it is soluble in water, but when it has been changed by hot acids it is insoluble, settles down upon the woody fibre, and remains with it. The reason why garancine made from washed madder, and garanceux also, are not dark colored is because their chlorogenine has been washed away; and the reason why some kinds of garancine are darker colored than others is because the original madder contained more chlorogenine. The quality of a madder does not, as far as I am aware, bear any relation to the amount of chlorogenine in it, and, therefore, the quality of a garancine cannot be correctly judged of by its shade of color. The formation of this dark green substance goes on step by step with the action of the acid in making garancine, but not apparently connected with it; it serves as a test to show the completeness of the action of the acid, for the change of the chlorogenine does not take place unless the garancine be made; but it is possible to make the garancine without using so much acid, or employing a temperature high enough to convert this principle into the green state.

With regard to the length of time during which the madder and acid must be kept boiling, it appears, from laboratory experiments, that five minutes boiling is quite as good as twelve hours; but large masses take much time to get thoroughly heated, and the risk in diminishing the period of heating would be that some parts would not be raised to the boiling point. I made a series of experiments upon making garancine at low temperatures, and have found that madder can be changed, at a temperature not exceeding 160° F., to garancine of the best quality, and with trifling quantities of acid. In using sulphuric acid, muriatic acid, and nitric acid, I have obtained results not distinguishable from each other. Other things being equal, muriatic acid gives the darkest colored product, and nitric acid the least colored. In the use of nitric acid there is not a very wide margin between the strength at which it acts, in the same manner as sulphuric acid or muriatic acid, and the strength at which its oxidizing properties come into play, and totally destroy all the coloring matter. If about half an ounce of good commercial nitric acid be mixed with 20 oz. of water, and 2 oz. madder be mixed up with it, and the whole gradually heated up in a water bath, with regular stirring, the madder looks unacted upon at first; but when the temperature approaches 150° it undergoes a remarkable change—it becomes of a brownish color, and contracts into a very much less space than it previously occupied. If the temperature be pushed a little higher, there is formation of the green substance spoken of, which floats in the liquor, and which, if not poured off, will settle upon the madder and adhere to it. If the heat be withdrawn, the mixture cooled, washed until all the acid is removed, and dried, it will be found that the madder has lost more than one-half its weight, and the product is found to dye up like good garancine. The same effects are produced by using sulphuric and muriatic acids. These experiments are interesting, as showing at what temperature the conversion into garancine takes place. By using a little more acid in the case of muriatic acid, a product can be obtained which is perfectly black when moist, and of a dull dark green when dried. It is, therefore, apparent that garancine can be procured without the dark colored body which generally accompanies it when made from unwashed madder.

A quality of garancine more especially intended for dyeing purples is made, it is frequently sold as commercial alizarine. (See ALIZARINE.)

Garanceux.—The manufacture of garanceux from spent madder is essentially the same as that of garancine, and the same general explanations apply to one equally as to the other. The difference between garancine and garanceux is not of a radical nature, and may be explained in a great measure by the different amount of coloring matter present in proportion to the amount of woody fibre and mineral matter, both of which are to a certain extent obstructives to the free solution of the coloring matter, and may produce those effects which distinguish these two preparations. The method of making garanceux is simple and well understood, and requires nothing but care and attention to obtain regular results. The spent madder from the dye becks and wince pits is collected in a sufficiently large receptacle, and allowed to settle, with the addition of a little weak vitriol, which both stops any inclination to fermentation and throws down a fine powder, which otherwise floats in the liquor. The wet mass should be well pressed until it does not contain more than two-thirds of its weight of water, and then broken up and mixed with the acid ; about a dozen pounds of brown vitriol to the hundred weight is a good proportion, mixed with about three gallons of water previously to degging the spent madder with it. It is an essential point that the acid touch all the particles of madder, and therefore such a system of turning over, sieving, etc., as will insure a perfect mixture of the acid must be strictly attended to. It is good to leave the degged madder for some days before steaming ; many weeks or months of storing does the mixture no harm, and allows the acid to thoroughly penetrate it. The steaming must be watched to see that all the madder gets a fair proportion of steam ; if the spent madder be too moist it sets in the steaming cistern, and forms chinks, through which the steam blows without permeating the whole mass. It should be so moist as to give water upon pressure between the fingers, but not wet enough to adhere in a lump when pressed in the hand. Garanceux can also be made equally well by boiling with acid, but the steaming process appears to be preferable. The washing of the acid stuff is important, and demands the greatest atten- tion. On account of its loose texture garanceux washes much faster and easier than garancine—the waters can be run off more quickly—so that two working days are enough to wash it in. The washing is known to be accomplished by the liquor not tasting acid, and by the springing out of a fine slimy pow- der, which never appears till the washing is almost completed. It is not to be recommended that any alkali should be used to neutralize the remainder of the acid as a usual thing : but it is sometimes done, and then crystals of soda or milk of lime may be added in proper quantity. The garancine keeps better when left a little acid, especially if it is to be kept in a moist state, as will be usually the case on printworks ; it can be neutralized in the dye-beck with safety. When the washing is complete it should be pressed again ; it might be used in a simply drained state, but it is better to have it regularly pressed, so as to know how much is being used, as well as for facility of storing. Well pressed garanceux contains between a third and a fourth of its weight of dry matter, the remainder being water. The strength of garanceux, in proportion to garancine of first quality, will vary according to the nature of the madder, and to the proportion in which it has been spent in the first dye- ing. Garanceux from mixed Turkey and French madder is equal, as it comes from the press, to from one-sixth to one-ninth of its weight of real good garancine, or, when dried, it is equal to about one-third. If the madder has been well managed in the dyehouse, making it go as far as possible, and that it is evidently the interest of the dyer, it will take nine parts of the pressed garanceux from it to be equal to one part of best garancine, and less as it has been used in excess of the requirements of the dyeing to begin with. It is a point of the utmost importance to see that garanceux is properly neutralized, either in the dye-beck or before it goes in; an error on either side—that is, an insufficient or an excessive quantity of the alkaline matter—is equally injurious. I have seen excellent garanceux condemned as worthless, and even sentenced to be thrown away, because too little carbonate of soda was used with it. The most useful matter to neutralize garanceux in the becks is the bicarbonate of soda, it is very regular in its composition and of a mild nature; no fixed proportion can be prescribed, because the acidity is various; the highest amount that should be required is about one pound to seventy of the garanceux; if the acidity passes this, good results will not be obtained—it should be washed over again. The lowest amount will be about one pound to a hundred and fifty pounds of garanceux, taking this last in the wet state, and containing about one-third of dry matter and two-thirds of water. The injurious effects of too much soda are perhaps even more marked than a slight deficiency; the colors are dull and cloudy, without brilliancy or solidity. A deficient quantity of soda is shown by a redness of the chocolates, grayness of the purples, poorness in the blacks, and the general bareness and poverty of the whole of the colors.

The colors which garancine and garanceux give to mordanted cloth are the same in kind as those from madder, but not exactly of the same quality; the whites are purer, on account of the absence of the soluble matters which stain them in madder dyeing. They will not stand a severe soaping, and cannot be brought to the same degree of brightness as madder colors. These products are not much employed in dyeing by themselves, but generally in combination with some of the cheaper dyewoods, as peachwood, sumac, and bark. Garancine dyeing cannot be understood until the action and nature of these woods are ascertained, since the simple garancine colors are modified and changed in a remarkable manner by the colors produced by these woods. Garancine is not simply a concentrated madder, to perform in less bulk what madder does or can do in larger bulk; it yields colors which cannot be obtained from madder. The existing garancine styles could never have been produced with madder, however much was employed in the dyeing. It is well known that a large quantity of madder must be used to obtain a chocolate, so much so that chocolates on account of their expense are nearly excluded from madder styles, and the best chocolate that can be obtained is of a chestnut color, and not of that deep brown shade which is required in the market. The coloring matter of madder is surrounded with so much foreign substance as to hinder its filling the strongest mordants, and producing the heaviest shades which the pure coloring matter can do. Garancine, being free from all these soluble matters, has a free power of combination to the utmost extent of its affinity and saturating power; and can produce those deep chocolates which no amount of madder could yield. It has the valuable property also of working well with other coloring matters, and admitting them to share in producing its effects; this madder does not do in the same manner, it is itself too powerful, and requires such energetic treatments to restore the whites to a good shade, that the weaker dyewoods, which may be used in combination with it, suffer to a great degree in themselves, and tend to deteriorate, to what seems a disproportionate extent, the shades which owe their chief part to madder.

Garancine Colors.—The chief colors for which garancine is employed in calico printing are black, red, chocolate, and brown; the latter being partly derived from catechu. Purples are also obtained from certain qualities of garancine; but what is called the *garancine style* or *brunette*, consists of deep colors of the chocolate, brown, and black species. The mordants used for obtaining the colors from garancine and garanceux are iron liquor, red liquor, and mixtures of the two. Garancine does not dye colors without mordants.

Red Colors are produced from red liquor or acetate of alumina —sometimes with addition of tin crystals.

Black Colors are produced from iron liquor or acetate of iron, at a strength from 8° to 12° Tw.

Purple Colors or *Lilac* are produced from iron liquor, at a strength of from 1° Tw. to 4° Tw.

Chocolate Colors are produced by a mixture of iron and red liquor—the darkest shades being made with, say, equal parts of iron liquor and red liquor at 12°. If a chocolate is desired of a redder shade, the proportion of red liquor to iron liquor is increased; when required of a lighter shade, the strength of both liquors is reduced.

Brown Colors are obtained by means of catechu. Red-browns are obtained by mixtures of catechu colors and red liquor.

Drab or *Gray Shades* are obtained from catechu colors, mixed with an iron salt.

In illustration of the proportions in which the various mordants are mixed I give a number of receipts for each color :—

Block Mordant for Garancine.

1 gallon iron liquor at 12°,
1 gallon water,
2½ lbs. flour,
¼ pint logwood liquor for sightening.

The flour in this receipt may be replaced by starch or gum; but, if by the latter, the iron liquor should be used somewhat stronger. Gallipolli oil, in the proportion of one-eighth of a pint per gallon of color, may be added to prevent frothing, etc.

Red Mordant for Garancine.

1 gallon red liquor, at 16° (p. 42),
1¼ lb. starch,
¼ pint peachwood liquor, for sightening.

Another Red for Garancine.

1 gallon red liquor, at 18°,
1¼ lb. starch,
¼ pint peachwood; boil, and then add
2 oz. crystals of tin.

This last red is capable of resisting light covers of chocolate; for resisting heavier covers the quantity may be increased to 8 oz. per gallon of mordant, or a proportionate quantity of liquid muriate of tin (about double the weight) may be employed instead. The addition of solution of tin gives a clearer and somewhat lighter red.

Light red shades are seldom required in garancine styles; if wanted, they are made by reducing the strength of red liquor.

Chocolate Mordant for Garancine.

8 gallons red liquor at 18°,
5 gallons iron liquor, at 8°,
20 lbs. flour,
1 quart logwood liquor; boil well.

This gives a dark chocolate of a medium shade, that is between red and black; by altering the proportions of red and iron liquor, about 30 well defined shades may be produced. For example, to produce a black chocolate the iron and red liquor may be in equal quantities and of equal strengths on the hydrometer; and, by maintaining the red liquor at one uniform strength and gradually decreasing the iron, the chocolates become less dark, and pass into the chestnut and red choco-late shades. The lowest chocolate—that is, the one containing the least amount of iron—usually worked contains about one measure of iron liquor at 24°, to 30 measures of red liquor at 18°.

Some of the brown colors applicable to garancine styles have been given under CATECHU;. a further selection is given here :—

Dark Brown for Garancine.

18 lbs. catechu,
2½ lbs. sal ammoniac,
2 gallons water; boil, strain, and add
3 gallons gum water,
2 quarts nitrate of copper, at 84°,
2 quarts acetate of copper.

This brown may serve as a standard from which to obtain the lighter shades by reduction with gum water; thus, instead of 3 gallons gum water, 4, 4½, 5, or 6 gallons may be employed. To modify the shade and to produce red browns, more correctly cinnamon shades of brown, a quantity of red liquor is added in proportion to the shade required. The above dark brown is converted into a red brown by addition of one quart of strong red liquor to 2½ gallons of color; a greater quantity of red liquor would make the brown too red. The proportion is re-duced for the lowest red shades to one part of red liquor to 30 parts of dark brown, as above.

Medium Brown for Garancine.

12 lbs. catechu,
2½ lbs. sal-ammoniac,
2 gallons water; boil, strain, and add
4 gallons gum water,
3 quarts nitrate of copper, at 90°,
3 pints acetate of copper.

This color serves to illustrate the reductiont o lighter shades, for it is essentially the same as the dark brown, only with less catechu and more gum water. From it the still lighter shades of brown may be produced by increasing the proportion of gum water.

The drab and gray shades are produced from catechu brown, modified by containing some iron salt. An example is given (p. 199) which may serve for all the drab colors in these styles; the following is a modification :—

Paste Catechu Drab for Garancine.

2½ lbs. catechu,
2½ lbs. sal-ammoniac,
3 quarts water ; boil, strain, and add
1 quart acetic acid,
1 pint acetate of copper.

The solution thus produced serves as a stock or standard liquor, from which the actual drab color is made as follows :—

No. 1, or Darkest Drab.

7 pints of the standard,
1½ lb. of starch ; boil, and add
1½ pint muriate of iron, at 8° Tw. ;

Diluted with water or gum water, and a variation of the quantity of muriate of iron, produces the numerous modifications of this color which enter into garancine styles.

An orange color for garancine styles may be produced by printing on a mordant of acetate of tin, as follows :—

` *Orange Mordant for Garancine.*

2 lbs. white acetate of lime,
3 lbs. crystals of tin,
1 gallon of water,
1¼ lb. of starch.

The garancine used to dye styles containing this color must be mixed with a considerable proportion of Persian berries or quercitron bark, to yield the yellow portion of the orange. The color thus obtained is so excessively fugitive that it has ceased to be produced as a regular shade in garancines. Another method of obtaining the same end consists in making a spirit yellow from bark liquor and muriate of tin, printing it with the usual colors, and dyeing in a mixture of garancine and quercitron bark.

Dyeing of Garancine Colors.—After being sufficiently aged, the goods are cleansed from excess of mordant and thickening matter, either in cow-dung or one of the dung substitutes at a high temperature. The washing after dunging must be of a very perfect nature, in order to remove every particle of un-combined mordant from the cloth. Great attention must be paid to this point, because the mordants are very concentrated and usually thickened with paste, which prevents their readily washing off. If any loose mordant gets into the dye-beck it is impossible to produce good work, because the parties 'exhaust the coloring matter and fix upon the whites completely, and beyond the power of cure, injuring the general appearance of the work. The dyeing may be commenced at a temperature of 100°, carried to 150° in 30 minutes, to 170° in 60 minutes, to 180° in 90 minutes, to 200° in 120 minutes, and then boiled for 20 minutes to finish; the whole time occupied being 2 hours and 20 minutes. The clearing generally consists of a good washing and a passage through chloride of lime and steam. (See CLEARING.)

The quantity of garancine or garanceux to be employed in the dye is of course variable, according to the design, nature of the colors, and quality of the garancine. No guide can be given upon that point, it being a matter for trial.

On account of the very deep shades required in the *brunette* styles, it has been found necessary to combine the use of garancine with some of the cheaper dyewoods to fill up the color. The woods so used are peachwood, quercitron bark, and sumac. If these substances be used in moderate proportions with regard to the garancine, the stability of the colors produced is not much affected; but in some instances they have been used in great excess, and the colors so dyed readily fade and wash out. It is difficult to say what proportion is allowable, but probably for best garancine, of three times the strength of madder, an equal weight or one and a half times the weight of woods may be used with advantage; that is to say, for each pound of the strongest garancine half pound each of peach-wood, bark, and sumac. For dry garanceux made from spent madder, which is from one-fourth to one-third of the strength of good garancine, not more than half its weight of woods should be used. The woods used in this quantity have a great influence upon the shade, and care should be taken to have them of regular quality. From a number of experiments I made to ascertain the influence of the woods upon the colors produced, I made the following conclusions:—

Peachwood darkens the chocolates, making them fuller; heightens the red, and, if purples be present, fills them up. The chocolate obtained with garancine alone is red; the addition of peachwood makes it fuller and heavier. When bark and sumac are used without peachwood the reds suffer in brightness, and the chocolate tends towards a cold clayey aspect, or cinnamon shade. If bark and sumac are used with a double portion of peachwood the chocolate is darker but suffers in brightness, the reds remaining about the same. The chief action of peachwood is, therefore, to fill up the garancine shades; the colors it yields not being much different from the garancine colors themselves.

Quercitron Bark itself gives yellow color with alumina mor-

dants, drab with weak iron, and greenish olive with chocolate
mordants. Garancine and it together differ from garancine
alone by a yellowish shade over the chocolate, taking it to the
clayey side; the reds are made decidedly orange, and the purples turned grayish. When peachwood and sumac are used
without the bark the chocolate is deficient in brightness; it has
a purplish cast, and looks heavy and flat; the reds also want
more fire and brilliancy. If a double quantity of bark be
employed, the chocolate turns to the yellowish clayey shade
and the reds look weak. All this time the black is not much
affected, the garancine itself appearing to have a strong possession of that color.

Sumac is the general darkener of all the colors. It improves
the reds, chocolates, and blacks; but destroys lilacs or purples
if they should happen to be in the piece. Without sumac, and
with peachwood and bark, the reds would be dull and heavy
and the chocolates not dark enough. It practically fills the
place of bark by its yellow coloring matter; but that is insignificant when compared with its action upon the iron mordants,
to which it communicates dark shades, giving depth and apparent solidity to the whole. Upon the red it has not much
influence; it turns it a little towards the orange, and thus
brightens it.

Size or glue is a very general addition in garancine dyeing.
Its chief use is in preserving the whites, which it effects by
forming an insoluble compound with the tannin matter of the
sumac, and thus preventing its fixing upon the unmordanted
parts of the cloth and constituting itself a species of mordant
for the other colors. The compound produced, which we may
call the *tannate of gelatine*, is, however, decomposed in the presence of mordants at a high temperature; the tannin matter
combines with them and the gelatine or glue becomes free
again.

Garancine is employed in dyeing Turkey reds and in imitation madder purples; it is also used as an ingredient in some
other styles, but not to any considerable extent.

Gardenia.—A genus of plants and trees, of which some
species appear to be capable of yielding valuable coloring matters. Bancroft and others have drawn attention to *Gardenia
aculeata*, or indigo berry, which is said to stain paper and linen
of a fine fixed blue color—it is found in Jamaica: the *Gardenia
florida*, which is said to be used by the Chinese for dyeing
scarlet, under the name of *unki;* and the *Gardenia genipa*, the
fruit of which contains a colorless juice, which goes blue immediately that it is exposed to the air. "It is universally employed," says Bancroft, "by the savage tribes of Guiana and

Brazil, to stain their skins with a variety of spots, lines, and figures, for the purpose of ornament at their feasts and dances, as well as to render themselves terrible to their enemies when going to war; as the *isatis* or *woad* was employed by the Britons in Cæsar's time." This blue coloring matter, though it attaches itself to the skin very permanently, does not appear to be useful in dyeing; it is certainly quite different to indigo, which, in some particulars, it appeared at first to have a resemblance to.

Glauber's Salts.—A name for sulphate of soda. (See SODA.)

Glucose.—This name is given to a sweetish matter, obtained by boiling starch from any source with acids, until it is decomposed and no longer colors iodine blue. The acid being neutralized, the solution is found to have a faint sweetness. It may be boiled down to a syrup, when, after standing a few days, it sets into a granular honey-like mass. As a chemical body it is distinguished by possessing high powers of deoxidation, especially in combination with alkalies. It has been applied to the reduction of indigo in one or two cases, which will be mentioned under INDIGO. This substance is sometimes called starch-sugar, grape-sugar, and—though this latter incorrectly—dextrine.

Glue, *Gelatine, Size.*—Glue, or size, is made from refuse animal matters, as bones, skins, etc. It is employed in printing and dyeing to a limited extent, but largely in finishing. In printing it is used as a thickener in some few cases, and enters into the composition of some resists. In dyeing it is employed in conjunction with what are called astringent substances, or substances that are in part, but not essentially, composed of astringent matter. It serves to protect the white and assist in the regularity of the dyeing. Size is used in cases where there is no astringent or tanning substance present, but in such cases I think it is unnecessary. It is a useful or essential ingredient where sumac or galls are used, as in ordinary garancine dyeing; but when garancine is used without sumac I do not think that size is of any use. Such was the conclusion made from careful experiments to ascertain if its addition improved the results or not. Glue is sometimes used on print works in the solid state, and dissolved as it is wanted; but generally it is used in the state of size or liquid glue, as being cheaper and more convenient. Of solid glues there are different qualities with regard to their power of dissolving in water and keeping in a fluid state. Some kinds, when dissolved at the rate of one pound per gallon of hot water, go firm and solid on cooling; others may be dissolved at the rate of three pounds per gallon of water, and only thick gummy fluids on cooling: the commercial size is of this nature. The best way to test the value of a liquid glue is to ascertain how much solid matter there is in a gallon of liquor. This can be partly estimated by its strengh, as shown on the glass; but more correctly by evaporating a known weight down to dryness, and weighing the solid matter. At the same time it should be examined for mineral matters, of which only a small amount are naturally present, but which might be added as an adulteration to deceive with regard to appearance of strength. I have found the following results:—

	WATER.	DRY GLUE.
No. 10 sample contained	68	32
No. 22 sample contained	70	30
No. 8 sample contained . , . .	65	35

Lightfoot patented a method of using glue for fixing colors; the coloring matter mixed with the glue was printed, aged, and, if necessary, steamed, then passed into a solution of some metallic salt, forming an insoluble compound with the glue—salts of mercury were preferred. I understand that the deficiency of brilliancy in colors so fixed is a radical objection to the process. In Germany, some of the printers fix pigment colors by means of glue; the process consists in mixing the ultramarine, or other pigment, with glue, printing, and then passing into hot solution of alum, which, forming a coagulum with the glue, fixed the color. I obtained very poor results by this method; but I was told that a peculiar quality of glue, only made in one manufactory, and that in Germany, was suitable; however, upon procuring samples of colors so fixed, or supposed to be fixed, it was evident that there was nothing in the process worth following.

To prevent fustians, and other goods finished with bone size, from becoming mildewed, the best addition appears to be purified coal-tar creasote, which is now an article of commerce, under the names of carbolic acid and phenyline. A comparatively small quantity suffices to prevent mildew under ordinary circumstances.

Gluten.—Gluten is the nitrogenous constituent of wheat flour, and appears to have some properties which bring it in relation to albumen and caseine or lactarine. It is easily obtained by mixing good flour to stiff dough, and kneading this under a slender stream of water: the starch is washed away, while the tenacious gluten collects into an India-rubber-like mass in the hands. In 1850 I made experiments upon this substance as a substitute for albumen, and obtained it easily—though, perhaps, wastefully by enclosing parcels of stiff dough in small coarse bags, and putting them into a dash-wheel, and so washing away the starch and soluble matters. This substance is soluble in weak alkalies, and has been proposed as a fixing vehicle for pigment colors; sometimes tolerably successful, at others an entire failure under unknown conditions, it is too irregular to be trusted. Recently efforts have been made to use it as an animalizing agent for the preparation of vegetable fibres, to receive colors, such as aniline and archil shades. Mr. Crum's patent claims the disintegration of gluten by allowing it to go into incipient putrefaction, then dissolving it in dilute alkalies, and impregnating cloth with it. I am not aware that this substance has yet been largely employed in this way.

Glycerine, *Sweet Principle of Oils.*—This substance, which has of late years become an article of extensive commerce, has been proposed for use both in color mixing and finishing on account of its never perfectly drying, and so keeping substances moist and soft. The advantages of its use are not, however, very apparent, while it is a comparatively expensive substance. It has some peculiar solvent powers, which may, probably, come in useful for coloring matters under some conditions.

Gold, *Gilding of Thread and Cloth.*—The application of gold and silver to textile materials is of the most ancient origin; and though the civilized taste of Europe does not admit of the gorgeous combinations of the Eastern peoples, there is still a regular, though small, demand for gilded fabrics; and no doubt it would be greater if the processes could be simplified and brought more in unison with the machinery and methods employed in fixing general colors and pigments. The following are some of the methods said to be employed for fixing gold or silver leaf upon fabrics: In the first method, the fabric is prepared with a solution of fish glue (isinglass) or gum tragacanth, and then the design printed on with an oil mordant; and when this has arrived at a proper state of dryness, the gold leaves are applied, and pressed down with a leather cushion, the fabric exposed some time to the air, and then washed to remove the gum or glue prepare. The oil mordant is prepared as follows: One pint of oil, which has been converted into drying oil by the usual processes, is mixed with two ounces of finely ground litharge and half an ounce of prepared acetate of lead—that is acetate of lead which has been dried and heated until it has been for some time in igneous fusion—the whole ingredients are very well ground together, and thinned with turpentine, if necessary.

A second method consists in giving the stuff to be gilded several coats of fish glue, until sufficient has been added to cause a moist hand to adhere rather strongly to the surface. The stuff is then hung up in a damp, cool place, until it is deemed in proper condition to receive the gold leaf, which is applied in designs, by placing the gold leaf loosely on the stuff, and then stamping them with blocks upon which the design is cut.   The gold leaf is prevented adhering to the blocks by the repeated action of talc, or French chalk, in fine powder.

The third process is applicable to moderately open fabrics, like muslin; a skin is stretched tight upon a table, lightly rubbed with tallow or suet, by means of a linen ball saturated with either of these fatty matters; the leaves of gold are then placed upon the skin in such positions as the design requires, and the muslin is then gently placed over without disturbing the gold.   The design is now printed upon the block with a hot solution of Flanders glue, containing one-eighth part of gum galbanum, or else with a very strong flour paste.   After a sufficient time has been allowed for drying, the piece is removed with the gold leaves adhering, and by a gentle brushing the gold not fixed by the paste or glue is removed, and the design becomes apparent.

A fourth process, for very open fabrics, as lace, crape, etc., is carried out as follows: The stuff is stretched, by means of pins, upon a well waxed cloth, thick flour paste is first printed and allowed to dry, and then paste colored with yellow ochre is applied; this paste, to make it adhesive and moist, is mixed with sugar; several applications must be made with the block, the first printing must be with pressure to force the paste into the meshes of the stuff, the second somewhat more lightly, and the latter ones without pressure, just allowing the paste to flow from the block.   When the color has acquired a proper consistency the leaves of gold are applied and the fabric dried, removed from the waxed cloth, which leaves it easily, and the superfluous gold brushed away.   The first method is the only one which will resist washing in water; it has the inconvenience of retaining the smell of the oil for a considerable period.

Other methods of applying gold and silver consist in reducing them to a very fine powder, mixing with some vehicle, as gum, and printing.   The metallic lustre is obtained by burnishing with a stone, or passing between heavy metallic callenders.

Many attempts have been made to deposit gold from its solutions upon fibrous matters, with but very little success.   One of the most promising methods consisted in saturating the fibre with solution of chloride of gold, and then bringing it into an atmosphere of phosphuretted hydrogen gas, by which the metallic lustre was developed; but the result is too irregular, and quite impracticable.

Golden Rod.—An American plant, *solidago canadensis*, which, according to several authorities, yields good yellows upon silk, wool, and cotton, with aluminous mordants. The attempts which have been made to introduce it into trade have not met with success. The shades of color it yields are similar to those obtained from weld.

Gray Colors.—Gray is a mixture of black and white; it also results from a mixture of the three elementary colors, red, yellow, and blue, in which the blue preponderates to a greater or less extent. There are almost an infinite number of gray shades, some named from fancied resemblances to natural ob jects, others with purely arbitrary names. In giving a selec- tion of receipts and processes for this color, I shall rather aim at laying down the principles upon which the results are ob- tained than pretend to include all possible modifications of it.

Grays upon Cotton.—These are practically dilute blacks; that is, the same materials are employed as for black, but in smaller quantity. The goods are worked in sumac and then raised with copperas; this gives a rather bluish gray, some- thing like diluted ink; to modify it to any particular hue, it is only necessary to add the coloring matter producing that shade. To make it more yellowish, a small amount of fustic and alum are employed; to make it fuller, peachwood and limawood with alum are used. Most of the colors called drab, the methods of producing which are given on page 199, are shades of gray, which, by varying the proportions of dyewood, may be made to assume various hues. Catechu is the basis of a great number of grays upon cotton goods, and is modified by logwood and copperas towards the stone grays, and by logwood and alum towards the slate grays. For the greenish or olive grays, logwood gives the blue part, and fustic the yellow, with alum for the mordant. Galls and copperas are sometimes used as the basis of gray, but not often, on account of the expense of the galls.

The gray colors obtained by printing on calico are generally compounded of the elementary colors, red, yellow, and blue, as in the following receipt:—

Pearl Gray.

3½ gallons gum water,
10 oz. yellow prussiate,
2 quarts of lilac standard (p. 200),
1 quart of bark liquor, at 20°.

The lilac standard quoted above is composed of logwood and red liquor chiefly, and yields what there is of the red part; the

17

prussiate, with the excess of acid contained in the lilac color, gives the blue, and the bark liquor gives the yellow. For all light shades it is customary, in color mixing, to employ standards which contain the elements of the color in a concentrated form. This is the case for several reasons: the colors are less bulky, can be mixed with greater certainty strong than weak, and, because these colors are mostly precipitates, do not print well or fix well if formed in dilute liquids, being apparently too irregularly diffused. This will explain the frequent reference to standards, which, though unsatisfactory in appearance, is the actual practical method of compounding these shades. For example, in making a stone gray, as in the following receipt, there are three standards employed:—

2 quarts pale purple standard (p. 189),
1 quart common blue standard (p. 104),
1 quart pink standard (see below).

Pink Standard.

1 gallon cochineal liquor, at 6°,
6 oz. alum,
8 oz. cream of tartar,
½ oz. oxalic acid.

The result of this receipt is itself a standard, from which the actual colors are made by dilution with gum water. Again, if we require a slate gray standard, we should find such a receipt as this:—

Standard for Slate Color.

2 quarts pale purple standard,
8 pints common steam blue (p. 104).

The colors are produced by diluting this composition with gum water. The following grays are based upon the lilac standard, given in the first column of page 200; they are actually for delaines, but applicable also to calico and to wool:—

Dark Gray for Grounds—Delaine.

4 quarts of gum water,
2 quarts lilac standard (p. 200),
14 oz. yellow prussiate of potash,
1 quart hot water,
1½ pint extract of indigo.

This gray, containing a considerable quantity of blue and no yellow, would be a dark mourning gray. It can be diluted by mixing with gum water. For calico, of course the extract of indigo would be left out, and, if the blue was deficient, the yellow prussiate would be increased.

Dark Pearl Gray—Delaine.

$3\frac{1}{2}$ quarts gum water,
1 pint lilac standard,
$2\frac{1}{4}$ oz. yellow prussiate,
$\frac{1}{2}$ pint of bark liquor, at 20°,
$\frac{1}{16}$ pint extract of indigo.

By varying the proportions here given, the gray may be modified to produce other shades, called silver gray, lavender gray, etc.

Gray from Galls—Calico.

22 lbs. gall nuts,
11 lbs. sulphate of iron,
10 gallons common vinegar,
10 gallons water; steeped for 15 days,

And the clear liquor strained ;

2 gallons gum water,
1 quart of gall and copperas liquor,
1 quart acetic acid.

A lilac gray color can also be obtained from logwood and mixed iron and red liquors, in the following proportion :—

Gray from Logwood.

1 quart logwood liquor, at 8°,
1 quart red liquor, at 14°,
1 pint iron liquor, at 21°,
2 gallons gum water.

Some agreeable gray colors have been obtained on calico by combining the chromium shades with red coloring matters ; an example of which is given under CHROMIUM.

Gray Colors upon Woollen.—For the light shade of mourning gray the coloring matters consist of blue in excess, modified by a slight quantity of red. The blue part is sulphate of indigo, and the red preferably ammoniacal cochineal ; the mordants being alum or sulphate of alumina and cream of tartar. For 10 lbs. of merino, or similar fabric, mordant as usual in $1\frac{1}{2}$ lb. alum and 1 lb. white tartar; then take $1\frac{1}{2}$ lb. of alum and 1 lb. of white tartar, dissolve, and add the sulphate of indigo and cochineal according to the shade required; about 45 minutes, at a temperature of 160°, is sufficient to dye. To obtain clear light shades the heat must not be pushed too high, nor the time protracted ; the quality of the sulphate of indigo will also considerably influence the nature of the shades produced. For the more neutral grays the three primary colors are blended;

the yellow being chiefly derived from fustic, the blue from sulphate of indigo, and the red from archil or cudbear, alum and tartar being as before the mordanting agents. According to the relative proportions of these coloring matters, the shade varies in an almost infinite variety of hues and tones, so that the quantities given below are merely suggestive.

Dark Chestnut Gray.—For 10 lbs. of woollen cloth mordant with 1 lb. of sulphate of alumina (or instead 2 lbs. of alum) and 1 lb. of tartar, and dye with ¼ lb. of extract of indigo, 1 lb. of archil, and ½ lb. of fustic, the two latter being in the state of decoction; work for an hour at the boil. One-half those quantities yield a lighter shade, and even one-fourth give a good color.

Brown Gray.—For the same quantity of wool mordanted as before, take 2 lbs. archil, 6 oz. extract of indigo, ½ lb. fustic, the archil or fustic as before being added in the state of clear liquor. This shade can be reduced again by taking one-half, one-third, or one-fourth of these quantities.

Yellow Grays.—For the same quantity of wool, and mordanted as before, take 2 lbs. of fustic, ¾ lb. of archil, and 3 oz. of extract of indigo, and dye as before. These quantities can also be reduced one-half and one-quarter for lighter shades.

In all complex shades of this kind the lighter styles are best dyed first, and the darkest colors dyed in the spent liquors, properly freshened up with coloring matters and mordant.

Besides archil, the red part of these shades can be given by cochineal, which, however, is expensive, or by madder and the common red woods. In some very few cases, where great stability is required, the wool receives its blue part by a dip in the indigo vat.

Another method of obtaining several shades of gray, on woollen cloth of a heavier kind, is performed by boiling the piece first in a small quantity of galls and sumac, and then passing in a dilute solution of green copperas, just warm; this gives a blue gray, which is afterwards converted to any desired hue by a fresh dyeing in a small quantity of madder, and with sulphate of indigo and decoction of fustic in quantities depending upon the shade required. Or the dyeing is performed at one operation, by taking, say for 100 lbs. of heavy woollen cloth:—

    1½ lb. fustic, ·
    ¾ lb. logwood,
    ⅐ lb. sumac,
    ¼ lb. alum ;

boil these for half an hour, or, what is better, using decoctions representing these quantities of dry woods, then adding ¼ lb.

madder and the woollen cloth, and boiling for thirty minutes. When the dye has well penetrated, the color is saddened by copperas and extract of indigo to the degree required.

A cheaper dye for merinoes than that given above consists in mordanting in alum and bichromate of potash and tartar, and then dyeing in a mixture of woods proportioned to the shade required; as, for example, for 10 lbs. wool, 1 oz. bichromate of potash, $\frac{1}{2}$ oz. alum, $\frac{1}{2}$ oz. tartar; boil thirty minutes, and dye in a mixture of fustic, peachwood, and logwood, saddening if necessary with copperas; or leave out the logwood, and use sulphate of indigo for giving the blue, and copperas to sadden. Madder frequently enters into these colors as a red constituent.

Gray colors for printing on wool are the same as those given for delaine, with some additional ones not applicable to delaine, as follows:—

Mourning Gray— Wool.

1 gallon crimson (p. 167),
4 oz. sulphate of indigo,
3 pints berry liquor, at 7°,
8 oz. alum; thickened according to requirement.

This is a dark color, which may be reduced by gum water, and addition of oxalic and tartaric acid.

Silver Gray for Wool.

3 oz. sulphate indigo,
1$\frac{1}{2}$ gallon hot water,
8 oz. tartaric acid,
8 oz. alum,
6 oz. ammoniacal cochineal,
2 gallons thick gum water,
6 oz. bichloride of tin.

Another Mourning Gray— Wool.

1 gallon logwood liquor, at 5°,
10 oz. nitrate of iron, at 80°,
1 gallon gum water.

This color—rather deep—can be reduced by gum water to any desired depth of shade.

Gray Colors on Silk.—Logwood and tin salts give a lavender gray; archil, dissolved in a soap lather, gives also shades of lavender gray; for the warmer grays a basis of anotta is dyed, and then the shade saddened by means of sumac and fustic with copperas and alum. Archil, cochineal, and sulphate of indigo also enter into the composition of these shades; in fact, any of the red, yellow, and blue coloring matters in combination yield grays, and, on account of their number, the methods of obtaining gray are so numerous as to render classification impossible.

The colors for printing on silk are mostly compound colors, obtained by mixing red, yellow, and blue standards, as in the case of calicoes. The following is from logwood :—

Mourning Gray—Silk.

2 quarts logwood,
2 oz. nitrate of iron, at 80°,
¼ oz. tartaric acid,
2 lbs. ground gum.

There are no colors so difficult to describe as those produced by a mixture of several dye-woods, and none in which the excellence of the results depends so greatly upon the care and tact of the colorist. There are many means of arriving at the same end, so that each dyer and color mixer has his own pet method, which he likes because he can manage best. The receipts and processes here given will, however, serve to extend and generalize the knowledge upon these shades of color.

Green Colors.—All the green colors in use—with the practically unimportant exceptions of Chinese green and oxide of chromium green—are compounded of blue and yellow; and though there is a considerable variety of processes, and some little choice of blue and yellow elements, the principles of the combination are very simple, and require no elucidation. It only remains therefore to give such a selection from receipts and processes as will illustrate the method of obtaining the chief greens at present in use upon cotton, wool, and silk.

Green Colors on Cotton.—The blue is generally dyed first for the faster kind of greens, and the yellow dyed upon it.

Vat Blue and Chrome Yellow.—Dye the cotton of a sky blue in the indigo vat, then work it in any of the lead preparations given for chrome colors (p. 142), and raise in bichromate. It requires some experience, and a good eye, to ascertain the right shade of blue for a given pattern of green. When the point has been missed, it is frequently compensated by sulphate of indigo; but, as this has not the slightest affinity for cotton, it washes out on the first contact with water, besides fading rapidly in light and air.

Vat Blue and Bark Yellow.—Dye a sky blue in the indigo vat, and then work the cloth or yarn in red liquor at 8° for a sufficient time; wash out, and dye with bark liquor, raising with alum or solution of tin.

Vat Blue and Fustic Yellow.—Precisely as the last receipt, only substituting the yellow coloring matter of fustic for that of quercitron bark.

Bark and Extract of Indigo.—This is a shamefully loose color, and quite unworthy of a record if it was not for the fact that a considerable quantity of cloth so dyed is regularly shipped abroad. The cotton is worked in red liquor at 6° Tw., then dyed yellow with bark, and the blue part obtained by working in extract of indigo and drying without the usual washing, which would, in this case, remove all the blue. The yellow may be obtained by fustic instead of bark.

Prussian Blue and Fustic.—Dye a light Prussian blue as given on page 102, then mordant by working in red liquor at 8°, and dye in fustic liquor, raising with alum. Bark can replace fustic in this receipt.

Greens Used in Calico Printing.—The best colors have Prussian blue for a basis, and either bark or berry liquor for the yellow part. Besides the receipts here given for steam greens most of those given for delaines are also applicable to calico.

Common Steam Green—Calico.

6 oz. starch,
2 pints berry liquor at 6°,
8 oz. yellow prussiate,
4 oz. alum,
½ pint acetic acid,
1 oz. oxalic acid,
4 oz. muriate of tin at 120°.

The following is a higher class of green:—

Green for Calico—Steam.

1½ lb. starch,
1 gallon bark liquor at 16°; boil, and add
6 oz. alum,
1 oz. oxalic acid,
2 oz. tin crystals; when nearly cool, add
1¼ lb. tartaric acid,
2¼ lbs. yellow prussiate of potash,
1 pint prussiate of tin pulp.

It is hardly necessary to indicate the *rationale* of these colors; but I may state that the bark or berry along with the alum, and partly also with the tin, if any be used, forms the yellow color; and the prussiate, acted upon by the free acid of tin salt, as well as by the tartaric and oxalic acids added, yields the blue part.

Green for Calico—Blotch.

2 gallons bark liquor at 10°,
3 lbs. starch; boil, and add
1¼ lb. alum, and, when cool,
3 lbs. yellow prussiate of potash,
1½ lb. tartaric acid,
6 oz. oxalic acid,
½ pint prussiate of tin.

Turpentine or gallipoli oil may be added in small quantities to assist in the working. Steam greens require raising in bichromate on account of the blue part.

Indigo and Chrome Greens.—These greens have nearly disappeared since the discovery of steam blue has enabled the printer to obtain more beautiful and regular results, and I shall only give a brief indication of the methods which were followed in obtaining them. For white objects on a green ground, a resist was printed similar to the resists used in indigo dipping, the pieces dipped for five minutes in a plombate of lime vat (see page 144), a quantity of lead would thus fix upon all the unresisted parts; the pieces then dipped in clear water, and then in an indigo vat until the desired blue shade was obtained. The yellow was then raised by passing in chromate.

The discharge white on these styles was obtained by first dipping a sky blue, then padding in lead salt and dyeing in chrome; this gave a uniform green, upon which one of the strong acid discharges of page 193 being printed, produced a white, provided that the blue was not too strong for the quantity of chromic acid present to discharge.

Another green was obtained by combining lead salts with pencil blue and raising the color in chrome.

Indigo and Bark Greens.—These greens were adapted to the calico printers' purposes by Thompson of Clitheroe, and met with great success about thirty years ago. A blue ground of the requisite depth was first dyed in the indigo vat, and the piece padded in red liquor containing a sufficient quantity of bichrome to discharge the blue. When dried, a white discharge (page 193) was printed on, which not only destroyed the blue but also cut out the alumina; the pieces were now dunged and dyed in bark, the yellow of which combining with the indigo blue produced the green. There have been other methods of combining mordants with China blue, etc., but the styles are now so seldom called for as not to require further notice.

Greens on Wool.—For all light woollen goods, merinoes, etc., the blue part is sulphate of indigo and the yellow chiefly fustic; but turmeric is largely used; weld and quercitron bark are also employed. In using turmeric, which is a very loose color, care must be taken not to dye at too high a temperature. To preserve the best color the turmeric is not added with the other ingredients—the extract of indigo and fustic—but towards the end of the dye. The wool is mordanted in alum and tartar as usual, lifted out, and the quantity of dyeing matter added. I give some examples showing the quantities used:—

Dark Green.—Decoction of 2½ lbs. of best fustic, 1¼ lb. extract of indigo, to 10 or 12 lbs. of wool; temperature not to exceed 170° nor time 60 minutes. If the exact shade is not hit, of course the quantities of blue and yellow part must be altered.

By diminishing these quantities of coloring matters by one-quarter, one-half, and three-quarters, various depths are obtained; by maintaining the strength of yellow and diminishing the blue, the green are more yellow; by taking the contrary course, the greens are bluer.

The best greens are dyed at twice, that is, the wool is mordanted separately; but most greens are dyed at one operation by adding the alum, tartar, fustic, turmeric, and extract of indigo, all together.

M. Théophile Grison, to whose "Teinturier au xix° Siècle" I have frequently referred for information upon woollen dyeing, proposes a modification of sulphate of indigo discovered by himself, for obtaining bright light shades of green. He dissolves one part of purified extract of indigo (*carmin d'indigo*) in four parts of ammonia, strains through a calico fent, and leaves in covered stoneware jars from six to twenty days; at the end of six days the green obtainable is of a bluish kind, and at twenty days it is very yellow. When the contact is judged to be sufficiently prolonged, the ammonia is neutralized by sulphuric acid. The green precipitate thus produced is drained on a filter and pressed to remove the greater quantity of water; it is then washed with a dilute solution of carbonate of soda, and is ready for use. To apply it, it is dissolved in water, containing its own weight of alum, and the solution employed fresh, without keeping. Acetate of alumina is the most suitable mordant for it, and a temperature of 140° answers best in dyeing.

A fast green is obtained by dyeing a blue in the indigo vat and then mordanting in alum, and adding fustic for the yellow part. Logwood is sometimes used, in the duller and more olive greens, as the blue part.

Picric acid has been proposed as the yellow part, but it is deficient in depth. Bark and weld are too much acted upon by the acidity of the bath to yield good colors ; fustic, although it has the inconvenience of making the stuffs rather harsh, is the only yellow color which works well with sulphate of indigo.

Greens on Wool by Printing.—These greens are chiefly sulphate of indigo for the blue part, and bark or berries for the yellow part. Fustic is seldom used, because the advantages it possesses in dyeing are not so conspicuous in printing, and it is much less rich in coloring matter than either bark or berries, and consequently more difficult to extract, nevertheless it is sometimes used.

Dark Green, all Wool—Blotch.

3 gallons bark liquor at 16°,
6 lbs. sulphate of indigo,
6 oz. sal ammoniac,
3 lbs. sulphate of alumina, ⎱
3 pints water, ⎰
1¼ lb. tartaric acid,
14 oz. oxalic acid; and when quite cold,
20 oz. bichloride of tin,
8 oz. yellow prussiate of potash,

thickened with gum substitute, according to the pattern. For bright light greens Persian berries appear the most suitable yellow part, turmeric next, but more fugitive, and bark last. The following dark green is somewhat different from the above:—

Deep Green—Wool.

1 gallon bark liquor at 15°,
12 oz. sulphate of indigo,
8 oz. alum,
3 lbs. gum in powder; heat, and then add
4 oz. tartaric acid, and, when cold,
1 oz. oxalic acid,
4 oz. bichloride of tin at 100°.

In the following green, which is of a lighter kind, fustic liquor is employed as the yellow part:—

Another Green for Wool.

2 quarts fustic liquor at 12°,
4 oz. sulphate of indigo,
14 oz. alum,
3 oz. tartaric acid,
1½ oz. oxalic acid,
1 gallon gum water,
4 oz. bichloride of tin at 100°.

In some receipts for green a small quantity of cochineal red is added to correct a grayness due to the sulphate of indigo. Out of at least fifty receipts for green on wool under my eyes I think it quite unnecessary to give any more, since the above are sufficiently representative for the purposes in view. Greens upon delaine, however, are somewhat different, and I give a few examples to illustrate them. It will be observed that, upon wool, the blue from the prussiates is not required, the extract of indigo fulfiling the blue part; but upon a fabric containing cotton the extract of indigo blue will not fix, so that the presence of the elements of Prussian blue is a necessity. Much care is required in delaine colors to obtain the wool and cotton threads of a similar shade, they will be usually found quite different. This is owing to want of care in the apportioning of the blue parts.

Dark Green—Delaine.

6 pints berry liquor at 15°,
4 pints red liquor at 20°,
1½ lb. starch ; boil, and add
14 oz. extract of indigo,
8 oz. alum,
2½ lbs. yellow prussiate of potash,
1 lb. tartaric acid,
4 oz. oxalic acid.

Another Dark Green—Delaine.

8 gallons bark liquor, at 16°,
13 lbs. starch,
4 lbs. gum substitute; boil, and add
5 lbs. alum,
1 lb. crystals of tin,
14 lbs. tartaric acid,
14 lbs. prussiate of potash,
1 lb. oxalic acid, ⎫
1 quart hot water, ⎬
½ gallon extract of indigo. ⎭

These dark greens may be reduced by starch or gum-water to yield the lighter shades. I give three more receipts for green, which in themselves would suffice for almost any number of shades. The No. 1 green is dark, the No. 2 is medium and blue, No. 3 is lighter or yellow. By combination of one with another, and by reduction with gum-water, the shades and hues can be modified at will :—

No. 1 *Green, for Delaine—Dark.*

3 quarts bark liquor, at 12°,
1 lb. starch; boil, and add
6 oz. alum; when cool, add
1¼ oz. yellow prussiate,
7 oz. tartaric acid,
1½ oz. oxalic acid,
5 oz. extract of indigo,
½ pint prussiate of tin.

No. 2 *Green, for Delaine—Blue Shade.*

1 gallon bark liquor, at 6°,
1¼ lb. starch; boil, and add
4 oz. alum; when cool, add
9 oz. yellow prussiate of potash,
3 oz. oxalic acid,
10 oz. extract of indigo,
¼ pint prussiate of tin.

No. 3 *Green, for Delaine—Yellowish.*

1 gallon gum-water,
2 quarts berry liquor, at 12°,
1 pint red liquor, at 15°,
4 oz. alum, dissolved in
3 pints hot water,
6 oz. yellow prussiate,
2 oz. oxalic acid,
4 oz. extract of indigo,
2 oz. prussiate of tin.

Green Colors on Silk.—These colors, whether dyed or printed, are very similar in their production and composition to greens on wool. The silk for dyeing is well alumed, and the yellow dyed first in fustic; then that purified kind of sulphate of indigo which has been described as DISTILLED BLUE is added, and the dyeing continued until the desired shade is obtained. Besides fustic, turmeric, weld, and bark are employed as the yellow part. Ebony wood chips are sometimes employed for the yellow; addition of logwood gives shades of olive-green. The following is a dark green color for printing on silk :—

2 quarts berry liquor, at 15°,
1 lb. gum; dissolve, and add
4 oz. extract of indigo,
2 oz. tartaric acid, at 40°,
1 oz. nitro-muriate of tin, at 80°.

If turmeric be used as the yellow part, a smaller quantity will suffice than of berries, on account of the greater intensity of its shade. (See also CHINESE GREEN.)

Gum.—Gum is a substance of the greatest importance to the printer, although hardly ever used by the simple dyer. Its study, therefore, deserves great attention, for success in obtaining good colors and good printing depends in a remarkable degree upon the nature of the gum or other thickening material employed.

The name of gum is given to any substance exuding from trees and vegetables, and drying up in the semi-transparent, globular-shaped masses we are accustomed to see in gum arabic. But another distinction must be made: some of these substances are of a resinous nature, and not acted upon by water. Gums may, therefore, be divided into gums proper and gum resins; the first being those soluble in water, the second including such substances as copal, animi, etc., which are called gums by the varnish makers, but which possess none of the properties which characterize gum arabic, the type and model of gums. A good gum is recognized by dissolving easily in water, to which it communicates a thickness or viscosity of a different nature to that given by flour or starch; the solution being fluid, easily poured from vessel to vessel, flowing with a long unbroken stream when poured from a height, and not going thicker by keeping, provided evaporation does not take place. The chemical composition of natural gums is nearly the same as that of starch, and their chemical relations to other bodies much of the same nature, that is to say very feeble and insignificant. It is this indifference which makes gum so valuable an agent in thickening colors; it does not interfere in the reaction of the various drugs upon one another, and interposes no obstruction to their combinations with the tissue, except such as are of a physical nature and inseparable from matter in its most inert form. There are only three or four compounds which cannot be thickened with gum in all the range of ordinary color mixing, and they are not often employed. There is the sulphate of chromium, which frequently coagulates the gum-water, but not always; I consider it is a deficiency of acid which causes it to coagulate; the basic acetate of lead forms a combination with gum of a curdy nature, quite unfit for printing, and the solution of protoxide of tin in soda the same. These, and one or two other mixtures must be thickened when required, either with sugar or some of the artificial gum substitutes to be mentioned afterwards. The principal natural gums in use among printers are as follows:—

Gum Arabic.—This is the most expensive and finest kind of

gum, seldom used in print works, but it may be taken as the sample of what gum ought to be—in color, solubility, and in keeping its fluid condition without any tendency to go sour. What is frequently sold as gum arabic is only picked gum senegal. It is employed principally in the fine arts, and for finishing crapes, silks, and similar goods.

Gum Senegal.—This is next in quality to gum arabic, and for all ordinary purposes, and certainly for calico printing purposes, it answers as well. It is a coarser looking gum, and when unpicked consists of masses of various sizes and colors. A portion is quite transparent and clear; this is usually picked out and sold at a higher price, on account of its yielding a colorless solution, applicable to many purposes where color is objectionable. Other pieces are colored in various degrees, from slight yellow to deep brownish-red, the whole intermixed with variable quantities of straw, bark, chips, sand, and gravel, depending upon the care used in gathering it. It dissolves easily in warm water, and gives a good strong gum water at about eight pounds to the gallon. It has a tendency to go sour upon standing, which may be partially corrected by the addition of crystals of soda; but this acidity is not, under most circumstances, objectionable.

East India Gum, Turkey Gum.—Other qualities of gum are in the market under the above and other titles. They are inferior to good gum senegal, but they are not constant or regular in their quality. Some samples which I have seen have scarcely been fitted at all for making gum water. Instead of dissolving easily in warm water, they swell into jelly, which is ropy and gluey, not flowing smooth, something like the white of egg before it is beaten up. It is possible to use this gum, for it has been used in large quantity, but it is not to be recommended for finer kinds of work. It does not give as good shades, and it does not wash off soft. This inferior quality is best recognized by steeping the lumps in cold water, and observing if they melt away into the liquor or swell up into toughish lumps, after some hours' standing. If they dissolve into clear gum water they may be accounted good; if not they may be useable, but certainly they are not worth so much as the other kinds.

Gum Tragacanth.—If gum arabic is a gum it is hard to know upon what kind of analogy tragacanth can be called a gum. It more resembles solidified paste, which resumes its bulk when it is wetted, but must be called a gum for want of a better name. It is an excellent gum for many purposes—smooth, firm, and solid—but unfortunately for the color mixer, so expensive that he can only employ it in very limited quantity.

The amount of water that it can thicken is very remarkable. At one pound per gallon it forms a mass which, in small vessels, may be inverted without running out, and at this thickness it can be used. It is dissolved by steeping it for several hours in lukewarm water, with occasional stirring. In some cases it is recommended to add nitric acid to the water, in small proportion, to be neutralized afterwards by crystals of soda; but I do not understand the benefit of it, and I question if it does not all reside in the nitrate of soda formed being slightly hygrometric, and keeping the color moister upon the cloth.

Any other gums which may be found in commerce will be similar to those mentioned, perhaps better for them, more likely worse. The only reliable test for the quality of gum is in making trial of it, dissolving it, seeing how it keeps, making color from it, noting the results, and observing if it washes off easily, leaving the cloth soft and fine. It is said that natural gums have been mixed with artificial gums ingeniously brought to resemble the lumps. I have never found anything of this kind, and doubt if it would pay to carry it out at the prices gum has commanded these few years past. For some of the properties of gum not mentioned here, reference must be made to the article treating on thickening matters generally.

Gum Substitutes.—Towards the end of the last century the natural gums, used by calico printers, became so dear as to seriously injure the trade. Many attempts were made to obtain some substitute, but with little success, the investigators chiefly looking to the gelatinous matters of linseed and mosses as the most probable sources of a substitute. The fact that heat converted starch into a gummy substance seems to have been accidentally discovered in the early part of the present century, but to whom the merit of this great boon to calico printing belongs is not generally known. There seems to be no doubt that it was a discovery belonging to these islands, and one which has done much to put calico printing in its present flourishing position. Foreign gums are now only exceptionally employed for purposes of thickening colors. The artificial gums have replaced them in the great majority of cases, for they can in general be better adapted to the requirements of any given color than the natural gums. I have had experience in the making and using of artificial gums, and believe that they can be made to answer every purpose of thickening as well and generally better than the natural gums; economy will in all usual circumstances be greatly in favor of the artificial gums. There are very few print works in Great Britain making gums; those who do find an advantage from it in several ways. In the United States nearly all printers make their own gums; but for the successful carrying out of this and other chemical manufactures, more knowledge of chemistry is necessary than is usually found on an English print works.

The general principle in gum making is to subject every particle of the starchy matter, whether wheaten starch, potato flour or farina, or sago flour, to the action of a high temperature. The effect of the heat is to cause a change in the structure of the farinaceous globules, so that, when put into water, they burst and dissolve, while previously they were unacted upon by it. Whether the change is chemical or physical is not well known. Under the microscope the globules seem unaltered except in size. They are become much smaller in roasting, but the envelope does not seem broken; the contact of water ruptures it directly, and all dissolves into a clear liquid of a pale yellow color. This yellow or brown color is an inevitable result of the heat used, and in some cases it is objectionable. A French chemist, M. Payen, found that if the starch or farina was mixed thoroughly with weak acids, it required a lower heat to make it into gum, and the product could be obtained nearly white. I found that if certain acid gases and vapors were passed over hot starch and farina, they were changed into gum at a comparatively low temperature without changing the color of the farina in a perceptible degree. These light colored gums are employed in many steam colors, and for finishing goods. The gums which are to be found in trade are very various in their properties; the quantity required to thicken a gallon of water; and the names which they bear. Each manufacturer has his own process of mixture and preparation, and has a reputation for producing a particular kind of gum, to which he usually gives any name that he likes. Of special trade gums I cannot here take full notice, but an account will be given of what may be called simple gums, that is, gums which are genuine results of roasting the pure starches. But, it must be borne in mind that most of the gums used by calico printers are mixtures made to suit the particular requirements of the style of work or the prejudices of the establishment.

Calcined Farina.—This is what its name indicates, farina or potato starch roasted or calcined to the required shade. It is the oldest form of artificial gum, and perhaps the most generally employed. Its color is usually much darker than is necessary; calcined farina of a light buff color dissolves as well and makes a better gum water than the very dark brown kind, but this latter is preferred more from prejudice than reason. There is a probability of a light-colored calcined farina going pasty on standing; in so far as this goes a dark-colored one is preferable, but to become pasty is a sign of care-lessness of manufacture. If a light colored calcined farina is quite soluble and remains gummy, I consider it preferable to a dark-colored one; it is more solid in the gum water, more tenacious, and gives a better mark and better shade of color, especially with iron mordants. There is less risk of burned particles, which remain in suspension in the gum water, and cause annoyance in the printing. Calcined farina thickens at from seven to ten pounds per gallon of water; some kinds are as thick at seven as others at nine; when ten pounds are re-quired for ordinary thickness, the calcined farina is said to be weak. What this weakness consists in, or is owing to, is not satisfactorily known; it is not in the calcining but in the raw farina, and may be attributable to the quality of the potatoes, or to the methods of extracting the farina from them. There are weak and strong farinas in the market, which yield gums of the same kind, and the gum maker is only indirectly re-sponsible for the difference in product. The objection to cal-cined farina is its expense, requiring, as it does, so great a quantity to thicken a gallon of water. It is a common cus-tom to mix thicker gums with it, and sometimes to mix flour with it, but it is not easy in this manner to produce a good gum water.

A method of testing a small sample of calcined farina is to mix it thoroughly with cold water, and place the liquid in a glass tube kept upright. In the course of a few hours all the insoluble or imperfectly calcined and raw particles fall towards the bottom of the glass, and from their bulk the quality of the sample is judged. When there is a sufficient quantity, the best test is to make a gallon or two of gum water from it, and examine when cold as to thickness, solidity, smoothness, and tendency to pastiness. For grit, pour off the top and feel the bottoms, if there are any, testing them between the teeth or between the point of a knife and a piece of glass; not all that seems to be grit is so. If the lumps yield to the teeth they will most probably be charred particles of the farina. It is one of the most valuable properties of calcined farina that when hot it is so thin that all the grit can sink through it to the bottom; other gums that are thick when boiled keep the grit in suspension, and it cannot be removed from them, causing much trouble in the printing. Good calcined farina keeps fluid for many weeks; only inferior qualities go thick within a month, but it is not unusual to have samples that set so hard and firm in about two days, that in color-shop phrase, "you might stand upon them"—a metaphorical expression for the most part, but sometimes literally true. There is some-

18

thing really remarkable in seeing a hot solution of bad gum as thin as water setting to a hard solid mass when cold, which can be broken in pieces and crumbled in the hand. Some farinas do not go hard, but thicken a little; these generally work thick in the color boxes; they are actually a mixture of real calcined farina and raw farina, or in an intermediate state of conversion. They are only mechanically intermixed, and the thin part appears to leave the color faster than the thick in the machine; the mass gets thicker and thicker until it becomes impossible to work it. It must then be either warmed up, mixed with thin, or, as frequently happens, thrown away.

Dark British Gum.—This name was originally given to pure wheaten starch, roasted to a dark brown color. Many gum makers sell a compound gum under this designation, containing only a small quantity of wheaten starch, or none at all. Others still furnish calcined starch under its old name; it will be taken here as made entirely from starch. It is a stronger gum than calcined farina, thickening at about six pounds to the gallon of water; it works and keeps very well when properly made; it is better adapted for alkaline colors than calcined farina, and generally with the same strength of mordant gives deeper and fuller shades. Owing to the costliness of wheaten starch, it is the most expensive of the artificial gums. For the same reason there is a great inducement to the maker to substitute cheaper original matters in its preparation, and sometimes to dispense with wheaten starch altogether.

Light British Gum.—This was originally pure wheaten starch, which had received a very slight roasting. It had in consequence a light color, and, from not being wholly converted into gum, had a pastier nature than the dark British; it is suitable for steam colors. It should thicken at from three and a half to four and a half pounds per gallon, and should keep in working order for ten days at least. Light British gum, like dark British, is now seldom made from starch alone, but is a compound gum, containing two or more different gums.

Gum Substitute.—This is usually the same as light British; but it is a name of so indefinite a nature that it can easily include any kind of gum, and does actually serve for many varieties.

Soluble Gum.—The gum sold as soluble gum, and known also as patent light gum, delaine gum, etc., is of a light color, and, as its name indicates, should be easily and largely soluble in cold water. It is made by the use of acids, and requiring more time and more care than most of the other gums, is more expensive. It should thicken at about five pounds to the gallon, but it is variable, and sometimes requires as much as eight pounds to give good gum water. The solution should be clear and but lightly colored. It is not so solid or adhesive a gum water as that made from natural gums at the same weight, nor would it answer well if it was.

Dextrine Gum.—In some establishments a species of gum is prepared in the fluid state by acting upon starch or farina, by means of the diastase or ground malt. It is said to possess some valuable properties for delaine colors in washing out easily, but has the defect of irregularity in body and proneness to fermentation.

Generalities upon Artificial Gums.—The goodness or badness of a gum cannot be tested upon small quantities with any trustworthy results, nor can it be predicated from the appearance of the gum water, except by one accustomed to judge of it; for the most part the quality of gum can only be decided upon after it has done fifty or sixty pieces. A good gum will print that number without requiring to be emptied out of the box, and without being thicker at the end than at first. Sixty pieces should be reckoned a sufficient test for any gum, and a gum deserves to be condemned that will not print more than thirty without changing. A first-rate gum for madder purples will keep fluid for a month in the color-house tubs, and, if going pasty then, will only require mixing with some hot gum water, or warming up by itself to be right again. Inferior gums go pasty in from three to seven days. It is very bad gum that will not stand good for three days; when gums go thick, the only remedy is to warm them up again and get them used without having to stand long. Such gums are wasteful, and should not be tolerated, for colors made with them become thick and unfit for working, and the frequent warming is injurious, so that a good deal is spoiled altogether and thrown away. Some gums froth much more than others in working, and in consequence give bad work. I could never satisfy myself as to what this froth-producing property could be attributed to with justice. It was always worse in gums made from inferior farina, that is, farina not perfectly freed from the pulpy matter of the tuber; but it existed in gums made from very good materials. A dose of linseed oil in the gum water as it is being boiled up is the best preventive of frothing, but it does not always stop it. Some colors are more liable to frothing than others—notably, buffs from acetate of iron, alkaline pinks, and light shades of madder purple. If a moderate quantity of oil does not prevent it, I know no other remedy. When the frothy color is left to stand, the froth collects on the top, and good color, which can be used again, settles down; but the time it requires to settle, and the quantity that will not settle at all, depends upon the kind of gum, and how much it has been worked. A Twaddle is sometimes employed to test gum water, but it is of very little good; if the gum is fluid enough to let it sink as far as it will, it marks higher in proportion as there is more gum to the gallon of water; thus calcined farina at eight pounds will mark more than dark British at six pounds, although this latter may be for all useful purposes a thicker gum than the former. What the hydrometer shows is density, not thickness; and it will be found to mark higher on a solution of common salt, at three pounds per gallon, than on any gum water, at the same quantity of gum per gallon, though one is as thin as water and the other of some thickness. An instrument has been devised, called a viscometer, for indicating the comparative thickness of gum waters. It consists of a tin tube, about an inch and a half in diameter, open at the top, and pierced with a hole about one-twelfth inch in diameter at the bottom end, which is closed with the exception of this hole; it is loaded with lead at the bottom. The manner of applying it is to place it gently on the surface of the gum water, and note how many seconds it takes to sink to a certain mark; the time depends upon the viscosity of the gum, measured by the rapidity with which it enters the small hole at the bottom. In water the viscometer sinks in a couple of seconds; in thick gum water it may take seventy or eighty seconds to sink under the surface. Another plan was to measure the time a gum water took to ascend a strip of bibulous paper, like blotting paper, or chemical filtering paper, and judge of its thickness by its resistance to capillary attraction. These, and other plans, will not give much assistance in forming a correct idea of the qualities of a gum water. An experienced machine printer or color mixer will inform himself of the value of the gum water he is working with, more effectively and truly, by simply putting his hand in the gum water or color, than by any elaborate scientific tests or apparatus yet invented. This is the result of long working among such things, which teaches how to arrive at conclusions by a kind of instinct, or at least intuition, which seems to have no steps. Artificial gums are sometimes overloaded with grit; as mentioned under calcined farina, it falls to the bottom when the gums are thin on boiling, but in gums thickening at five and six pounds to the gallon, and possessing some degree of thickness when boiling, the grit remains in suspension throughout the mass of gum water. To ascertain whether a certain gum contains grit, and how much it contains, the following plan may be adopted: Take any convenient quantity of the gum (for a color-shop experiment about five pounds will be required), mix this quantity with a gallon of water, in a clean copper pan, and add a gill of vitriol and a gill of spirits of salts; raise to the boil, and boil half an hour; the acids take all the thickness out of the gum, making it as thin as water: let it stand for half an hour and draw off all but the bottoms; wash the bottoms two or three times with warm water, letting them settle properly before drawing off the liquor, then dry in the pan and take out the dry bottoms for examination. If there be grit in the gum it will show here, and can be felt by the teeth. Besides grit, there will be mostly a quantity of vegetable matter present, rather disguising the appearance of it: to get rid of this the whole must be made red hot on a clean slip of metal; the organic substances will burn away, leaving the grit behind, which may be now more satisfactorily examined and its quantity weighed. It will be found exceedingly fine, passing readily through the finest sieve, actually running through dressing silk as rapidly as if it were quicksilver. At first it does not seem as if this were the grit which destroyed the face of the roller, necessitating frequent polishings in order to get clean work, and roughening the doctor edge; but if it be tried upon a polished plate of copper it soon shows how quickly it will roughen the surface. I have never found gum totally free from grit; the very best I have ever tested contained about one grain to a thousand of gum, and this quantity never gave rise to any complaints on the part of the printers. The worst I ever tested contained about sixteen grains to the thousand of gum, and was very bad indeed, spoiling the roller before twenty pieces were printed. This grit existed in the farina and sago flour from which the gum was made, and whatever blame there was really rested upon the maker of the farina. Upon analyzing the grit, I found it to be similar in composition to the stone from which mill-stones are usually made, and suspected it came from the stones used in grinding the potato pulp. Subsequent inquiry supported this assumption; but some is derived from other sources. The amount of grit which is unavoidable, I estimate at two parts in one thousand, or one five-hundredth. Even the largest usual amount of grit found does not seem much in itself, about sixteen pounds weight in half a ton of gum; but if it is reckoned for gum water it arises to a serious amount. Suppose calcined farina at ten pounds per gallon, there is two and a half ounces of grit in every gallon or five quarts of gum water; and in gum at five pounds per gallon there will be an ounce and a quarter. In the case of a gum containing five grains per thousand, a gallon of ten

pound farina water would contain nearly an ounce of grit. In
open patterns the grit passes with the color on to the piece ; it
is in closer styles that it is troublesome. There is no method
of removing the grit when it is once in the gum ; the most
careful straining has very little influence in taking it out.

A good gum has a very indifferent or neutral chemical character, having little or no affinity for the various chemical substances employed in printing. It is observed that different
gums give different shades of colors when the drugs are quite
the same; this is usually owing to the physical, and not the
chemical, nature of the gum. The shade of color from two
different qualities of gum is usually in relation to the quantity
of gum required to thicken a gallon—the less gum the darker
shade—and also it is influenced by the structure of the gum
when dried; if the gum dries hard and flinty, the colors are
not usually so good as when it dries porous and soft. It is for
this reason that compound or mixed gums are in many cases
preferable to pure ones; a pure gum, like calcined farina, is
too dense or close in its texture, when dried it is too much like
varnish. A good gum will partly partake of the properties of
paste, drying up porous, not so close and hard as a natural
gum. A gum may contain sugar, or saccharine matters, which
will interfere with the fixing of the color and mordant, obstruct-
ing the deposition of mineral matter upon the fibre, and injuriously affecting the shade of color produced, especially in the
case of iron mordants. The gums most likely to be bad from
this cause are the soluble gums, which are made by means of
acids, and the light gum substitutes, which often contain soluble
gum. As before stated, acids have a tendency to convert starch
or farina into sugar; and it frequently happens that in the
making of these light gums sugar is formed from the raw material. Its presence can be detected by the taste. Though sugar
is mentioned in particular, it must be understood as including
other substances, not correctly coming under this designation,
but producing the same effect, and having nearly the same composition; compounds possibly intermediate between gum and
sugar, possibly produced by the decomposition of the sugar
itself into some more developed compound. The existence of
something similar to caramel, or burnt sugar, in calcined farina
and dark gums, may be easily conjectured from its color and
taste. Gums which enter into fermentation in warm weather
generally contain saccharine matters, and are also generally
made from inferior material. That fermentation can reach an
extent capable of injuring the gum is very probable, but a
slight fermentation does not do so. In fermenting, the sugar
is destroyed and acid produced; the acid is the acetic acid,

which is not injurious, in moderate quantities, to mordants or the majority of colors. According to my observation, a little alcohol exists at the same time as the acid, but it appears to soon pass away, and be changed to acid ; if alcohol was present, even in larger quantities than is possible, it would do no harm.

A gum will either dissolve totally in cold water or only dissolve in part. The first class is distinguished by becoming thin on boiling, the second by going thick. A soluble gum has some advantages, and some defects, as a working gum, and can only be advantageously applied to certain classes of work. Dissolving off the cloth easily, it should be used for steam and spirit colors; and generally, in all cases where the pieces are washed off cold, less washing is required to take the gum out, and the color is, of course, less injured. Beyond these cases, I do not see any advantage in having a gum perfectly soluble ; it ought not to be a required condition in a gum which is employed, or intended to be applied, in making colors for dyeing. The hot dunging or cleansing liquor clears off a thickening equally well, whether it be paste or gum; and though a gum thickening will be more speedily cleansed than a paste thickening, and, consequently, a soluble gum more quickly than one only partly soluble, the time required in the hot dunging liquors is greatly in excess of that required to take off the most resistant of ordinary gum thickenings.

Gum Thus.—A resinous substance which is employed in bleaching, in the same way as RESIN. Its properties are very similar to those of common pine resin, and it is preferred to it by some bleachers. I had an opportunity of testing the relative merits of gum thus and resin upon the large scale, but I could not perceive any difference when both were equally well prepared. Gum thus is usually higher priced than resin.

H.

Hachrout.—The Hindoo name for a plant yielding the same colors as madder; only known in Europe from the accounts of travellers.

Harmaline.—This name is properly given to the red coloring principle of the seeds of *peganum harmala ;* but this is not a commercial article, and the name has been improperly used to distinguish some qualities of the aniline colors. The seeds containing the real harmaline have been tried for dyeing, but the results were not encouraging.

Hartshorn. *Spirits of Hartshorn.*—An old name, and still in common use, for ammonia.

Hellebore.—It is said that the Canadian Indians made use of the three leaved hellebore (*helleborus trifolius*) for communicating a fine yellow color to skins and wool.

Hematine, *Hemateine.*—Names of the pure coloring principle of Logwood, which see.

Hematosin, *Hœmatosine.*—The coloring matter of blood. In the cases in which blood is used in dyeing operations, it is supposed by some authorities that its coloring matter fixes upon the fabric. For a method of preparing and preserving blood for dyeing purposes, reference may be made to a patent granted to James Pillans, December 29th, 1854.

Hematoxyline.—The more appropriate name for the pure coloring matter of logwood. (See Logwood.)

Hemlock Spruce.—The bark of this tree (*pinus abies Americana*) is employed in America for tanning, and, like most other barks, gives some colors to mordanted goods, but nothing worthy of special notice.

Hiccory.—Bancroft patented the application of the bark and fruit of the American hiccory, or walnut tree, as a dyeing material for producing yellow and green colors. It was not found practically economical.

Hydrate.—In chemistry, this name indicates a compound of a substance with water. Thus lime, in the freshly burnt state, is dry or anhydrous, but, if left in a damp place, it falls to powder, at the same time absorbing water. The powder, though dry to the touch, contains about one-fourth of its weight of water, and is the hydrate of lime. If this hydrate of lime be heated red hot the water is expelled, and the lime is said to be dehydrated.

Hydrochloric Acid, *Muriatic Acid, Spirits of Salts.*—This acid, commonly known as muriatic acid or spirits of salt, is a compound of chlorine with hydrogen; it is a gas, and the liquid commercial acid is a solution of it in water, the strongest acid containing about one third part of its weight of dry acid gas. The strength of muriatic acid can be very well ascertained by means of the hydrometer; it is too cheap to make it worth while to adulterate it; such impurities as it contains are what it takes up in the course of its manufacture, these are principally iron, to which the yellow color of the acid is owing, and sulphurous acid, or some sulphuretted compound derived from the oil of vitriol, used in the manufacture of it; some other substances are likely to be contained in it, in small quantity, but don't interfere with its use. There are two kinds of acid known in commerce, the one called "cylinder salts," because it is derived from the manufacture of salt cake in cylinders, the other called "tower salts," because made by the method known as the tower method; preference is usually given by the consumers to the former, and a slightly higher price can be commanded; but in what the difference consists, or if there is any real difference in them, the author is not informed. Spirits of salt is used for several purposes in the arts; in bleaching it is often used instead of vitriol to make the sours which follow the chemic or bleaching powder solution, and it is thought to give better results than vitriol, especially for calico intended for garancine work; it is the best acid for taking iron mould spots out of the cloth, applied at the full strength and washed as soon as it is seen to have dissolved the iron mould. It does not destroy cloth immediately as strong vitriol does, in fact calico will stand the strong acid for a long time, provided it be quite immersed in it; but if calico is dipped in even very weak spirits of salt, and left in the air so as to get dry, it will become quite tender; if a piece of calico quite dry be passed up into a jar of the dry acid gas no immediate effect is visible, but if before passing it into the gas it be breathed upon or held over boiling water, the gas is immediately absorbed and the fibrous texture destroyed, as much, probably, by the effect of the heat developed as by the greater concentration of the liquid acid formed. The usual commercial spirits of salt has a specific gravity of 1.170 or 34° Tw., and contains from 30° to 33° per cent. of the dry acid. The amount of real acid can be ascertained by the quantity of alkali which it can neutralize, as ascertained by the method given p. 52.

The following table shows the amount of real acid in 100 parts of the liquid acid, at the strengths set down:—

Twaddle.	Per cent.	Twaddle.	Per cent.	Twaddle.	Per cent.
40	40.77	26	26.50	12	12.23
38	38.4	24	24.48	10	10.19
36	36.30	22	22.42	8	8.15
34	34.2	20	20.28	6	6.11
32	32.2	18	18.34	4	4.07
30	30.17	16	16.31	2	2.03
28	28.30	14	14.27	1	1.00

Hydrometer, *Twaddle.*—Hydrometer is the scientific name for an instrument for readily ascertaining the density or specific gravity of solutions. The hydrometer in use in this country was originally graduated by a manufacturer named Twaddle, and bore his name; hence the instrument itself is called a Twaddle. The construction of the hydrometer is too familiar

to all who use it to need any description, and its application is
so simple as to be acquired without any trouble or instruction ;
a few general hints however may be useful.   •

The Twaddle should not be used in warm liquids; for, besides
the risk of breakage, it will not show the same degree upon
the same liquor when hot and when cold. In warm liquors the
Twaddle sinks lower than in cold, to the extent of one or two
degrees. In testing liquids, therefore, care should be taken to
have them at nearly one uniform temperature.

The Twaddle shows the weight or density of liquids, and
nothing else. It does not show the thickness for example, and
can only be a fallacious guide in testing gum waters. Neither
does it show the purity of any liquid, though its indications
upon this point are frequently accepted without question. It
is quite common, for example, to strengthen liquors with common salt, as in the case of bleaching liquor, which cannot very
well be made stronger than 7° or 8° on Twaddle, yet it is frequently found marking 24° or 28°. The ill-informed bleacher
would not be satisfied with a liquor at 7° or 8°, and the manufacturer is in a manner compelled to bring up the density by
adding common salt, which of course contributes nothing to the
bleaching power of the liquid, though it causes it to mark 15°
or 16° higher on the hydrometer. Thus a philosopical instrument, in ignorant hands, becomes a delusion instead of a protection. An instance came under my notice when a cunning
manufacturer attempted to take an advantage of the half knowledge of the hydrometer possessed by some purchasers. He
offered what he called a double stannate of soda, much stronger
than ordinary, for he stated "if a pound of ordinary stannate
of soda be dissolved in a gallon of water it will only mark 12°,
but a pound of mine will mark 16°;" and such was the fact,
but, nevertheless, it was not worth a fraction more than the
common stannate. Ordinary stannate contains about one fourth
its weight of water, the double stannate simply consisted of
ordinary stannate, from which the water had been expelled, and
common salt put in to make up the weight. If manufacturing
chemists were honest, and all the larger ones, at any rate, I find
to be so, the hydrometer would serve most of the purposes of
analytic chemistry on a works; in such liquids for example as
muriate of iron, muriate of tin, nitrate of iron, nitric acid, etc.,
there should not be any other test required; but, unfortunately,
there are unprincipled traders who, under cover of the hydrometer, perpetrate the most impudent frauds upon purchasers
—frauds which can only be detected by chemical analysis.

An hydrometer is most delicate when the bulb is large and
the stem out of the water thin.

The degrees of Twaddle's hydrometer are easily turned into specific gravity numbers—a quality which makes it preferable to any other hydrometer in use. The rule is to multiply the indicated degree by 5, and add 1000 to the product; for example, 9° Tw. equals specific gravity 1045; 25° Tw. equals specific gravity 1125; 100° Tw. equals specific gravity 1500; and so on. To bring specific gravity numbers to degrees of Twaddle, subtract 1000, and divide the remainder by 5; for example, specific gravity 1100 equals 20° Tw.

An instrument so much in use should be correct in its indications, but there are some very false ones offered for sale, and purchasers should only trust to an instrument made by some well known and established house.

Hygrometer.—This is an instrument for indicating the degree of moisture in the air; its use on a print works is to enable the stoves to be kept at a regular degree of moistness, for, if it begins to indicate too much moisture, the steam is cut off, if it indicates too little moisture more steam is admitted, and so on. The form of the instrument generally used is the modification called Mason's hygrometer. It consists of two exactly similar thermometers fixed to a stand ; the bulb of one thermometer is kept constantly wet, by a thread communicating with a reservoir of water, while the bulb of the other is freely exposed to the air. Mason's hygrometer is from this fact frequently called the wet and dry bulb hygrometer. The dry bulb always indicates a little higher temperature than the wet bulb, because the evaporation of water from the wet bulb absorbs heat from the mercury and glass; but the difference between the two bulbs is in ratio with the dryness of the air, the dryer the air the more rapid the evaporation, and the lower the temperature of the wet bulb, and the greater difference between it and the dry bulb. Moisture in the air prevents rapid evaporation from the wet bulb, and there is a less difference between the two thermometers. If the air is perfectly full of moisture there will be no evaporation from the wet bulb, consequently no cooling, and the temperature of the two thermometers will be equal. Such an event rarely occurs in the open air or in rooms, but is not uncommon in close dyehouses or in stoves containing damp goods.

Hypochlorite of Lime.—A name sometimes applied in scientific works to bleaching powder, or chloride of lime, or rather to that part of common bleaching powder in which, according to theory, all its active properties reside.

Hyposulphites.—A series of salts so called, the only one of which in common occurrence is the hyposulphite of soda. This salt has received some applications, and, from its peculiar properties, it is anticipated it may yet be largely employed. It is used as an antichlore, or substance for neutralizing any excess of chlorine which may be left in fabrics after bleaching.

Another application was invented by Kopp, who found that the hyposulphite of alumina produceable by its means deposited its alumina by simply drying, and constituted a red mordant, which did not require any ageing. The patent in which this novelty is included bears date July 10, 1855. The following are some working receipts in which the hyposulphite red was applied. The first step is to prepare a muriate of alumina liquor as follows:—

Muriate of Alumina Standard.

6 lbs. alum,
3 quarts hot water,
10 oz. ground chalk; dissolve, and add
5 pints muriate of lime at 35° Tw.

These ingredients were well stirred, an abundant deposit of sulphate of lime formed, and the impure solution of muriate of alumina strained off. The color (mordant) was prepared as follows:—

Dark Hypo Red.

3 quarts standard above,
1½ lb. starch; boil, and when cool, add
2 lbs. hyposulphite of soda.

Stir well and sighten, preferably, with ground indigo. For light reds or pinks a slightly modified process was followed:—

Hypo Standard for Pink or Light Red.

2 gallons hot water,
5 lbs. alum,
8 oz. ground chalk; dissolve, and add
2 quarts of muriate of lime at 34°.

The color for printing was prepared as follows:—

Hypo Pink or Light Red.

3 quarts water,
1 quart of standard above,
1½ lb. starch.
½ lb. of gum substitute; boil, cool, and add
4 oz. hyposulphite of soda.

The muriate of alumina presents some difficulties in thickening, on account of its acidity; and it is imperative that the hyposulphite should be added to the already thickened and cold color. I have seen a good deal of this style worked; but it was subject to irregularity, and did not present any conspicuous advantages over common red liquor mordants. The property of the hyposulphites, which is taken advantage of in this application, is that of their ready decomposition in the presence of acid salts, and heat. As soon as the hyposulphite of alumina (formed by interchange of acid and base between the salts) was drawn over the hot tins or steam chests it was decomposed, sulphurous acid was given off, sulphur deposited and the alumina free from acid remained attached to the cloth.

The hyposulphite was applied in a nearly similar manner to iron or tin mordants.

Another application has been made of the hyposulphite, in which the atom of sulphur it holds in loose combination has been taken advantage of to produce metallic sulphides upon fabrics. This application was patented 1855, Nov. 20, and consists essentially in steeping dyed fabrics, to which the metallic lustre is to be imparted, in solution of sulphate of copper, and then in a strong solution of the hyposulphite; a precipitate of the sulphide of the metal takes place, which, having a metallic reflection, produces the intended effect. A similar effect was previously obtained by Schischkar and Calvert, by submitting goods impregnated with metallic solutions to the combined action of steam and sulphuretted hydrogen gas. See patent, dated 1854, January 5.

I.

Indigo.—This most important dyeing material is contained as a colorless juice in a genus of plants, to which the general name of indigofera is applied. The method of extracting it consists in steeping the plant in water; it enters into a state of fermentation, and the coloring matter dissolves in the water in the yellow state; the water is drawn off, and by agitating it so as to bring it freely into contact with the air, the indigo acquires the blue state, and settles down as a blue precipitate, which is drained, pressed, and dried into the form under which it is sold to the consumers. India, and the islands of the Indian Archipelago, produce nearly four-fifths of all the indigo consumed in the civilized world; of the remainder, Central America furnishes the greater portion; the quantities received from Egypt and other African countries is very small.

There is a considerable difference in the value of different sorts of indigo, the best quality commanding from three to four

times the price of the lowest quality, there being many intermediate qualities and prices. The brokers' of indigo believe
that they are enabled to fix the true value of a sample by its
external characters alone, and I am inclined to think they are
seldom far from the truth; nevertheless, it would be extremely
desirable that their judgment should be controlled by chemical
analysis; but at the present time this is not possible for two
reasons—first, the method of selling indigo does not permit of
any testing; and, secondly, there is no really trustworthy
method of analyzing it. Gross frauds, which are said to be
sometimes attempted by the indigo makers, could be easily
detected by chemical analysis, such as the admixture of ground
slate, black sand, plumbago, lead powder, starch, etc. It is
stated that indigo is sometimes adulterated with alumina colored
with logwood. It is doubtful whether such adulterations are
frequent, and whether they could deceive an experienced purchaser; on the other hand, it is extremely doubtful if any
method of chemical analysis as yet known can be depended
upon to give results true to five per cent. of the pure indigo in
the sample. The best method I know of testing indigo is by
making an imitation blue vat, which may be done with the
following quantities: Take a fair sample of the indigo, and,
having ground it very finely, weigh 75 grains, and in order
still further to soften and disintegrate it, boil it for a short time
with weak caustic soda, and then, if there be any soft lumps or
clots, strain through calico; mix this with three quarts of water
in a narrow-necked bottle, which it will nearly fill, and add 400
grains of quicklime which has been slacked as perfectly as
possible: shake well up, and add 1000 grains measure of solution green copperas at 30°; cork the bottle closely, and leave
it for three days, frequently shaking it in the interval. The
indigo will be dissolved by this time; one quart of the clear
solution is drawn off, shaken up in a bottle to oxidize it, acidified with acetic acid, and the pure indigo collected upon a filter,
dried, and weighed. Four times the weight of the pure indigo
is the percentage of indigo in the sample. This method is
tedious, requires great skill and extreme care, and even then
is liable to failure from several causes; for it is based upon the
assumptions that all the indigo is dissolved by the lime and
copperas, and that nothing but pure indigo is precipitated from
the solution when oxidized and treated with acetic acid. Very
careful experiments throw doubt upon the truth of either of
these hypotheses, and tend to prove that samples of different
origin behave quite differently in the deoxidizing fluid.
Another method, recommended by a great number of experimenters, is more remarkable for its simplicity than for the

accuracy of its indications. A weighed sample is converted into sulphate of indigo, which is dissolved in water and decolorized by solution of bleaching powder, chlorine, or chlorate of potash; according to the quantity of decolorizing solution required so is the quality of the indigo. There are some difficulties in this method which render it untrustworthy, for no two experimenters can obtain accordant results by it; the process may probably be available for the comparison of two or more qualities of indigo by one operator, but it will not answer for general application.

Of the indigo sold in the English market, fine violet paste Bengal commands the highest price; Kishnighur ranks next; good qualities Kurpah and Madras Bimlipotam afterwards; while lower qualities of Madras, fig indigo, and sweepings occupy the lowest place. For silk and fine woollen dyeing, the best qualities are taken; for calico dyeing, medium qualities are the most economical; while the lower qualities answer for coarse woollens.

Refined indigo, obtained by dissolving ordinary indigo with lime and copperas, and then precipitating by acid, is used for blueing finished goods; it may be considered as nearly pure indigo.

Indigotine is chemically pure indigo, obtained by applying heat to refined indigo; a purple vapor rises which partly condenses into needle-like crystals of indigotine.

Commercial indigo yields from 10 to 80 per cent. of pure indigotine, the remainder being earthy matter or vegetable impurities either purposely added or resulting from defective processes of manufacture.

Pure indigo is not soluble in water, nor in weak acids or alkalies; it dissolves to a very small extent in alcohol and turpentine, to a somewhat larger extent in aniline; but for practical purposes blue indigo is insoluble in all liquids. It is dissolved by oil of vitriol, but becomes radically changed during its solution, and cannot be brought back to its primitive state. This is not, therefore, properly speaking, a case of solution. It is also dissolved by alkalies, when these are mixed with copperas, tin, sugar, and other substances, which exert a particular chemical action upon the indigo, depriving it of its blue color and a portion of its oxygen. From this state of solution the indigo can recover both its oxygen and its color by exposure to the air; and it is entirely through the agency of alkalies and these substances that indigo is applied as a dyeing matter.

Chemists distinguish two kinds of indigo, called, respectively, blue indigo and white indigo. The white indigo is obtained from the blue by depriving it of an atom of oxygen; it is insoluble in pure water, but soluble in lime-water and alkaline liquids generally. If it comes into contact with air or oxygen gas, it absorbs the latter and becomes converted into blue indigo again. The yellow colored fluid, which forms an indigo vat, contains this white indigo, dissolved by means of lime, soda, potash, or ammonia; the blue scum on the surface is blue indigo, which has been formed by contact with the air. The change from blue to white, and back again to blue, appears capable of being made any number of times without destroying the indigo.

The multifarious processes for applying indigo, should, from the foregoing explanations, be rendered intelligible. Indigo cannot enter into the fibre until it is dissolved; it cannot be dissolved so long as it is in the blue state; when reduced by deoxidation to the white state, it is easily dissolved, and can enter the pores; upon exposure to air it returns to the blue state, and, being insoluble, cannot again be washed away from the fabric. This is the general theory of fixing indigo, and applicable to all the particular cases given below, as used in practice.

Applications of Indigo.—The greatest consumption of indigo is for forming the blue vats, in which woollen or cotton goods are dyed by simply immersing them in the solution of white indigo. The same vat is not equally adapted for wool and calico, and, as will be seen in the following details, there is a wide difference in their composition:—

Blue Vat for Wool.—According to the general accounts, the lime and copperas vat is not well adapted for woollen goods; yet, in the most recent French treatise on woollen dyeing (Grison's) there is no mention of any other kind of vat. He gives the following proportions and directions for setting a vat for dark blue:—

> 1200 gallons water,
> 34 lbs. quicklime,
> 24 lbs. green copperas,
> 12 lbs. ground indigo,
> 4 quarts caustic potash at 34°.

The indigo is in every case ground excessively fine by trituration in properly constructed mills for several days; this is a point of the utmost importance. In the above receipt the potash is mixed with five gallons of water in an iron pan, and the indigo added; the mixture is gradually heated to ebullition, and kept boiling for two hours, with uninterrupted stirring; this softens and prepares the indigo for dissolving. The lime is very well slacked, so as to have it very fine, and passed through a sieve in the liquid state; it is then mixed with the indigo and potash; the copperas, previously dissolved, is added to the vat, and well stirred; then the mixture of lime, potash, and indigo, and the whole well stirred for half an hour; if the proportions have been well kept, the vat will be fit for working in twelve hours. If, however, it looks blue under the scum, it is a sign that the indigo is not wholly dissolved, and more lime and copperas must be added, and the vat left for another twelve hours. It is worked at a temperature of from 70° to 80°. This is the common composition of a vat for dyeing cotton, but I have never seen it before prescribed for woollen.

The usual blue vats for wool contain neither copperas nor lime, or but little of the latter, as seen in the following examples:—

> 500 gallons water,
> 20 lbs. indigo,
> 30 lbs. potash,
> 9 lbs. bran,
> 9 lbs. madder.

The water is heated to just below the boil; the potash, bran, and madder first introduced, and then the indigo, previously very finely ground. Cold water is added to bring the heat down to about 90°, and kept at that temperature all through; the ingredients are very well stirred every twelve hours, and the vat should be ready for use in forty-eight hours after setting. The vat does not work more than a month, and is somewhat expensive on account of the loss of potash. Another vat, called in France the German vat, is much more manageable, and may be worked for two years without emptying, being freshened up as required. It is put together as follows: 2000 gallons of water are heated to 130° F., 20 lbs. of crystals of soda, 2½ pecks of bran, and 12 lbs. of ground indigo are then added, and well raked up. In twelve hours a fermentation sets in, bubbles of gas arise, the liquid has a sweetish smell, and has become greenish; 2 lbs. slacked lime are now added, well stirred, the vat heated again, and covered up for twelve hours, when a similar quantity of bran, indigo, and soda, along with a little lime is again added. In about forty-eight hours from the setting it may be worked; but as the reducing powers of the bran are somewhat feeble, an addition of six pounds of molasses is made. If the fermentation becomes too active, it is repressed by addition of lime; if too sluggish, it is stimulated by addition of bran and molasses. Like all the other blue vats for wool, it is worked hot.

In the following vat, which may be called a woad vat, be-

cause a considerable quantity of that plant is employed, there
is also a large proportion of madder; whether the madder is
useful on account of its coloring principles, or whether it is the
saccharine and fermentable portion of it which is useful, is not
clearly known.   It is thought to produce darker colors, and to
give them a violet tint; but this may be only fancy, for it seems
very improbable that there can be any notable quantity of its
coloring matter fixed under the conditions of indigo dyeing in
the hot vat.   The proportions employed in one case are:—

> 1 lb. ground indigo,
> 4 lbs. madder,
> 7 lbs. slacked lime,

boiled together with water, and poured upon the woad in the
vat; after a few hours fermentation sets in, and fresh indigo is
added according to the depth of color required to be dyed.

The pastel vat is set with a variety of woad, which grows
in France, and which is richer in coloring matter than the plant
commonly called woad.   Its coloring matter no doubt adds to
the effect, but it is probably only used as the remnant of a
prejudice and because it furnishes fermentescible matters useful
in promoting the solution of the indigo.

The rationale of these vats, so far as chemistry can perceive
any, rests upon the fact that fermentation is in many cases
accompanied by the formation of nascent hydrogen, which
either hydrogenizes the indigo, as M. Dumas has it, or deoxidizes
it, according to the most usual view.   Bran is one of the agents
which is most active in setting up this species of fermentation,
but it takes place in other substances as well, especially nitrogenized substances.   In the case of madder, sugar, and molasses, the reducing action is no doubt on similar principles to
that which is taken advantage of in some cases of calico printing, where glucose or grape sugar is employed as the reducing
agent.

The method of dyeing is very simple.   The wool, thoroughly wetted out, is suspended on frames and dipped in the
vat for an hour and a half or two hours, being agitated all
the time to insure regularity.   The pieces are then carried to
water and washed, and then treated with weak muriatic or sulphuric acid sours to remove the alkali retained by them.

Blue Vat for Cotton Dyeing.—In some exceptional cases the
same kind of vat described as the German vat is used for cotton
piece dyeing, that is the one containing carbonate of soda, bran,
and indigo, sometimes with madder and molasses.   This is principally for thick and heavy goods, into which the cold lime and
copperas vat would not penetrate sufficiently well.   But, gene-

rally speaking, all calicoes are dyed blue by means of the cold lime and copperas vat, especially those in which any design enters by way of resist. The materials used are lime, green copperas or sulphate of iron, indigo and water. The chemical action consists in the formation of sulphate of lime and prot-oxide of iron in the first instance—the latter body, having a considerable affinity for oxygen, removes an atom of it from the blue indigo, converting it into white, which dissolves in the excess of lime and is ready for dyeing. The proportions are about as follows:—

Strong Vat.

900 gallons water,
60 lbs. green copperas,
36 lbs. ground indigo,
80 to 90 lbs. dry slacked lime,

stirred up every half hour for three or four hours, then left twelve hours to settle; well raked up again and as soon as settled it is ready for dyeing in. It is usual to work vats in sets of 8 or 10; in such a set this vat would be the best or strongest, and the pieces would get their last dip in it; after one day's use it would become the second best—another one being freshly set as best vat; in two days it would be third best, and so on until it became the eighth or tenth, at which stage it is supposed to have lost 34 out of the 36 lbs. of indigo with which it was set, only retaining two pounds, and only capable of dyeing up very light shades, little more in fact than wetting out the pieces; after a day or two's use it is run off, and, being again fresh set, becomes once more the best vat. It is recommended to add the copperas first in a dissolved state, then the indigo, and lastly the lime; but the order of adding the materials is not absolute and is frequently varied, neither are the qualities fixed but actually differ very much in different works, and as an illustration I give four other receipts:—

		No. 1	No. 2	No. 3	No. 4
Water, gallons	. .	1000	1000	1000	1000
Indigo	℔ . .	45	34	12	2½
Green copperas,	℔ .	38	80	22	7
Quicklime,	℔ . .	45	90	34	12

Nos. 1 and 2 are for dyeing dark blues, No. 3 medium, and No. 4 only for light grounds. The proportions of ingredients it will be seen are without any rule; but, unless we knew the percentage of pure indigo in the samples, we could not tell with what reason these differences had in their origin. The high price of indigo will, of course, stimulate watchfulness that none is lost, and each dyer or manager flatters himself that he is working closest and exhausting his vats best; but I have known instances of extraordinary loss of indigo through want of skill and knowledge. As far as the effect of the vat upon the whole body of the indigo is concerned, it appears that if the quality in use contained 50 per cent. of pure color, the whole of it should be dissolved, except a portion retained by some kind of attraction by the bottoms, that out of the remaining 50 per cent. of impurity scarcely a trace dissolved—in ordinary qualities it remaining with the bottoms, so that an impure indigo dyes up as pure shades as pure indigo would. In certain contingencies, which are but imperfectly understood, there is formation of an insoluble compound of white indigo and lime, which goes with the bottoms, and is lost to the dyer; it is important to prevent this; all that is known, however, is that the formation of this compound appears to be owing to the presence of too much lime.

Dyeing Dip Blue or Navy Blue.—The method of dyeing is simple; the only skill required being in the management of the vats, which sometimes get out of order, and require additions of lime or copperas, or both. Practical men can discern the state of a vat by its external appearance, or experience has taught them, in the majority of cases, how to apply the proper remedy. The pieces to be dyed are stretched on frames, and either at once plunged into the first weak vat or into water, or, if the cloth contains strong resists, into lime water, to fix them. The piece is gently moved to detach air bubbles, and left in about seven minutes and a half, then the frame raised, and the pieces left exposed to the air for the same length of time, to take green off them, that is, to oxidize the white indigo into blue; then plunged in again, but this time into the next stronger vat, for seven minutes and a half, oxidized for another seven minutes and a half, and so on until the eighth vat, requiring two hours, of which one hour has been spent in the dye, and the other hour in the air. This length of time is sufficient to give the darkest shades. When lighter shades are required less time suffices, or the process called "skying" is had recourse to. This term, derived from "sky blue," expresses a method of dyeing somewhat different from dipping. The indigo is dissolved by means of the same ingredients, but in a vat provided with a double system of rollers, by which a piece of cloth is made to traverse through the liquid, nearly the same as in the cleansing dolly of a madder dyehouse. It acquires sufficient indigo in its passage to be dyed a light sky blue.

The pieces after the last dip, are washed over rollers, by the process known as "bowling," then passed into weak sours, and finally washed by the fly wince.

By preparing the pieces before dyeing with sulphate of copper the time of dyeing is lessened and a higher color is obtained, along, it is supposed, with some slight economy of indigo, but this is very doubtful. This preparation may be made as follows:—

10 gallons water,
1½ lb. sulphate copper,
3 lbs. starch, boiled well, and mixed with
1 gallon of glue size.

The piece is padded in this, dried, and dipped in lime before dyeing. Very little, if any, advantage attends the use of this prepare, and it is seldom employed. Sulphate of manganese has the same effect.

The bottoms of spent vats still contain some indigo, which, in well-conducted establishments, is extracted. The lime water is neutralized so as to precipitate any indigo in solution, the bottoms collected and treated with caustic soda and orpiment, which dissolves out the indigo; the solution is mixed with water, and allowed to precipitate in large sunk cisterns, which will hold a week's collection, from which it is collected and used over again.

Dip Blue Styles.—I give here some receipts, etc., used in combination with the vat dyeing, with brief accounts of the processes in use in obtaining some of the styles of work prepared for the market.

Light Resist for Azure Style.

3 gallons water,
16 lbs. British gum,
4 lbs. soft soap,
10 lbs. sulphate of zinc,
1½ pint nitrate of copper.

The white design being printed with this resist, the color is obtained by the "sky vat," the rate of passing through is regulated by the shade required: after skying, the pieces are winced or bowled, then soured and washed off. The principal difficulty in this style results from a dragging of the resist from the design on to the cloth, disfiguring the pattern; this accident is known technically as "tailing," and is owing to the piece being too tight on the rollers, and the rollers not moving with equal velocity.

Strong Resist for Navy Blue.

10 gallons water,
38 lbs. flour,
8 lbs. British gum,
36 lbs. sulphate of copper,
8 lbs. brown sugar of lead.

This is a very thick, rough paste, and must be printed with deeply engraved rollers. It will stand an hour's dipping, or sufficient to dye up the darkest usual colors, but is only safe for rather small masses of white. The following stronger resist is adapted for bold designs :—

Strong Resist for Stripes, etc.

8 gallons water,
21 lbs. flour,
7 lbs. calcined farina,
32 lbs. sulphate of copper,
7 lbs. brown sugar of lead,
10 lbs. sulphate of lead pulp.

Both these resists contain sulphate of lead and mixed acetate and sulphur of copper; the resist acts partly chemically and partly mechanically; the copper salts oxidize the indigo and prevent its deposition on the fibre, and the sulphate of lead and paste form a ground to receive any indigo which is not rendered inactive by the copper. For excessively large designs the pieces are dipped first in lime to fix the lead and copper; but usually an extra dip in the entering vat suffices, especially if the vats are strong in lime, or, as the dyers technically term it, " very hard."

Besides the ingredients mentioned in the above receipts we find lard, oil, and pipeclay as being occasionally used. The receipts admit of trifling modifications, but do not depart widely from the above examples.

Blue, Orange, and White Styles.—In this style the orange color is chromate of lead, the lead basis being printed on with the white resist, and going through all the dyeing, and afterwards raised in chrome. It is evident that in this style there must be no lead in the white resist, or it will become yellow or orange when passed in chrome. The following are receipts suitable for this style of work :—

Orange Paste.

6 gallons water,
21 lbs. flour,
3 lbs. calcined farina,
50 lbs. sulphate lead pulp,
28 lbs. nitrate of lead,
30 lbs. sulphate of copper.

This is a very strong color, and requires great care in printing and drying, on account of its dusty and friable nature. The copper is put in to help it as a resist. It is evident that the whole of the lead of the nitrate of lead is converted into sul-phate, so that there is no soluble lead salt in the resist; by the dipping in lime, which precedes the dyeing of this style of work, the sulphate of lead is enabled to adhere to the fibre in some curious way.

White Resists for Chrome Styles.

3 gallons of water,
11 lbs. flour,
2 lbs. British gum,
13 lbs. sulphate of copper,
2 lbs. sulphate of zinc,
1 lb. acetate of copper,
1 pint nitrate or copper, at 80°.

This is not a very strong resist, on account of the absence of solid mechanical matter like pipeclay, but it resists well enough for the depth of color required.

After the pieces have been dipped in the usual way they are winced and soured as usual, then well washed, and the orange color raised in neutral chromate of potash mixed with lime and kept at the boil. It would not answer to use bichromate of potash, on account of the injurious action of this salt upon the indigo blue; consequently, the orange is raised at one opera-tion. A standard chrome liquor may be made by taking 90 lbs. of chromate of soda or chrome salts, dissolving them in 50 gal-lons of water, and adding 30 lbs. of slacked lime: the raising vat is made to stand at 4° Tw., and freshened up every six pieces.

If the chrome liquor be only neutral, and not alkaline, a yellow is produced instead of orange; but the preferable way of obtaining a pure yellow is to first raise the orange and then cut it down to yellow with a wince in weak nitric acid sours; the acid and the liberated chromic acid act upon any blue which may be mixed with the orange and destroy it, so puri-fying the hue of the resulting yellow.

Two Blues and Green.—Dye a light blue by skying, print on a white resist for pale blue and an orange resist for green ; dye up as usual ; wash sour and chrome, then pass through very weak nitric acid. The white resist keeps the sky blue from becoming any darker ; the orange paste gives the lead basis from the chrome yellow, which, upon the blue ground, shows as green. Unless the nitric acid be very dilute there will be a risk of discharging the blue and obtaining only a yellow.

Besides the styles indicated above a variety of others are in existence, produced by modified treatments and by combination of others with the blue. Besides the colors which may be blocked in, mordants for madder, garancine, and other dye woods may be blocked or machined ; indigo being almost the only coloring matter which will stand a madder dye.

All the styles of work produced by dipping are cheap and low class ; the blue thus fixed possesses extraordinary stability, but very little brilliancy ; its chief consumption is consequently among the poorer classes, and it never enters into high class work. Some of the indigo colors in the following processes have a lighter and more pleasant appearance ; but it is remarkable that no attempts to communicate brightness to indigo colors have yet been successful :—

Methods of Fixing Indigo by Printing.—There are several very ingenious methods of preparing indigo, by which it can be printed in designs upon white calico without the necessity of having recourse to the expensive and clumsy system of resists, which is, in fact, only tolerated in any style because of the want of a method of directly applying indigo which will yield the deepest shades. The chief methods of printing indigo are here enumerated.

China Blue.—This blue derives its name from having a resemblance to the shade of color upon old china ware. It is produced by a process which is so extraordinary in its results that it is impossible to conceive how it originated. The indigo in its natural state is very finely ground, and mixed with deoxidizing bodies, such as sulphate of iron, acetate of iron, orpiment, and protochloride of tin. In old receipts a great number of apparently useless substances are prescribed. Good results can be obtained by the addition of sulphate of iron alone to the indigo. As thus applied to the cloth, the indigo could be removed by washing, because the deoxidizing agents are in the inactive state, and the indigo is not brought into solution. It is necessary that it should be deoxidized or dissolved in order to contract an intimate union with the fibre ; for this purpose it undergoes a treatment somewhat analogous to that employed in dyeing from the same coloring matter.

After printing and ageing, to bring the color into a proper condition, the piece is first dipped in clear lime water; this serves to wet it out and to form an insoluble or difficultly soluble compound of the gum, paste, or starch of the thickening with the lime. Since my experiments have convinced me of the existence of such a compound, the researches of M. Kuhlmann, of Lille, have demonstrated that such compounds are always formed under proper circumstances between starch and the similarly formed bodies and the alkaline earths. It is to the existence of this coating, pervious to water, but holding like a fine net the indigo particles in their place, that the first portion of the fixation of the china blue is owing. The piece is next placed in the copperas vat for ten minutes; the lime water which adheres to the cloth precipitates a little oxide of iron over its whole surface, but it does not appear that the slightest dissolution or deoxidation takes place. The piece is now moved to the lime vat which has been raked up, and being plunged in is moved about with a gentle motion. What takes place here is at first a precipitation of all the oxide of iron of the copperas upon the cloth, along with sulphate of lime together forming a thicker coat of slime; as soon as the whole of the iron is precipitated, the excess of lime begins to act (in conjunction with the protoxide of iron) to deoxidize and dissolve the indigo. The dissolved indigo has no tendency to spread beyond the design, for the reason that it is surrounded with fibres saturated with water, containing also a species of coagulum of gum and lime, and everywhere filled up with the slime of gypsum and oxide of iron—no distant capillary motion is possible, and it is absorbed by the fibres in close contact with it; another reason is that an excess of lime prevents the solubility of the indigo to a great extent, and as an excess is present, the dissolved indigo cannot pass away from the spot where it was formed; in addition to this, there is the positive attraction of the vegetable fibre, which is strong enough to take away indigo from lime, and keep it intact even in the presence of the agents which could dissolve free indigo if presented to them. These considerations are, I believe, amply sufficient to account for the retention of the dissolved indigo on the spot where the undissolved blue indigo was placed in the printing. To complete the process, the piece is again dipped into the copperas, and again into the lime several times, the number of dips depending upon the depth of the color, the last dip is a long one in the lime. The pieces at the end of the process are covered with slime to the thickness of nearly half an inch; this is partly removed by a wincing in water, and then the pieces are turned over into sours, and left for several hours, so that all the irons may be removed from the cloth; they are then washed and cleared in weak soap and warm sours.

Very dark shades cannot be obtained by the China blue process. It is a fast color, but expensive on account of the time and labor it requires. In the old process of dipping on frames, which gives the best results, it takes about two hours to accomplish the dipping for the light shades, and twice that time for the darker colors. One man can only dip two long pieces at a time, or four short ones, so that it is one of the most tedious and costly processes in use. A more modern method of raising is carried out in many places, in which the lime and copperas vats are supplied with rollers and the pieces passed through quickly; time is saved, but it is at the expense of quality.

A good quality of indigo should be used for a China blue, because in inferior qualities the resinous matters interfere with the regularity of the shade, but it is not necessary to have a first-rate quality.

Dark China Blue Color.

24 lbs. indigo,
5 gallons iron liquor,
24 lbs. green copperas,
6 lbs. orpiment.

The indigo and iron liquor are ground together for three days in an indigo mill, and then the copperas and orpiment added and the grinding continued for three days more, and then three gallons of gum water added and the grinding continued until perfect mixture is accomplished. This gives the dark color; the lighter shades are obtained by reducing with gum water.

Another China Blue—Dark.

2 quarts water,
2 quarts honey,
4 lbs. green copperas,
2 lbs. starch.

Other receipts include nothing but muriate of iron, indigo, and gum water. Protoxide of iron appears to be the only really necessary agent in addition to the indigo; but the other substances used may be useful under particular circumstances. The copperas vat is made to stand between 1 and 2° Tw., and is freshened up with 5 lbs. copperas after every piece dipped; the lime vat is set with about 2 cwt. of lime, and freshened up with a gallon or two of thick milk of lime every piece or two pieces dipped. For light patterns the following is the time and order of the dippings:—

First dip in clear lime water . . . . 15 min.
Second dip in copperas vat . . . . 15 "
Third dip in clear lime water . . . 10 "
Fourth dip in copperas vat . . . . 5 "
Fifth dip in lime vat, well raked up . .10 "
Sixth dip in copperas vat . . . . . 5 "
Seventh dip in lime vat, well raked up 10 "
Eighth dip in copperas vat . . . . 10 "
Final dip in lime vat, well raked up . 25 "

Dark patterns require about twice the time in each vat. A resist for China blue can be obtained by thickening a mixture of sulphate of copper, acetate of lead, and nitrate of copper, and adding lime juice. The following proportions may be employed:—

Resist for China Blue.

1 gallon water,
2 lbs. sulphate copper,
2 lbs. sugar lead,
2½ lbs. flour; boil, and when nearly cool, add
5 lbs. nitrate copper crystals,
1 quart strong lime juice.

Pencil Blue.—This blue receives its name from the manner in which it was formerly applied to the cloth, viz., by means of a fibrous matter like an artist's pencil. It is a tolerably old color, and no improvements have been made upon its preparation since the earliest account we possess of it; it is subject to the same difficulties in the application and requires the same precautions. Pencil blue consists of indigo in the deoxidized and dissolved state. It is made by heating a mixture of finely ground indigo, orpiment, and potash. The orpiment and potash together form a powerful deoxidizing mixture, which speedily reduces and dissolves the indigo. In a short time after mixing, the blue of the indigo will have disappeared, and given way to a yellow color, except at the surface, where the oxygen of the air continually revives the indigo, and causes it to assume a coppery blue color. The avidity of this mixture for oxygen, which restores the indigo to its blue insoluble state, is so great that it cannot be exposed a moment to the air without being covered with a scum or pellicle of blue indigo. It is this property which makes it so difficult to apply pencil blue in a regular and satisfactory manner. As soon as the block or roller leaves the color and enters the air the surface of the color is covered with a scum of indigo, which, being insoluble itself, cannot enter into the fibre of the cloth, and, being on the top of the soluble color, is a hindrance to its entering into the fibre. Peculiar arrangements have to be made in applying this color to prevent contact with the air; they are all more or less defective, and the results are seldom regular. In the old method of applying it with a pencil the pressure upon the fibres of the pencil containing the blue could drive the film of the indigo aside at the very moment when the pure color beneath could enter into the cloth and unite with the fibre; but, in either block or roller printing, the cloth and design are perpendicular to each other, and the oxidized face of the color comes in flat contact with the cloth; the insoluble particles being deposited first hinder and prevent the fixation of the others. Many ingenious constructions of reservoirs and sieves for block-printing have been made specially for this color, and, with the exercise of great care, it has been possible to print with some of them and obtain tolerable uniform results; but they have not admitted of general application. The difficulties in the way of roller printing are greater; and, though it is possible to print it like any other color, and obtain fast blues, it can only serve for styles of work where there is no particular demand for uniformity of shade. It is not possible to print five pieces of one shade in such a manner; and generally a single piece will be found to have two or three shades in it—an irregularity which would utterly condemn it for trade purposes.

The following receipts are adapted for pencil blue, for roller or block :—

Dark Pencil Blue.

2 lbs. ground indigo,
2 lbs. orpiment,
1 gallon caustic potash at 36° Tw.,
2 lbs. gum Senegal.

The orpiment, indigo, and potash are boiled together in an iron pan (a copper pan would be rapidly destroyed by the sulphur), until the blue of the indigo has entirely disappeared; the gum is then added.

Light Pencil Blue.

1 lb. indigo,
1 lb. orpiment,
1 gallon caustic potash at 26° Tw.,
2½ lbs. gum Senegal.

Treated as above. Lighter shades can be obtained by diluting with gum water mixed with caustic potash.

Another Dark Pencil Blue.

8 gallons water,
6 lbs. carbonate of potash.
5 lbs. quick lime,
5 lbs. indigo,
6 lbs. orpiment.

This mixture kept hot two hours until all the blue color has gone, then allowed to settle, the liquor drawn off the sediment and thickened with gum.

Another Pencil Blue.

1 lb. ground indigo,
2 lbs. granulated tin,
1 gallon caustic potash at 30°; boil
    for two hours, strain, and add
2½ lbs. gum.

Another Pencil Blue.

1 lb. grape sugar or glucose,
1 lb. ground indigo,
1 gallon caustic potash at 26°.

Heat to 120°, and allow it to stand for a few days in a warm place until ready, then thicken with 2½ lbs. gum.

Gas Blue.—The economical advantages of being able to apply a good dark blue by the roller are so apparent, that it is not surprising that many efforts have been made to over-come the difficulties and obstacles in the way of the pencil blue. One of the most ingenious was made by Woodcroft, and patented in 1846 (June 22); acting under the knowledge that it was the oxygen of the air which, both on the roller and on the color, was the obstacle to its neat and proper printing, he proposed to construct covered apparatus to surround the color box and roller, and to receive the piece when printed; and in all these spaces where the air was in contact with the color, he proposed to expel it by introducing common coal gas from the gaspipe—this gas, not containing any oxygen, could not act upon the color, which it was supposed would then have a fair chance of entering the cloth. The idea was good, but, as may be imagined, the application was difficult in the extreme. Large sums of money were expended in giving it a trial, and all the resources of an extensive concern, com-bined with the best chemical information, brought to bear upon it, but without avail. The exact point where it failed is not known, but it was deficient in many respects. It necessitated expensive alterations, and prevented the machine printer having that constant eye upon the color, roller, and cloth, so necessary to success; a large escape of gas into the machine room was unavoidable, it annoyed the workmen, and rendered them either unable or unwilling to pay close attention to the printing; it was necessary to wind the pieces on a roll soon after leaving the roller, and in a vessel filled with gas. It was extremely difficult to prevent them marking off upon one another. Beyond these difficulties, which seem only to be of a practical nature and surmountable by perseverance, there were others which were more discouraging because they were unaccountable. The shades given by the same color at the same printing were irregular; one piece would be several shades lighter than another, and the same piece a good color at one end and bad at the other; there were streaks and cloudy work from bad cleaning of the roller by the doctor, and so many mishaps, that it was dropped in despair, and has never since been worked.

Glucose Blue.—A new method of applying indigo was patented, December 8, 1857, by Ward. The idea, in this method, is to accomplish a kind of China blue dyeing without the vats. The indigo in the blue insoluble state is mixed with a deoxidizing matter (preferably, the substance known to chemists as grape-sugar or glucose) and alkali, as soda and lime. Here are all the elements necessary to the deoxidation and solution of the indigo, but heat is required to bring them into operation. The color is applied cold to the cloth, and as soon as the piece leaves the printing machine, and without drying, it is passed into steam for about a minute. The heat of the steam causes the chemical action to take place; the indigo is deoxidized and dissolved, and enters the fibre of the cloth. It then requires washing and oxidizing to bring up the blue color. The plan has been put in practice in two or three print works, but it has not been successful. The only use to which it can be advantageously applied is when it is desired to combine madder, or other dye colors, with the blue. It might enter into competition with the dip blue in resist styles, having a possible advantage in saving the cost of the resist and the printing. I have seen dark blues from this process, but they are of a rather coarse nature, resembling dip blue more than any other style. I am informed that there are great practical difficulties in the way of applying other colors and mordants at the same time as the blue, and the steaming of the pieces before they are dry leaves them open to many serious accidents.

The published process will have to be much modified before it can be expected to take its place as a regular plan in calico printing.

Fast Blue or Precipitated Blue.—Under the name of fast blue, a color is obtained from indigo upon principles differing from any of the previously given processes. The indigo is applied upon the cloth in the white deoxidized state, but not in the soluble state. It differs from China blue by the indigo being deoxidized, and from pencil blue by its being insoluble. It is prepared in several manners: indigo, soda, and granulated tin are boiled together until all the blue of the indigo has disappeared; or ground indigo is mixed up with crystals of tin and soda in the same manner until it is all dissolved. When all dissolved, it is precipitated by the addition of an acid or a salt of tin; the white indigo settles down, mixed with oxide of tin, as a grayish pulp. This is thickened and printed. The pieces are then passed through ash (either potash or soda); it requires the ash vat to be strong in order to get good results. The alkali of the ash vat renders the white indigo momentarily soluble, in which state it is absorbed by the fibre, and constitutes the blue color when oxidized by wincing in clear water. Although custom has given the title of *fast* to this color, it is really the loosest of all the indigo colors. This may be owing to the shortness of time given it to enter the fibre, and the excessive alkalinity of the fluid, in presence of which the indigo becomes soluble, and has to contract its adhesion to the fibre. The depth of color which can be applied in this way is limited. It is principally used in chintzes and furnitures as an addition to the other colors. If the other colors are liable to be affected by the alkaline nature of the vat they have to be protected by a paste; but, as the fast blue is mostly used as a cover, the same paste which resists the blue will resist for a sufficiently long time the action of the alkaline raising vat.

The following receipts illustrate this application of indigo in printing:—

Precipitate or Fast Blue.

1 gallon caustic potash at 20°,
1½ lb. indigo,
1¼ lb. crystals of tin,
½ gallon water; boil down to one gallon, and add
2 gallons thick gum-water,
½ gallon muriate of tin at 120°,
1 quart muriatic acid.

Another Method.

4 lbs. ground indigo,
8 lbs. copperas,
12 lbs. lime,
30 gallons water.

These materials are put together as a blue vat, stirred every half hour for twelve hours, then allowed to settle for twelve hours. Fifteen gallons of the clear liquor are drawn off and precipitated by addition of two quarts of muriate of tin at 120°, the precipitate drained to a pulp, one part of this mixed with two part of thick gum-water, and 2 oz. of crystals of tin per gallon added.

Pencil blue, if precipitated by muriate of tin, will also answer for this style.

Indigo Sulphate, *Extract of Indigo, Saxony Blue, Soluble Indigo, &c.*—This substance is obtained by treating ground indigo with strong sulphuric acid; a chemical action takes place which entirely alters the constitution of the indigo, but without destroying its color; the sulphuric acid or its elements, or some of them, combine with the indigo and form compounds, which are soluble in water. The coloring matter so produced has affinity for woollen and silk, with or without mordant, but none for cotton; it is chiefly used in woollen dyeing and printing as the most convenient blue part for all compound shades. It does not produce fast colors like indigo; in fact, though directly derived from indigo, it is as different from it as any other coloring matter can be, nor can it by any known means be restored to its original state of indigo. There are many modifications in the method of preparing it, but the following receipts will be sufficiently illustrative:—

Extract of Indigo.

10 lbs. of rectified oil of vitriol,
2 lbs. of ground indigo.

Place the oil of vitriol, which must be the very strongest, in a mug provided with a close cover, and add the indigo about two ounces at a time, stirring well up at every addition until the whole is added, then put the mug in a warm place; if the heat is about 80° it will require 48 hours or thereabouts to effect the action, but if kept at 150° it will be accomplished in 12 hours. Practically, the completion of the operation is ascertained by rubbing a little of the paste upon a piece of glass and viewing it by transmitted light; if all is dissolved and the color is transparent, the action is over and the solution made.

The liquid or paste thus produced is intensely acid, and not generally applicable in dyeing until the excess of acid is removed; this is best done for common purposes by diluting with three or four gallons of water and adding common salt; the blue at first dissolved is precipitated by the salt, the whole thrown upon a woollen filter and drained to a paste; if again dissolved, and again precipitated and filtered, it is obtained still more neutral. To purify it still further it is customary to filter it before adding the salt, by which the impurities contained in the indigo are removed. In woollen dyeing excessively acid liquors are sometimes used, and but little attention paid to purifying the preparation; but for printing and the fine-dyed colors a more careful preparation is required, which is called refined extract, neutral extract, or carmine of indigo. This can be prepared by the method given under ACETATE OF INDIGO, and the solution can be still further refined by the process given under BLUE, DISTILLED. The preparation of the sulphate of indigo admits of many variations in the quantities and methods employed; but the differences are more apparent than real. The following method, considered as yielding a very suitable extract for combining with the bark or weld yellow to produce a green is taken from Persoz iii. 395, and has some peculiarities. One pound of Nordhausen or smoking oil of vitriol is placed in an earthenware pot and heated by being placed in boiling water twenty minutes; half a pound of finely ground indigo is then mixed with it and carefully stirred for ten minutes, so as to thoroughly incorporate it; then three quarts of boiling water are carefully added in small portions, the stirring being continued, and lastly three pounds of acetate of lead are added, by which sulphate of lead is formed and the product, called acetate of indigo, remains in solution.

Within a short time Bolley has proposed the manufacture of a species of sulphate of indigo, by fusing bisulphate of soda with ground indigo instead of using sulphuric acid, and patented the process. From some specimens I have seen, it appears very well adapted for woollens.

Iodine.—This is one of the chemical elements which, though capable of producing several colored compounds, has not received any application in printing or dyeing on account of the excessive sensibility of its combinations to the actions of air and light.

Iron.—Iron is extensively employed in printing and dyeing for purposes in which it comes into contact with many chemical substances. The pure metal has no action upon most ordinary materials, but it is easily oxidized, and then it becomes chemically active, producing well-marked phenomena. The use of

iron vessels for dyeing was long considered impossible, but it
is now known to be the best material in all cases where no
acid liquors are employed. Alkaline fluids have no action
upon clean iron, but acids, even in a very diluted state, attack
and dissolve it. It is evident, therefore, that iron vessels cannot be safely employed in any operation which requires the
use of acids, or acid salts, which term may be taken as including all the salts of the metals proper. Iron vessels cannot be
safely used in color-mixing, or for storing colors, not even iron
mordants; but iron pans can be employed for alkaline colors,
such as pencil blue or aluminate of potash. Wrought iron is
with difficulty kept from rusting, on account of its lamellar
structure; cast iron is easily kept from rusting because of its
homogeneity, and is to be preferred for the making of dye-
becks and similar vessels. But cast iron soon rusts in the state
in which it leaves the foundry. To prepare it for the purposes
of the dyer the surface must be covered with some kind of pro-
tective coating. Ordinary paint would not answer well. Experience has shown that the best method of giving this pro-
tective coat is to boil water, containing some organic substance,
in the iron vessels. Cow-dung is most generally employed.
Madder, logwood, and other coloring matters answer the same
purpose. The effect seems to be that a combination between
the firmly attached oxide and some of the principles of the
organic substance employed is formed, which effectually cuts
off any further action of the oxygen of the air, and of course
any possibility of rusting. When, through long disuse an iron
beck has become rusty, it is necessary to scrape it well, and
boil it for some time with cow-dung or madder; by so doing
any free oxide is combined, and prevented from acting injuriously upon the subsequent materials used. Cast-iron vessels
used in bleaching are not considered safe unless the metal is
covered with a film of lime or carbonate of lime; this is readily
accomplished, in most cases, by scraping the metal and painting
it over with milk of lime. The contact of cotton goods with
rusting iron causes iron-moulds, or stains, resulting from the
partial fixing of the oxide of iron. A drop of strong hydrochloric acid suffices to remove a single spot of iron mould,
and may be applied to the cloth without injury to its strength.

Iron combines with oxygen in two proportions to form sali-
fiable oxides; the one called protoxide, the other peroxide of
iron. There are, consequently, two series of salts of iron,
which differ from each other so much in their properties that
they might be salts of different metals. For the sake of brevity,
the salts are called respectively proto and persalts of iron.
Sulphate of iron, or green copperas, is a proto-salt; and com-

mercial nitrate of iron is a per-salt; by using dilute solutions of these two salts the most conspicuous characters of the two classes of iron compounds may be studied. Yellow prussiate gives a light blue precipitate with the sulphate of iron, but a dark blue with the nitrate; the red prussiate of potash gives a dark blue precipitate with the sulphate, but no precipitate with the nitrate. Caustic soda produces a greenish olive precipitate with the sulphate, but a red precipitate with the nitrate. These are the respective oxides of iron. But the protoxide, when precipitated under favorable circumstances, is white; it readily combines with more oxygen, changing to green, olive, and eventually to the well-known rust colored oxide. When the buff color from acetate or sulphate of iron is being raised in lime the protoxide is precipitated, and the cloth has only a greenish color, but by exposing to the air, or acting upon it with oxidizing agents, it absorbs oxygen, and becomes the buff peroxide. The protosalts have a continual tendency to pass into the state of persalts, absorbing the necessary oxygen from the air or other substances; and there are cases, on the other hand, where the persalts pass, by losing oxygen, into the state of protosalts, but this is less usual than the contrary. The use of sulphate of iron in indigo dipping, and in China blues, depends upon the affinity of its oxide for more oxygen; it deprives the indigo of oxygen, thus altering it, and putting it into a state favorable for solution. M. Kuhlmann has drawn attention to some cases of what he considers the oxidizing effects of the peroxide upon calico. A rust spot is generally observed, upon dissolving the iron out, to be greatly weaker and thinner than the rest of the cloth. Calico strongly impregnated with buff is, upon the oxide of iron being removed, found to be more tender than is usual. These effects are attributed to an oxidation or slow combustion of the cloth, the oxide of iron acting as a carrier of oxygen to the organic matter. These points are of much importance in dyeing and printing, and deserve every attention.

Sulphate of Iron.—This salt has been used in dyeing from very early times, under the names of vitriol, green vitriol, and green copperas. It is a plentiful secondary product in some chemical manufactories; it is cheap, and not liable to be adulterated. It may contain salts of alumina and salts of copper to a limited extent, which would probably be prejudicial in some of its applications. A simple inspection is usually sufficient to know if a sample is good or not. It should not be wet or dirty; if dry, and with signs of rust, it is usually esteemed good, because such appearances indicate an old-made and well saturated copperas; but it is also possible for that character to be fraudu-

lently given to it. Practical dyers form opinions from other
appearances of the fitness of copperas for their uses, but it is
doubtful if they are of any value. The points to be attended
to in a chemical analysis are the acidity, which may be too
great; the amount of water which it contains; and the absence
or otherwise of alumina salts, which are liable to injure certain
colors for which copperas is used. Copperas is much used in
the dip blue and China blue styles, in dyeing black on cotton
goods, and for numerous shades of gray, drab, and olive upon
heavy cotton goods. It may be used for the preparation of
acetate of iron by double decomposition. In calico printing
it is very little used. Some receipts prescribed "calcined copperas," that is, sulphate of iron dried in an iron pan, and heated
pretty strongly, with occasional stirring of the mass. If sulphate of iron be calcined at a very strong heat, only peroxide
of iron remains. A gallon of cold water can dissolve about
four pounds of sulphate of iron, and it is much more soluble
in boiling water.

Muriate of Iron.—A solution of iron in muriatic acid, marking
about 80°, is sold under this name. It can be made by dissolving iron in hydrochloric acid in a mug, or similar vessel, having
an excess of the metal present. When the solution is concentrated by boiling, it deposits crystals of chloride of iron, resembling the sulphate in appearance, but much more oxidizable.
The crystals are sometimes preferred to the solution, they are
likely to be purer and more neutral. Muriate of iron is only
sparingly used in printing and dyeing; it serves to obtain some
shades of slate and drab by means of catechu for madder and
garancine dyeing, and is used in a few combinations with salts
of manganese.

Nitrate of Iron.—There is a nitrate of the protoxide of iron,
but the commercial nitrate of iron is always a persalt. It is
made by dissolving old iron hoops, or smithy scales, in moderately strong nitric acid. It requires some experience to make
nitrate of iron successfully; if too much iron be added at once,
if the liquid becomes heated, if the acid be too strong or not
strong enough, there are a number of bye products formed,
and sometimes the whole spoiled. The chief points are gradual
addition of the metal in tolerably sized pieces, not to work
upon more than a carboy of acid at once, and to have it so
situated as to keep cool. Nitrate of iron is a dark red liquid,
marking about 90° Tw. When diluted with water, it should
remain clear without any addition, and should not give a blue
precipitate with red prussiate of potash; if the nitrate of iron
is over saturated it becomes turbid upon dilution, some oxide
or basic nitrate falling out, which may cause irregularity in

dyeing; if, on the contrary, it is too acid, it does not yield up its base in sufficient quantity or sufficiently rapid; the addition of a little acid to the water in the first case, and some alkali in the second, will prove of advantage. Nitrate of iron is extensively used in dyeing and printing; in the former it is the preferable iron mordant for all varieties of Prussian blue, and is used as the basis for black and gray. In calico and delaine printing it is used in a few steam and spirit colors. The only adulteration in nitrate of iron likely to occur is the substitution of hydrochloric acid for a portion of nitric acid. All samples that I have tested contain some chloride, but not more than five per cent. of the iron iu solution should be allowed to be in this state. The perchloride of iron, is not decomposable by fibrous matters to nearly the same extent as the pernitrate, and is consequently not worth so much.

Alkaline Solutions of Iron.—There are some compounds of iron soluble in alkalies; they have been proposed as mordants and said to answer that purpose. I have tried them all, but found no good practical result. I believe they are not in use. The pyrophosphate of iron, formed by mixing a very neutral persalt of iron with pyrophosphate of soda, is a white powder, insoluble in water but soluble in ammonia, forming a feeble mordant. M. Persoz states that this mordant will dye up colors in a bath of madder spent to ordinary mordants. I did not succeed in obtaining so desirable a result. Some arseniates of iron are soluble in alkalies, but do not yield anything of value as a mordant. Concentrated solutions of commercial nitrate of iron and carbonate of potash, when mixed, give a precipitate which re dissolves in excess of the alkali, forming a clear dark red solution.. This property of nitrate of iron was known to Scheele, and has been used in medicine, but not yet applied to dyeing or printing; it is a curious and unexplained reaction.

. *Ferric Acid.*—Under certain circumstances iron assumes a superior degree of oxidation, and seems to act the part of an acid, forming highly-colored compounds with alkalies. Ferric acid has not yet been isolated; it is easily decomposed, and does not seem likely to have any application at present.

Iron Liquor.—See ACETATE OF IRON.

Isatis Tinctoria.—The botanical name of the woad plant. (See WOAD.)

Isopurpuric Acid.—An interesting result of the action of cyanide of potassium upon picric acid, by which a brown or chocolate coloring matter is produced of considerable tinctorial power—dyeing wool and silk in deep rich colors without the aid of any mordant. The colors are, however very fugitive, and do not resist the action of steam. They have not yet been applied.

Ivory Black, *Bone Black, Animal Charcoal.*—This substance is distinguished by its powers of withdrawing coloring matter from solutions; and though not directly used in dyeing, it receives some applications in the arts which may probably be capable of extension. ·

Jamaica Wood.—A name for logwood, a portion of which comes, or did at one time come, from Jamaica.

Juice, Lime.—See CITRIC ACID, page 148.

K.

Kermes.—This ancient coloring matter is so little known at present in England that when some parcels of it were received in London a short time ago, not one of the brokers recognized it. It is more interesting, therefore, in an historical than in a practical point of view; although it appears to be much more extensively used in France than was supposed, since in 1856 about twenty tons of it were imported. Its coloring matter is similar, if not identical, with that of the cochineal insect; but it is poorer in tinctorial power, requiring twelve times as much to produce shades of equal fulness. Its principal employment appears to be in dyeing the turbans or fez of the Persians, and other oriental people; it produces a crimson, with a rich purplish hue, very much admired: the color is more stable than that obtained from cochineal, not being so readily stained or faded.

No coloring substance, or other material used in dyeing, possesses so great an antiquity, or has so many scriptural, classical, and historical associations as kermes. It is proved to have been known in the time of Moses, and mentioned by its Hebrew name of *tola* in the Pentateuch: its name, *coccus*, frequently occurs in the Greek and Latin writers; and from the use of this material in dyeing the imperial purple, the adjective *coccinus*, or *coccineus*, arose, applied to those who were entitled to wear such colors. From the ancient Greek and Latin versions of the New Testament, it appears évident that the robe in which the soldiers clothed and mocked our Lord was one dyed with kermes. Kermes, like cochineal, were supposed to be berries or grains, and colors dyed with them were said to be *grained*, or *engrained;* and, as the kermes colors were fast and durable colors, the term grained expanded in its signification, and meant a fast color whether dyed with kermes or not, and is even used in that signification to this day. But kermes are insects, and the word is Arabic, signifying "little worm;" and in the middle ages they were called *vermiculi*, and the cloth dyed with them, *vermiculata*, whence, through the French, we have vermilion, which is now employed as the name for one of the compounds of sulphur and mercury. The term crimson is also derived from kermes, through the Italian and French.

Knoppern, *Valonia Nuts?*—An excrescence upon the oak, something similar to gall-nuts, but more irregular in shape. They are used in Germany as a substitute for galls and sumac in saddened colors: they have also been tried in England, but have not met with general approval. They contain some as-tringent matter, but appear more suitable for tanning than for dyeing purposes.

L.

Lac-Dye, *Lac-Lake, Lac.*—This is an East Indian product, prepared from a resinous substance, which covers the branches of certain trees and shrubs. It is derived from a variety of the cochineal insect, which settles upon the branches in such num-bers as to entirely cover them ; a resinous exudation from the tree is excited by their punctures of the bark, and cements them to the branch and to one another, when having performed the functions of their existence, they die, and become so incor-porated with the resin, that it is difficult, if not impossible, to distinguish the insect. This resinous substance is called *stick-lac*, and it is from stick-lac that the lac-lake is obtained by dissolving out the coloring matter with water and alum, and precipitating it by alkalies. There appears to be a considerable difference in the value of lac-dye as imported, some qualities being worth at least twice as much as others ; the higher priced varieties are taken for home consumption—the conti-nental consumers believe that the medium and lower priced sorts are for their prices more economical in use.

Lac-dye is only employed in woollen and silk dyeing; it yields the same colors as cochineal, for its pure coloring matter is chemically identical with that of cochineal ; but on account of the various processes to which it is subjected in extraction, the colors it gives are not quite so brilliant as cochineal, but, on the other hand, they appear somewhat more durable and better fitted for rough usage. Its tinctorial power is variable, according to quality, but the better qualities are from one-half to one third as strong as cochineal. On account of the coloring matter being in a state of combination with alumina in the lac-dye, it is not available for dyeing without some preparation, which essentially consists in acting upon it by an acid or a strongly acid salt, which attacks the alumina, and thus isolates the coloring matter, and renders it capable of combining with the mordanted goods. Sulphuric acid is used for this purpose, but more generally muriate of tin, containing an excess of acid, is employed.

The method of applying lac-dye in practice is as follows:—

Preparation of the Lac.—It is ground in a coffee mill or pounded in a mortar, and passed through a fine sieve; then, for every 10 lbs. of it, one gallon of water, and one gallon of oxymuriate of tin are added, and the whole carefully and completely incorporated into a homogeneous mass. The mixture should stand not less than 24 hours, but preferably should remain for a week before using, and should not be kept longer than a fortnight. Instead of the above process, about a gallon and a half of water and half a gallon of sulphuric acid may be employed; or muriatic acid may be employed at full strength. The only object of this treatment is, as before stated, to liberate the coloring matter from its combination with alumina, and though tin salts are chiefly in use, the simple acids are quite sufficient for the purpose. After a sufficiently long digestion, the pasty mass is mixed with hot water, and the clear liquor used in dyeing.

Dyeing with Lac-dye.—The mordanting is exactly the same as for cochineal scarlet, taking for a piece of merino of 10 lbs. about $1\frac{1}{2}$ lb. of tartar and $1\frac{1}{2}$ lb. of oxymuriate of tin or lac spirits; the piece is worked in this for half an hour, and then a quantity of prepared lac liquor, equivalent to $1\frac{1}{2}$ lb. of lac-dye, added to the dye, the piece again entered and kept near the boil for three-quarters of an hour. The cloth is afterwards carefully treated with warm water to remove from it any adhering particles of resin.

Practically, lac-dye is scarcely ever used singly for bright colors, but generally in combination with cochineal. The cloth, after being dyed as above in lac, is "topped" with cochineal, by giving it a few minutes in a fresh beck, or by adding cochineal to the same dye beck. Scarlets, scarcely inferior to those obtained from pure cochineal, may be thus obtained at a reduced expenditure.

Although some attempts have been made to prepare lac in a condition suitable for the use of the printer, I am not aware that they have been successful; some of these preparations, under the name of "cochineal substitutes," have been examined by me. I found them unfit for the best work, the shade being considerably inferior to cochineal, and not presenting any considerable advantage in cost for lower styles.

Lactarine.—This is the name employed in England to designate the curd of milk prepared in the dry state for the use of the calico printer. It was introduced by Pattison, and patented November 2, 1848, and is now extensively used for fixing pigment and aniline colors, as a substitute for albumen.

Lactarine is dissolved by means of ammonia or other weak alkalies, but preferably ammonia; the coloring matter or pigment is mixed with it, printed, and fixed by steaming. There are some points in the application of lactarine which do not appear to be very well understood, so that most contradictory reports of its fitness as a vehicle have been made. It does not appear to be equal to albumen under the best circumstances, and is particularly liable to coagulation when dissolved, by which the color made with it is rendered completely useless. For fixing the new aniline colors, it appears to be not only cheaper, but better than albumen, working more easily, and finishing off softer, but not fastening them so completely. In order to preserve the dissolved lactarine fit for use it should be kept as cool as possible, the mug containing it being placed in cold water. A friend of mine, employed on a print works in a warm climate, was much embarrassed by the spontaneous coagulation of his lactarine solution when standing in the color shop, but completely remedied this defect by keeping it in a box packed with ice. However, even with this assistance, it is not advisable to dissolve more at once than is required for the day.

Cheese which does not contain much fat, when digested with ammonia, produces a solution capable of replacing lactarine, and is employed in some print works as a substitute.

Lake.—A lake is a colored body produced by the combination of a coloring matter with an earthy or metallic basis. The only ordinary bases of lakes are alumina and tin; but some lakes have an iron basis, and others may have the oxides of lead, zinc, antimony, etc., as bases. Although nearly all steam and spirit colors actually consist of lakes of tin and alumina diffused, suspended, or temporarily dissolved in the thickened color, the separate manufacture of lakes for calico-printing purposes has not yet met with much success. For paper hangings, where penetration of the color is not an object, prepared lakes are extensively employed; but for printing fabrics, where the coloring body must be in so finely a divided state as to easily penetrate the fibres, the application of ready-made lakes presents considerable difficulty, on account of the impossibility of diffusing them with sufficient uniformity through the thickening. When tin crystals or alum are stirred into a thickened extract of logwood, cochineal, or other coloring matter, the tin or alumina do undoubtedly form a lake with the coloring matter, but of such excessive tenuity or fineness as to differ very little from a solution, and the colored compound is readily absorbed by the fibres of the cloth. Coëz obtained a patent, March 28, 1854, for the preparation of lakes for the use of printers. His process of manufacture is understood to consist in adding alumina in the gelatinous state to a decoction of the dyewood, and keeping warm until a combination has taken place between the coloring matter and alumina; the lake produced, settling out, drained, and kept in a state of pulp. For use as colors, these lakes simply required mixing with gum water and tartaric acid, oxalic acid, or crystals of tin, the object of these additions being to act upon the alumina and partly dissolve it, so as to facilitate the further division or perhaps solution of the lake; for the rest, the colors were treated just as ordinary steam colors. I have seen and employed the prepared lakes of M. Coëz, and obtained fair results, but nothing that seemed to render them preferable to the ordinary methods of color mixing, while they were not so regular, nor so much under control, as colors prepared on the spot. It is very doubtful if, under ordinary circumstances of care and skill in the color house of a print works, there would be any pecuniary advantage in the use of prepared lakes; it should be always cheaper to employ the raw material at first hand.

The most ordinary method of preparing lakes consists in adding alum to the solution of coloring matter, and then adding crystals of soda and heating; the alumina is precipitated and combines with the coloring matter. By boiling a decoction of coloring matter with acetate of alumina a lake is also produced. The tin lakes are prepared by using tin salts instead of alumina or the gelatinous oxide of tin. Some of the lakes I have had occasion to employ in my experience are the following:—

Sapan Wood Lake or Pulp.

56 lbs. rasped sapan wood,
10 gallons water boiled three times,
2½ lbs. alum,
1 lb. sulphate of copper.

The clear decoction of sapan wood being mixed with the salts and heated, precipitated a lake which contained both copper and alumina. It was used in the preparation of a brown color for delaine.

Logwood Lake.

10 gallons logwood liquor at 9°,
1½ lb. sulphate of copper }
1 gallon hot water, }
¼ lb. bichromate of potash, }
¾ lb. crystals of soda, }
1 gallon water, }

Mixed together, and the resulting pulp drained until it measured four gallons. This was used as an ingredient in preparing a black color for delaine.

Fustic Lake for Brown.

56 lbs. fustic made into decoction,
2½ lbs. alum,
1 lb. acetate of lead.

The alum dissolved first, then the acetate of lead added and heated, the pulp drained and kept for use.

Lamp Black.—This pigment is the soot obtained from the imperfect combustion of resinous or oily bodies; finer qualities are obtained by burning turpentine, and it is said that the black used in the preparation of artists' Indian ink is derived from the combustion of camphor. Spanish black, drop black, and some other kinds used as pigments, are obtained by burning peculiar substances, not for the smoke, but for the charcoal they leave. Lamp black intended for printing purposes requires a preparatory treatment to remove certain gritty particles it contains, and which would produce abrasions of the doctor and scratches on the roller. In the first place, it must be calcined at a moderate red heat in close vessels, to destroy any volatile matter present. When cool, it is mixed with strong sulphuric acid till it forms a thin paste, left in contact with the acid for twenty-four hours or longer, then mixed with water, and well washed until the acid is removed. This treatment leaves the black soft and fine; it gives a good black color with drying oils; but in calico printing it is used with albumen for shades of gray and drab, which are very pretty in combination with other pigment colors. It is best adapted for furniture styles, hangings, etc.; it does not resist washing very well, but never fades in the light, a desirable quality for certain classes of colors.

Lawsonia Inermis.—The botanical name for a plant, the leaves of which have been used from the most ancient times amongst the oriental nations for the purpose of dyeing the finger nails of a reddish color: it is the Henna of the Arabians. A French chemist has recently taken out a patent for the application of this substance in dyeing blacks; but, from all the experiments with which I am acquainted, it is not likely to be much used for that purpose on account of its price, and from the want of my distinct advantage in the color it produces. The turks have long employed this substance for dyeing horse hair and leather.

Lavender Colors.—The color of the flowers of the lavender plant—a bluish lilac. This shade is obtained on cotton goods by mixing blue with the lilac color from logwood. On silk, lavender shades are obtained from archil and cudbear, as well as from logwood. By dyeing a safflower pink on the top of a Prussian blue, very agreeable lilac and lavender shades are obtainable—the hue depending upon the relative depths of the blue and red employed. The lilac, or light purple colors, are taken as standards from which to obtain lavender colors.

Lead.—Lead, as a metal, is more indifferent to the action of chemical materials than copper; but, owing to its softness, it cannot be applied well except as a stationary fixed apparatus. It is well adapted for hot acids, holding tight, and lasting for years, where wood and stone have both yielded. It must be well supported on account of its softness, and, of course, without solder, the different joints being made by fusion, or burning, as it is technically termed. It does not withstand the action of aquafortis, but both vitriol and spirits of salts are without action upon it.

Lead has three combinations with oxygen, but only one of these forms salts, and this the protoxide, composed of single atoms of lead and oxygen. When pure it is white; with some little impurity, it exists in litharge, as a commercial article. Litharge has some few applications in dyeing and printing. In dyeing it is used for obtaining chrome oranges: in printing, it serves to purify caustic potash from sulphur, and to prepare basic acetate of lead.

Nitrate of Lead is made by dissolving litharge in hot aquafortis to saturation: it is not easily prepared, except on a large scale, in a state of purity. It forms white crystals, very soluble in water, in which they should dissolve without leaving any residue. They are very pure as sold by respectable drysalters, and, not as far as I am aware, subject to any regular adulteration. Nitrate of lead is employed as a mordant for chrome yellow and orange; for the purpose of preparing a number of soluble nitrates from their sulphates, and in mixing of the murexide purple, etc.

Sulphate of Lead is an insoluble salt, and precipitates whenever nitrate of lead or sugar of lead is mixed with any liquor containing sulphuric acid. It forms the bottoms from the making of red liquor when the acetates of lead are used, and from buff liquor in the same way. It is because the sulphate of lead is insoluble in water that sulphuric acid is used to pass pieces in, which are mordanted for chrome orange; the lead does not then wash off, and the pieces can be entered clean and free from gum, etc., into the chrome liquor. Sulphate of soda is sometimes used, but this is when there is some other color on the cloth that the acidity of the vitriol would injure; the effect produced is the same, viz., the formation of a sulphate of lead. Although sulphate of lead is insoluble in water, it is a remarkable fact that, if the pulp be applied to calico, it enters into an intimate connection with the fibre, and after drying and hanging some time it cannot be removed by washing in water. Good solid chrome oranges can be raised in such a manner. It is used extensively as a resist in indigo dipping.

Chromates of Lead.—There are two chromates of lead—the dichromate, or basic chromate, which is of an orange red, and the mono-chromate, which is yellow. The latter can be transformed into the former by heating it with an alkali, when it loses chromic acid. It is worthy of note that that chromate of lead which is the most highly colored contains the least amount of the coloring acid, and that the nature of the shades are exactly opposed to those of the chromates of potash. Both salts are in extensive use as pigments; the red chromate is a good deal used as a dye, while the yellow has not much use, but is frequently required, especially in printed indigo styles. In dyeing chrome orange the yellow chromate is generally produced first, by passing the mordanted cloth through bichromate, until it has well taken up the chromic acid; it is then changed into orange, by adding a proper amount of caustic soda to the liquor, and keeping in the pieces till they have taken the right shade. The orange could be as easily produced at once, and very frequently it is done so, by using yellow chrome salts, or converting the red chromate into the yellow, by adding a sufficient amount of caustic alkali. Or the yellow chromate of lead, instead of being changed into the orange in the same beck it is dyed in, is taken out, washed, and passed into boiling lime water, which changes it. In the case of employing caustic, care must be taken that no excess is used, for if there is more than necessary it robs the color, and if the excess is considerable it discharges the color altogether. The action of the lime water and soda are similar; they deprive the yellow chromate of lead of part of the chromic acid, reducing it to a compound containing less chromic acid in proportion to the lead; and if the action is carried too far the alkali will remove the whole of the chromic acid, leaving upon the cloth only oxide of lead, which is white. It depends upon practical considerations whether it is better to use the one or the other method of obtaining the chrome orange, whether the orange should be obtained at once or by conversion of the yellow; in piece dyeing the former is the usual plan, while in calico printing the latter is the method usually followed.

Red Lead is a mixed oxide of lead. Though a bright-looking color, it is not fit for printing on cloth; probably, if there was any means of forming or producing it on the fibre of the cloth it might be valuable, but at present the only method known of making is to roast the litharge at high temperatures; it cannot be made in the wet way.

The Peroxide of Lead has a deep puce or chocolate color, and can be fixed on cotton fibre. The process consists in mordanting in acetate of lead, fixing in lime water, and then passing in chloride of lime or soda until the color is raised. This color is never worked now, because similar shades are more easily and economically obtained by garancine.

Lead Acetate.—See ACETATE OF LEAD.

The oxides of lead have at various times been tried as mordants, but they appear unfitted for the purpose. Messrs. Perkin and Gray patented, May 21, 1859, a process for fixing the aniline purple upon a lead mordant, obtained by printing acetate of lead, and fixing in a mixture of soda crystals and ammonia; the shades obtained are good, but the process as a whole was not successful. All colors which contain lead as a constituent part are liable to injury from sulphurous gases in the air; the browning or blackening of chrome yellows and oranges at the edges of the piece is due to this cause; the whites of indigo styles, from which the lead has not been wholly removed, are frequently discolored from the same cause, which operates continually against the use of lead in printing.

Leiocome.—A species of gum substitute of French origin. It is analogous to the soluble gum of the British printers, being made by the action of heat and acids upon farinaceous matters.

Lemon Juice, *Lime Juice.*—The uses of this article depend upon the citric acid which it contains. (See CITRIC ACID, page 148.)

Libi-Davi, *Livi-Dibi, Divi-Davi.*—An astringent substance, suggested as a substitute for sumac and gall nuts in dyeing, but found to be too poor in tannic acid. It is used in tanning.

Lichens.—The tinctorial lichens are very widely diffused over the surface of the globe; the greater portion of those used in England are collected in the Canary Islands, on the shores of the Mediterranean, or in the mountainous districts of Spain and France. Until within the last few years the methods of obtaining color from them were rude and but little understood; already some successful attempts have been made to improve both the hue and fastness of the colors yielded by lichens, and it may be hoped that, with the aid of science, some considerable improvements will yet be made. It is remarkable that there is no coloring matter ready formed in the lichens, but it is produced by the combined action of air and ammonia upon some colorless principles contained in them. Archil, cudbear, and litmus, are the coloring substances obtained from the lichens.

Light.—Light seems to be a chemically active agent, inducing decompositions and changes in salts and neutral substances. It either acts itself, or its presence is the cause of action in numerous cases interesting to the dyer. I do not know what credit should be given to the assumed importance of bright light in dyeing colors of the finest quality. It is pretended that if all other things are equal, the brightness or dulness of the daylight influences the product of the dyeing. It is well known that plants which grow in dark places are pale colored, and that the most sunny climates produce the brightest colors in the vegetable and animal kingdom. But there is nothing in the case of a growing plant or animal comparable to dyeing. In the one case the color is being formed, in the other it is only being transferred. It is within the bounds of possibility that a formation of coloring matter may take place sometimes in dyeing, but I know of no case where this is effected by light. In nearly all coloring matters used in the arts, the color is fully developed before the dyer uses it, and it is interesting to consider in how many cases without the access of light; the heart of logwood cannot be supposed to have been influenced directly by the rays of light; we know that the color of indigo is only developed after the death of the plant; and in madder root the coloring matter must have been formed in perfect darkness. Yet practical dyers insist that the finest colors can only be produced in good clear weather; the most beautiful dyed silks and velvets of France are, I am told, produced by small dyers, who perform the final operations in clear bright weather and out of doors, turning the goods over and over in the dye, lifting them to meet the sun's rays, and, regardless of prescribed times and quantities of material, handling them till they find them finished.

Whatever doubts may be entertained upon the beneficial effect of light in certain circumstances, there can be none as to its destructive action upon coloring matters, under almost any circumstances, when prolonged beyond a short time. In most cases of rapidly fading colors, such as safflower pink on cotton, or archil and cudbear shades upon silk, it is the light and not the air which destroys them, or at least the light is the more rapidly destructive of the two; the direct rays of the sun, being the most concentrated form of light, are the most active, but a bright diffused light is very energetic. A safflower pink on muslin may become nearly white in three or four hours' sunshine, and a peach-colored silk ribbon, dyed with archil, is destroyed in about the same length of time. That it is the light, is demonstrated by the preservation of the color in the folds, or protected parts of knots and bows formed by the material; the air could have access to these almost as readily as to the exterior portions, but they are nearly uninjured. Light is an imponderable body, and as such, leaves no marks behind it detectable by the balance. It is not clear whether these transformations are simply a change in the chemical or molecular structure of the coloring matter, or whether the light induces an oxidation or deoxidation of the coloring principles. The first view seems the most probable, from the knowledge we have of the action of light upon chemical substances, though the second is not without probability. The action of light upon substances in general may be profitably studied by the colorist; it seems probable that some means of protecting the easily alterable colors may be devised from a knowledge of the laws and properties of light. The action of light upon the salts of gold and silver is the foundation of photographical art; and many curious particulars have been discovered in it with regard to sensitivizing, and, if the expression is allowable, desensitivizing, the layer of silver salt. The deposited iodide or chloride of silver is so easily acted upon by light, as to necessitate the greatest precautions in keeping out a single ray from the closet in which the processes are conducted; but if the light be made to pass through a yellow medium, such as stained glass, it loses all its active chemical properties, and the prepared plates may be exposed to it and handled with the greatest ease and safety. There are other cases in which an additional film of another material renders the sensitive one insensible to the action of light. It is possible that the very delicate and fugitive vegetable colors may, by some practical process, be similarly desensitivized. This is a line of research very difficult, no doubt, but in which there is both hope of success and rich reward.

Photography has not yet any application in calico printing; but it may be interesting to know, that it is possible to print pictures by photography and dye them up in madder and other dyewoods. I have done this several times and in several ways with iron salts. Calico may be prepared with the ammonia-citrate of iron, dried in the shade, and then covered with the object or picture, and exposed to the rays of the sun in a photographer's copying press; the iron is partially fixed upon those parts exposed to light, the calico may then be passed in yellow prussiate, which forms Prussian blue with the iron fixed; this can be decomposed by dilute caustic alkali, leaving the oxide of iron, which, by boiling in cow dung, becomes fit for taking up dyes, and with madder gives a lilac or purple. The bichromate of potash is decomposed by light in contact with calico, washing in water removes the unchanged bichromate, the remainder may be fixed by dilute alkali, and forms a weak mordant for several dyewoods. Indigo blues, padded in bichromate of potash for discharging with acid, must be kept from strong direct light, on account of its decomposing action upon the bichromate.

Lilac Color.—The color of the flower of the purple lilac; lilac is understood in printing and dyeing as being a low toned purple; violet is nearly if not quite synonymous, but scarcely ever used in the English trade.

The lilac generally employed in delaine and calico printing is the one given p. 189 as a constituent of dahlia color, and consists essentially of a solution of the coloring matter of logwood in red liquor made by digesting rasped logwood for 24 hours in that liquid in the cold. Other receipts are as follows:—

Dark Lilac Delaine.

4 gallons logwood liquor at 11°,
3½ lbs. alum,
12 lbs. gum.

This color is reduced by addition of gum water.

Lilac for Wool.

1 quart ammoniacal cochineal liquor,
1 quart vinegar,
5 oz. alum,
4 oz. oxalic acid,
1 lb. bichloride of tin,
3 oz. extract of indigo,
1 gallon of gum water.

This lilac is a direct mixture of the red and blue parts represented in this case by sulphate of indigo and cochineal.

In woollen dyeing various shades of lilac are obtained by first mordanting in alum and tartar, and then dyeing in logwood, with addition of sulphate of indigo. Logwood alone gives a reddish lilac, which can be brought to the blue shade, in any required degree, by properly apportioning the sulphate of indigo.

Lilacs on wool are also obtained by dyeing in a mixture of ammoniacal cochineal and sulphate of indigo.

Archil in combination with sulphate of indigo, also yields rich shades of lilac. The mauve color from aniline is also a species of lilac.

Alkanet, with alumina mordants, gives lilac colors; they are difficult to work and very fugitive.

For madder lilac, see MADDER.

21

Lima Wood.—One of the woods yielding red colors;—similar, or identical with BRAZIL WOOD, which see.

Lime.—Quick lime, prepared by expelling the carbonic acid from the carbonate of lime, is the oxide of the metal calcium. Its uses in bleaching and dyeing are dependent upon its alkaline properties. Presenting some analogy with potash and soda, it differs from them in being much less soluble in water, and consequently in many cases much less energetic in its action; but there are conditions under which it may act with even greater power than the more soluble alkalies.

Slacked lime is a combination of water with lime; very considerable heat is evolved in the combining of water with lime. Two cases have come under my notice where the accidental admission of water to lime in wooden vessels has caused sufficient heat to set fire to the wood. Lime is often kept in old hogsheads on print and dye works, and care should be taken of the possible occurrence of such an accident. Lime, mixed with an additional quantity of water, forms what is known as milk of lime; it consists of particles of hydrate of lime suspended in lime water. When milk of lime is allowed to stand quietly, the particles of lime subside, and a clear liquid is left, which is lime water. Lime water has alkaline characters, but very weak on account of the small quantity of the lime it contains: a gallon of lime water will not contain more than a quarter of an ounce of lime, nor can the strength be increased by concentration of the liquid. Hot water dissolves less lime than cold water, which is contrary to the usual law of solution; the most reliable experiments show that it would require a gallon and a half of boiling water to dissolve as much lime as a gallon of cold water. The first water obtained from lime is usually stronger than the subsequent ones; this arises from a minute quantity of the alkalies, potash and soda, being present and being all dissolved at once: the second and third waters from the lime are pure lime water. It is a question how many waters can be obtained from lime bottoms. That depends upon the quality of the water used. Pure water would continue to dissolve lime and yield good lime water many times; but water containing bicarbonate of lime will not yield above three or four good lime waters, and that only with active stirring and raking up of the lime bottoms. Water containing organic matters does not yield many lime waters; in both cases the lime is coated with a pellicle of insoluble precipitated matters which prevent the access of the water to it. The fact that cold water is a better solvent of lime than boiling water has induced some scientific men to advise that it should be always used cold, as then a greater quantity of the active material is in solution. But this advice is not founded on scientific principles, for it is well known that heat gives an energy to the action of chemical substances, the absence of which could not be compensated for by the use of a tenfold quantity of the material. In the general applications of lime in dyeing and bleaching, it is used in the milky state, that is, containing undissolved lime; and though it is contrary to theory to suppose that the undissolved lime is chemically active, there can be no doubt that, besides acting as a reserve for maintaining the water saturated with lime, the finely divided particles have an action which is at present not to be distinguished from what is considered purely chemical. It is known that lime can disorganize vegetable textures, and that some cases of tender cloth in bleaching are attributable to the action of milk of lime while it cannot be shown that clear lime water produces such effects. Lime combines with all acids, neutralizing them and forming salts of lime or calcium; the film or crust which forms upon lime water exposed to the air is carbonate of lime, the carbonic acid being derived from the air. Lime is a powerful base, and can displace the oxides of the metals proper from their combinations, itself combining with their acids. Upon this property depend the uses of lime in raising colors, in indigo dyeing, or other cases. In raising or fixing the buff from salts of iron, for example, the cloth containing acetate or sulphate of oxide of iron in its pores is passed into milk of lime, the lime combines with the acid, forming acetate or sulphate of lime, while the oxide of iron, deprived of the acids which made it soluble in water, rests upon the fibre. The action of lime in bleaching depends also upon its powerful basic properties.

Carbonate of Lime.—The only form of carbonate of lime familiar to the dyer and printer is chalk, which, being ground, is used in some few cases as an anti-acid. It is very suitable for this purpose, especially when an excess of alkali would be injurious. Chalk does not completely neutralize diluted acid liquors. A beck or cistern of dye liquor can have an acid reaction to test paper, though an excess of chalk be present; and this acidity, though small, would be too much for some styles. In such cases carbonate, or bicarbonate of soda, may be employed, or even lime water, if cautiously used. Ground chalk, though cheap, is liable to adulteration with sand; I have found ten per cent. of coarse sand in ground chalk; it could not be observed by inspection, but was easily shown by treating the chalk with muriatic acid, which left the sand undissolved. The quality of chalk is liable to variation, and all kinds are not equally suitable for the calico printer's use; some varieties contain magnesian salts, others a good deal of silicates. Chalk is frequently used in madder dyeing, and care should be taken that it is tolerably pure. The lighter variety appears better adapted for general use than that which is dense and heavy; good qualities do not contain more than five per cent. of moisture. Carbonate of lime is insoluble in water, but forms a soluble combination with another atom of carbonic acid, which exists in many natural waters. The extra atom of carbonic acid is so loosely held, that it escapes by simple agitation of the liquid, or exposure to the air, leaving the ordinary carbonate of lime in the insoluble form. Some spring waters are so saturated with this solution of carbonate of lime, and let it fall out so easily, that it collects in stony masses about the source, deposits in boilers fed with it, forming incrustations, and is productive of many inconveniences in application. Some idea of the amount of carbonate of lime dissolved in water may be formed from the statement of Bischof, that a single small stream in Germany carries away each year as much of this salt as would be equal to a cube of building limestone of one hundred feet, in lateral dimensions.

Sulphate of Lime.—This salt is an abundant natural product. It is known as gypsum, and when deprived of its water by roasting or calcining, forms plaster of Paris. It exists in most spring and river waters, and affects the dyeing of certain colors, as alluded to. It is only slightly soluble in water; it is produced when sulphuric acid and a soluble salt of lime are mixed together, and is the precipitate which forms when sulphate of alumina or alum is mixed with acetate of lime in the making of red liquor. It is sometimes used in finishing to give the appearance of body to inferior qualities of calico.

The Nitrate and Muriate of Lime are not generally known in trade. They are both deliquescent salts, and have been sometimes used in color mixing on account of that property.

Phosphate of Lime.—It was long ago known that the bones of animals which ate madder were tinged red; the conclusion that it was the phosphate of lime which attracted the color was too hasty; it now seems probable that some of the animal matters also present in the bones had more to do with it than the mineral matter. Nevertheless, phosphate of lime has some attraction for coloring matters. If an excess of ivory black or calcined bones be digested with citric acid, phosphate of lime is dissolved, which being properly applied to calico, can be shown to form an intimate connection with it, and to have also an affinity for coloring matters. It does not attract so much coloring matter as to be of value as a mordant, for the shades it yields are poor in depth and of a dry absorbent character.

In repeating this experiment care must be taken that there is no iron dissolved by the acid, or this will entirely change the nature of the colors produced; if iron exists in the liquor, it can only be precipitated by the prussiate of potash.

The best test for lime in solution is oxalate of ammonia, which gives a white precipitate provided the solution be neutral; in moderately strong solutions sulphuric acid gives a bulky precipitate of sulphate of lime.

Litmus.—This is a coloring matter similar to archil and cudbear, and capable of yielding blue and violet colors upon silks, but the colors are so excessively fugitive that they are never employed except in the extreme fancy styles.

Litre.—The standard French liquid measure: is equal to a kilogramme of water, that is nearly $2\frac{1}{4}$ lbs. English. An English imperial quart contains $2\frac{1}{2}$ lbs. of water, and is therefore nearly equivalent to a litre. The two measures may consequently in many cases be reckoned as equal, but in particular cases the difference would lead to considerable errors; for the exact relation of the litre and parts of a litre (taken in decimal parts) the following table may be consulted. The equivalents are given in ounces of water, and as color mixers are provided with ounce measures, they will be enabled to follow as closely as necessary the receipts given by French authorities.

Table showing the Value of a Litre and Decimal Parts of a Litre, in measure ounces. (Pint equals 20 ounces.)

Litre.	Ounces.	Litre.	Ounces.
1	35	0.1	$3\frac{1}{2}$
0.9	$31\frac{1}{2}$	0.09	3
0.8	28	0.08	$2\frac{3}{4}$
0.7	$24\frac{1}{2}$	00.7	$2\frac{1}{2}$
0.6	21	0.06	2
0.5	$17\frac{1}{2}$	0.05	$1\frac{3}{4}$
0.4	14	0 04	$1\frac{1}{2}$
0.3	$10\frac{1}{2}$	0.03	1
0.2	7	0 02	$\frac{3}{4}$

The weights are within a quarter of an ounce of the exact equivalent, which could not be given without employing larger fractions. As a good many color mixers are not familiar with decimals, I may explain that such a figure as 0.9 is the same in meaning as $\frac{9}{10}$, and indicates nine-tenths of a litre, 0.5 equals five-tenths, 0.09 is the same as $\frac{9}{100}$, and means nine-hundredths of a litre, and so on with the remainder.

Logwood, *Campeachy.*—Logwood is the wood of a tree flourishing chiefly in Mexico and the adjacent parts of America

It arrives in Europe in large pieces, and is rasped by machinery
into small fragments fit for dyeing or extracting the color from.
The coloring matter requires a large quantity of water to dissolve it, but when dissolved, can be concentrated or boiled down
to any degree of concentration. During the boiling down of
logwood extracts, and especially during the cooling, a considerable quantity of tarry matter is deposited, the nature of which
is not well known—probably it is similar to the resinous substances which exist in many species of woods. A weak solution of logwood in pure water has a yellow color; when strong
it has a reddish color, a sweetish astringent taste, and a peculiar
odor. Chemists consider it contains either two coloring matters,
or one coloring matter in two distinct states of oxidation. Like
indigo, it is supposed to contain a colorless body which, by the
absorption of air or ammonia, becomes colored; but this statement is by no means so well proved as to be taken for a fact.
The wood is very hard and dense, and as before stated does not
yield its color quickly to water. The rasped logwood is usually
damped and kept in that state for some weeks before it is used,
being turned over when it shows any inclination to heat.
Instead of degging it with pure water, sometimes lant or stale
urine is used, either alone or mixed with water or lime, and
sometimes soda is dissolved in the water. It is considered that
logwood is improved in dyeing power to the extent of fifty per
cent. by this process; or that ten parts of it thus treated are
equal to fifteen taken in the dry state from the rasping mill.
It has been attempted to prove that some chemical change takes
place in the coloring matter of the logwood; that the colorless
principle, supposed to exist in the wood, absorbs oxygen and
ammonia, and becomes colored, and thus its dyeing power is
increased. I consider that there is no real foundation for this
belief, and that the action of steeping and ageing logwood may
be simply and sufficiently explained on physical grounds. The
water may be supposed to soak gradually into the hard fibres,
swell them out, soften the mass, and render the coloring particles accessible to the action of liquids, and so readily soluble
in them. The change of color from the dusty yellow rasped
wood to the reddish hue of damped wood may be due to the
simple effect of water dissolving the coloring matter, and covering the fibrous part with it; but water always heightens the
hue of coloring matters. The use of urine—if it has really any
use upon the wood beyond that of increasing its depth of
color—may be looked for in the action of the ammonia it contains upon the resinous matters of the wool; it would dissolve
them in part, and set at liberty the coloring matters which it
may be supposed they would otherwise prevent from coming

into contact with the water. Solution of logwood has an inclination to form blue compounds with mineral substances, such as lime, baryta, copper, alumina, iron, etc., but if in large quantity the blue becomes so intense as to be considered a black. No good blues can be obtained from logwood, the best of them are dull and absorbent, and inclined to go brown or black; it is principally employed in dark colors, black, chocolate, etc. The pure coloring matters which may be extracted from it have received the names of *hematoxyline, hematine,* and *hemateine.*

Logwood is very extensively used in the black dye for silk, woollen, and calico; its cost comes considerably under that of galls, which give the best and firmest colors. Upon calico the mordant may be either alumina alone—but that gives a black too much on the purplish side—or iron alone, but the black from this will be too brown and dull. Perhaps the best mordant is a mixture of the two, in which the alumina predominates; other dyewoods are used to modify the shade of the black, according to the requirements of trade. Wool is dyed black in various ways, mostly with a blue foundation, which, for fast colors, is from the hot vat, but more frequently from the sulphate of indigo. Logwood black withstands washings pretty well, and is not much injured by a moderate soaping; it does not withstand the action of the air and light, but soon loses its lustre, becoming brown and faded. This change takes place more quickly upon cotton than upon silk, and sooner upon silk than upon wool. A logwood black can be discovered by the action of weak spirits of salts upon it; a drop turns it to a bright red. If the black be from galls, it changes the color, but does not give a red. The substitution of logwood for the astringent substances in dyeing black has been injurious to the general character of the black dye. For printing blacks on calico or wool the logwood is mixed with nitrate of iron. Logwood liquor is much employed in steam and spirit colors, for other colors besides black. With tin mordants it yields shades of purple, lilac, and violet; mixed with other colored extracts it helps to produce chocolates and similar colors. Logwood produces an intense black with chromate of potash under certain circumstances, and this salt, is occasionally employed in logwood colors, but it has several difficulties attending it. It coagulates the solution, and then it is impossible to work it; the only way in which, at present, it can be used is to pass the pieces printed with logwood color through it. It gives a harshness to the pieces, and is seldom employed; but it deserves the attention of dyers, because the blacks thus made seem faster than blacks produced from iron and alumina mordants.

Logwood is a very rich coloring matter, and under chemical treatment can be made to assume several different and valuable shades of color; but they are very unstable, and peculiarly susceptible to the destructive action of air and light, while they withstand washing with tolerable firmness. Chemistry does not at present give the slightest clue to the reason why one coloring matter is fast and another fugitive, why one can resist the detergent action of soap and the other cannot, why one is not particularly affected by exposure to the air and another is almost destroyed. There is, doubtless, some general law governing these things. It may reasonably be expected that, when the principles of the fixation of coloring matters are better understood, something may be done to communicate to the fugitive coloring matters some of that permanency which distinguishes the majority of the substances employed by the dyer and printer. Logwood seems to be one of those substances most likely to reward the labor which may be spent upon it in the endeavor to improve its permanency.

The coloring matter of logwood is distinguished from that of the red woods of the caesalpinia tribe by giving blue-colored precipitates with the alkaline earths and several metallic solutions, while the red woods give precipitates of a crimson hue. The fixed alkalies, in contact with air, appear to have the power of developing a red color from the yellow hematoxyline; this property is possessed by lime water, and also by bicarbonate of lime.

Lustres.—This is a trade term which was applied a few years ago to a style of delaine work in which the delaine was dyed before printing, but so dyed that the woollen threads were of a different color from the cotton threads, an effect being produced something like that seen in shot silks. (See MIXED FABRICS.)

Luteoline.—The chemical name of the pure coloring matter of weld, or dyers' weed.

M.

Madder.—The madder plant grows in many parts of the world, and seems to yield a nearly equal product in very various climates and situations. It is not cultivated for commercial purposes in this country, but is largely grown in France and Holland, whence the bulk of that used in England is obtained. Madder roots in the unground state are imported from the Levant, and called Turkey roots; small quantities are obtained from other countries in Europe, and some from the

East Indies, generally known as Bombay roots. The madders from these places are very similar in chemical and tinctorial characters—their differences do not appear to be of an essential nature. Some are preferred for one style of work and some for another, while some are not so rich in coloring matter as others. The market price is a correct evidence of the goodness of a known kind of madder; the tests of its quality being perfectly practical in their nature, are for existing methods of using this coloring matter unmistakable, and the true commercial value of a madder is soon ascertained. The French madders are in a state of very fine powder, packed tight in large casks, where, owing to a gelatinous or gummy substance which all madder contains, the whole powder is sometimes firmly cemented together, requiring to be cut with a pickaxe; when placed in water the gummy matter immediately dissolves, and the madder falls to powder. The roots which are imported are generally ground by the consumer. I have made many experiments as to the fineness to which madder should be ground, so as to give out all its available color; and working upon Turkey roots, I have found, as a general result, that there is no advantage in bringing it to an exclusively fine powder; if it passes through a sieve of about twelve wires to the inch, it is fine enough. The French cask madder, when dry, will pass through a sieve of eighty wires to the inch. That degree of fineness is perhaps necessary to make it a commercially saleable article; but for a manufacturer to grind his own roots so fine would be a great loss of labor without any corresponding advantage. The reason for this lies in the texture of madder; it is not like a hard wood, enclosing its coloring particles in walls of insoluble ligneous fibre, which must be torn up by mechanical force, in order to set them at liberty in the dye beck; it is, on the contrary, soft, it contains one-half its weight of gum, sugar, salts, and other soluble matters, which the water speedily dissolves, and reduces the remainder into a porous, spongy state, so that water has easy access to the coloring matter. The finer particles which all ground madder contains, if separated by a sieve from the coarse parts, will be found to be no better for dyeing, but frequently rather worse, because they generally contain the sand, stones, and dirt, which grind to dust sooner than the tough fibre of the root.

It is a general opinion that madder, when well kept, improves for some years after it has been gathered. A kind of slow fermentation appears to go on, the madder swells, often bulging out the casks, and even bursting them. I consider that the question of the improvement of madder by age is not a general one, it is confined to peculiar qualities; the contradictory

results of many experiments, and conflicting statements of
those who have compared fresh and old madders, can only be
explained or reconciled by this supposition. With regard to
Turkey roots, I feel certain they are old enough when they
arrive in England for all useful purposes. Madder, under
ordinary circumstances, is not deteriorated by age, nor even
sensibly altered, if it has been kept dry and out of strong
light. I have had an opportunity of trying madder forty years
old, which was very little different in its behavior from fresh
madder of the same kind. Its solution in the beck was blacker
and of a different taste to new madder, but the colors it yielded
were as good in every respect. It has been proposed to moisten
madder with water, or expose it to a damp atmosphere to absorb
moisture, with a view to extracting its coloring matter more
quickly or in greater quanity. This method, though found
useful with regard to many dye-woods, is no good in the case
of madder. I have tried it under many different circumstances. If madder be exposed to a thoroughly moist atmosphere it will absorb about twenty per cent. more water than it
usually contains; this will make it feel damp, it will adhere
together when pressed, and its color is heightened. If degged
with water its gummy character makes it adhere in lumps, and
it cannot be turned over and exposed to the air in the same
way that logwood or fustic can. Either of these treatments
leaves the madder about the same for dyeing; there is no perceptible improvement in it.

Action of Solvents upon Madder.—If ground madder is stirred
up with a large quantity of cold water, and strained off clear,
without standing more than an hour, a good deal of the coloring matter will be removed by the water. But it is not
possible in this manner to extract all the coloring matter from
madder, and the portion extracted is mixed up with so much
sugary and gummy matters that it is inapplicable as an
extract of coloring matter. If the madder be left to stand
for twenty-four hours before straining, it will be found to
have assumed a gelatinous state, more or less apparent according to the quantity of water used, and if the liquor be
pressed out it will be found not to contain any appreciable
quantity of the coloring matter of the root. This is a very
curious property of the coloring matter of madder; it will dissolve if it does not stand; if it stands it becomes insoluble,
and only the soluble and useless part of the root is washed
away. In this manner madder is often treated. It loses nearly
one-half its weight; when dry has a peculiar smell, somewhat
resembling sour milk; it is lighter colored than before the
treatment, and does not injure the whites so much in dyeing.

It is called in trade "Fluer de garance," or, in English, " Flowers of madder." It is suitable for some styles of work, but it presents no advantage on the score of economy; it goes about twice as far as madder, and is about twice as dear; not tinging the whites as much as the madder, it saves soap, and can even be cleared without soap, but in such case the colors are dull.

Hot water, if poured upon madder and strained off, dissolves some coloring matter, but less than cold water; if left to stand until cold, it acts nearly the same as cold water. Both hot and cold water, when not left standing on madder, injure the un-dissolved residue to an extent which seems disproportioned to the amount of coloring matter to be found dissolved by the water.

The other liquids which are generally used as solvents for coloring matters do not make any satisfactory extraction of this root. The coloring principle of madder is an anomaly in its behavior to different substances. In some respects one of the strongest of colors, it is at other times injured or destroyed by the slightest chemical action. It appears to be at the same time soluble and insoluble in water. How are these conflict-ing phenomena to be explained? Either by supposing that there is more than one coloring matter in the root, or that, if it is a simple principle which yields all the shades, it must be capable of assuming different forms and properties, passing from one condition to another, as from a soluble to an insoluble one, and so on. Indeed, it seems probable that madder does not contain a really isoluble coloring matter, but it contains something, or perhaps several things, which, under the influ-ence of water, air, warmth, etc., become colors. By several complicated chemical processes a substance can be obtained from madder, in small quantity which is crystallized in beau-tiful orange-red colored needles; it is named *alizarine*, and is the reputed pure coloring principle of madder. There can be no doubt that it is so in a most important manner, since all the colors which madder gives can be obtained from these crystals, by simply dyeing mordanted cloth in them. But it may not be the only coloring principle in the root, or it may not be in the root at all, but produced by the chemical operations performed upon it. That is of no practical consequence; it is either in the root, or something else which forms it is there. It does not appear necessary to look for other coloring matters while this one is capable of yielding all the shades which madder itself gives. There may be, indeed, a separate principle for the red color, and one for the purple, and, if so, also one for the black and chocolate; but it is apparent, if this be the case, that they are respectively convertible, under the influence of mordants, one into the other; and for all practical comprehension of the properties of madder, the assumption that there is one coloring principle, and that one alizarine, may be held as true. The peculiar behavior of water upon madder may be considered as attributable to the alizarine existing partly formed and partly unformed; the completely formed portion not being sensibly acted upon by cold water, and very little by hot, while the unformed alizarine is dissoluble by water, but has a constant tendency to pass into its complete state of alizarine, in which it is not dissolved by water. The jellifying of the madder has probably no necessary connection with this alteration of the coloring matter, for the gelatinous substance can be separated, and is found to possess no dyeing properties at all. What the water takes away is vegetable extractive matter, partly of a gummy, partly of a saccharine nature, and some earthy matters which are soluble in water; what it leaves behind is woolly fibre, earthy matter, this peculiar jelly-like matter, called by the chemists pectine and pectic acid, and also the alizarine.

A question of the highest importance is yet pending with regard to the coloring matter of madder; to extract it in a pure state from the other matters that are with it in the root, or if not in a pure state, yet sufficiently strong and pure to resemble other extracts, as logwood liquor, sapan liquor, etc. If this could be accomplished, it seems probable that it would be more economical to use it as a steam color than to dye with; it is likely it would yield better shades and more regular results, and be in many cases extremely preferable to the mixture of a small amount of pure coloring matter and large amount of useless encumbering matter which constitutes madder. The Industrial Society of Mulhouse has at various times offered large premiums and honors to any one who should resolve this question in a satisfactory manner, having regard to commercial requirements. These inducements, and the certainty of otherwise making large profits, have led some of the ablest colorists in France and England to turn their attention to this topic, but without the slightest success so far. Nothing, perhaps, could so well illustrate the immense difficulty of dealing with this matter than the failure of so many attempts to accomplish something, so well defined as a solution of the coloring matter of this root. Solutions and extracts have been made, but they did not fulfil the required conditions; they were either too impure and contaminated with foreign matters, or else were too expensive on account of the use of spirits of wine, or such solvents, in extracting the color. Some of these extracts have been employed on a small scale, and pieces done with them which, if not perfect, at any rate seemed to indicate the strong probability of success awaiting continued efforts in the same direction. To judge by the numerous patents taken out in this country and in France with reference to this subject, one would be led to conclude that not only was the matter not difficult to accomplish, but that actually a great number of persons had succeeded in accomplishing it. There could not be a greater mistake; either the enrolled specifications of those patents are fraudulently deceptive in concealing the real method of accomplishing the end, or else the patentees are most lamentably ignorant of what has been done in the same direction. Processes are patented which are perfectly impossible, and one might imagine that the parties did not actually know what madder was, or had never worked upon it for an hour in their lives. Much information upon the behavior of madder, under various circumstances and with various chemical re-agents, has been acquired, but nothing of any practical utility has yet resulted from experiments to concentrate it by extraction. It is a subject yet inviting to research, and seems to promise great things to the discoverer. A necessary condition in preparing an extract of madder is, that it must not, at least, give inferior results to those at present obtainable by dyeing. No facilities of application would compensate for inferior results, and no extract can hope to meet with extended employment which does not enable the printer or dyer to produce as good or better results than he can now obtain. There is a sufficient margin in the amount of coloring matter, which is never obtained from the roots, to make an effective extraction pay for its expenses, if these are not very heavy, by the use of solvents, which are dear in themselves, or get lost in the process of extraction.

Madder does not give up all its coloring matter to water, however long it may be boiled with it, even in the presence of mordants, which remove it from solution as fast as it is dissolved. This may not be true when very large quantities of water are used for comparatively small amounts of madder, but in all practical dyeing operations it is the case. The spent madder contains nearly as much coloring matter as has been extracted from it, but it must be in some different form, or else it would come out upon treating with fresh water. The only way in which it can be obtained is by treating the spent madder with acids; the acids liberate the coloring matter, and, when they are washed away again, the spent madder is able to dye up almost as before. (See GARANCINE.)

The coloring matter of madder is dissolved in alcohol in larger quantity than by water, but it does not extract it from the root in a state of purity, but mixed with several other substances. It is soluble also in acids, when heated, and generally falls out again when cold. Boiling alum water dissolves it in comparatively large quantity, and lets it settle out upon cooling as a yellow-colored sediment. The caustic alkalies, ammonia, potash, and soda, dissolve it to a large extent, but not in its integrity. They cause it to undergo some change, which either injures or destroys it, and this is more especially the case when they are left in contact for any length of time. Even the milder forms of alkali, as the carbonates, bicarbonates, and the borates, injure it very much indeed if left in contact with it.

Action of Heat upon Madder.—Madder is so complex a substance that the influence of any chemical agent upon it resolves itself into the action of the agent upon several matters mixed together, and produces results which defy all attempts to say where the principal action has been exerted, and how far secondary actions have influenced the final result. So it is with heat. Madder is injured by heating; if steamed it is much deteriorated, and more by low pressure than by high pressure steam. Dry heat above the boiling point of water does not injure it nearly so much as moist heat. Madder may be exposed to a temperature of 300° F. without being much injured, but if the temperature be pushed a little higher it begins to be seriously injured, and at a roasting point it is destroyed. The pure alizarine can be sublimed by heat in crystal; it is not destroyed by it, but, as it exists in madder, mixed with many other substances, it does not sublime but is destroyed. M. Camille Kœchlin, and others more recently, have proposed a plan for extracting the coloring matter from madder by heat and currents of gases, but it could not be practically applied. Messrs. Pincoffs and Schunck have patented a process for submitting washed madder to the action of high pressure steam in the production of a species of garancine, sold under the name of alizarine. If an acid mixture of madder and sulphuric acid be gently dried, and then exposed on an iron plate to a heat about sufficient to brown flour, there will be formed in a little time on the surface a number of small crystals of a brilliant orange and red color—these are alizarine, pure so far as they can be obtained without being soiled with the residue to which they are attached. Much alizarine is destroyed compared with what is obtained by this method; but modifications of this plan might be devised by which the coloring matter could be sublimed away with the worthless residue.

M. Schlumberger estimated the amount of pure coloring matter, or alizarine, in good qualities of madder at about four per cent.; inferior qualities only yielded him from two to two

and a half per cent. I have found good qualities of Turkey
madder to contain about sixty per cent. of attractive matters,
which term includes everything removable by water and dilute
alkalies; the woody fibre was therefore about forty per cent.
I never could pretend to say either by direct or indirect means
how much real available coloring matter there was in a given
quantity of madder. French madder, upon an average, contains
also about forty per cent. of a ligneous matter. The amount
of mineral matter in madder is tolerably constant at about
ten per cent.; that is, for ground madder which contains the
attached mineral matters of the roots besides the contained
mineral salts. Madder root cleansed from all adhering soil
and sand will give from six to eight per cent. of residue upon
calcination.

There is no method known by which the value of a sample
of madder can be determined in a direct manner from the
amount of coloring matter which it contains. The only reliable
test is that of dyeing either pieces or fents with the sample in
question. A quantity of madder as small as twenty grains
will suffice for a laboratory experiment, from which an experienced manipulator can obtain tolerably reliable results. But
it requires very great care in placing all the samples under
examination in perfectly equal circumstances; the precautions
can only be learned by experience.

Madder is said to be adulterated with mineral matters and
valueless vegetable substances. As madder admits of no additions of this kind, without an immediate injury to its dyeing
properties, the dyeing test will suffice to point out the presence
of such foreign matter. Various qualities of madder may be
mixed together, or even dried spent madder may be ground
up with the roots. These falsifications will all show in the dye-
beck. The admixture of other dyewoods with madder for
adulterating purposes would for the most part destroy their
object, by so injuring the madder as to render it quite unfit for
its usual applications. The same is true of garancine; although
very few articles are more subject to falsification, it is entirely
in the addition of neutral and inert matters which increase the
weight. That such an adulteration should be possible for
lengthened periods can only be owing to gross ignorance on
the side of the purchaser, or a guilty collusion on the part of
his servants.

With regard to the scientific investigation of madder I shall
confine myself to the published accounts of Dr. Edward
Schunck, who may be considered the best authority upon all
that concerns the chemical action of this root. He considers
alizarine to be actually contained in madder, since it can be

obtained from it without having recourse to sublimation.
When acted upon by nitric acid it gives a peculiar acid, which
was at first announced as a new acid, but afterwards proved to
be identical with the phthalic acid, which Laurent had obtained
by the oxidation of chloro-naphthalic acid by nitric acid. Because nitric acid acting upon two separate substances has given
rise to the same acid, Strecker and other chemists have been
led to believe a similarity in composition of the original bodies;
nothing, however, proves this, and the hopes which have been
raised of producing alizarine from any of the compounds of
naphthaline seem delusive. A French chemist, M. Roussin,
actually announced to the world, about a year ago, that he had
obtained alizarine by acting upon a naphthaline compound with
sulphuric acid and zinc; a little examination proved that he
had allowed himself to be deceived upon very insufficient
grounds.

Rubiacine is a yellow coloring matter, crystallizing in
greenish yellow needles of much lustre; it is believed not to
be the only yellow coloring matter in madder; when treated
with perchloride of iron it is oxidized, and forms an acid—
Rubiacic acid, which yields crystalline compounds. *Chlorogenine* or *rubichloric acid* is a substance which, under the influence
of strong acids, is converted into a dark-green powder—whence
its name; it is supposed to be this principle which stains the
unmordanted parts of calico in the dye-bath, and which, being
partially removed and destroyed in garancine making, permits
this latter to dye without staining the whites so much. *Pectic
acid* or *pectine, sugar,* and *gum* are also found in madder. Mr.
Schunck points out the existence of a ferment in madder,
which he calls *erythrozym,* and which, in confirmation of Hig-
gins, he looks upon as capable of transforming some of the
uncolored principles into alizarine. He goes further, and says
that fresh madder root contains no red coloring matter at all,
only a yellowish color, which is of no technical value, but it
contains this ferment, which changes the yellow into alizarine.
The principle from which all the color is supposed to spring is
called *rubian.* Schunck claims to have isolated this principle
in a pure state, and to have satisfied himself that it is capable
of producing alizarine by the action of the pure ferment in
madder, and also by the influence of acids and alkalies.

Here is a list of some of the bodies which are found to result
from the decomposition of rubian by ferments or acids:—

Rubiretine,	Rubiadine,
Verantine,	Rubiafine,
Rubianine,	Rubiagine.

It would be useless to enter into details upon the manner in which the discoverer connects these bodies with one another and the original rubian. He does not look upon them as necessary to the reaction, and the probability is that they are accidental and indefinite mixtures of secondary products. Rubian should yield about 80 per cent. of its weight of alizarine, but, owing to the formation of bye-products, not more than from ten to twenty per cent. can be obtained. Besides the above-mentioned compounds, Mr. Schunck describes the following, of which only the names are here given :—

> Rubianic acid,
> Chlororubian,
> Chlororubiadine,
> Perchlororubian,
> Purpurine.

The latter named substance he looks upon as a species of coloring matter, distinct from, and inferior to, alizarine, which fixes upon mordants, but is removed in the subsequent treatments to which the better class of madder colors are subjected.

Coloring Matters similar to Madder.—Besides the various qualities of madder received from different parts of the world, and possessing general characters of resemblance, although not identical, there are dyeing matters which come from plants of a similar nature, and possess some of the characters of madder, while they are deficient in others. Several of these are known to be in use in Hindostan, and two or three of them have been imported in rather large quantities into this country. Perhaps the chief of them is the substance known as munjeet. If looked upon as a species of madder, it must be considered a very inferior one, producing the same colors but requiring much larger quantities, and not giving them so bright or so fast. The soluble parts of madder are nearly absent in munjeet. It is a dry, dusty, reedy substance, very different in general appearance, taste, and character to madder. It has been used in Lancashire by some printers, and I suppose that it may be assumed it was found as cheap as madder although much poorer in coloring matter. It seems to answer best when made into a kind of garancine ; on account of its woody nature it does not lose much weight in this process, for while six hundredweight of French madder only gives about seven hundredweight of pressed garancine, the same weight of munjeet gives about nine hundredweight. A French writer, who resided some time in India, M. Gonfreville, states that the dyers there produce very fine reds from some roots not known in Europe, but all belonging to the same botanical class as madder ; and MM. Persoz and

E. Schwartz declare them all to contain coloring principles
identical with the coloring principles of European madder, and
that with using proper precaution in the dyeing they can be
made to yield good colors. Among others of this class of
plants, roots, or products, occur the names of *nona, chayaver,
ouonkoudou* and *hachrout.* Some of these are probably syno-
nymes, and there may not be that number of separate red
dyeing materials of the madder kind which would appear from
the list of names. None of them seem to be so good as our
own madders; but the native Hindoos can produce dyes from
them which are declared superior in stability and brilliancy
to those produced in Europe. But their processes are long,
tedious, and expensive, and inapplicable to general calico
printing and dyeing.

Madder Colors.—The chief colors obtained from madder
are the madder purple and pink. Turkey red may also be considered here as a madder color. Madder yields also a dark
purplish black, strong reds, and, with catechu, various shades of
brown. Madder chocolate, obtained from a mixture of alumina
and iron mordants, is not much used.

Madder Purple or Lilac.—This is probably the most important color produced by the art of calico printing, always
beautiful, and the very type of a fast and permanent dye; it
withstands all the accidents of wear without fading, and permits
the fabric to be washed an almost unlimited number of times
without deteriorating in shade, provided ordinary care is employed. The madder styles are .produced in great perfection
by many printers in Lancashire, and especially by Messrs.
Thomas Hoyle and Sons, who have for two or three generations
been looked upon as the leading house for these particular
colors. For some years that the author was engaged in that
establishment he had excellent opportunities of studying this
style of work. There is nothing easier than the production
of second-class madders; but the very highest-class work requires an infinite number of precautions in every step, from
the singeing or shearing to the starch used in finishing, and
success entirely depends upon the close attention and intelligent
observation which is bestowed upon each of the processes. If
any one or two of the processes can be named as those upon
which success more particularly depends, perhaps the thickening
of the mordants, and the soaping after dyeing, might be selected
as the most important, or rather those in which the use of
inferior articles, or neglect in manipulation, would operate most
strongly against obtaining the best result. But, in fact, the
whole of the processes hang together like a chain, in which if
there is one faulty link the whole is bad.

The mordants used are very simple, consisting of nothing but commercial iron liquor, suitably thickened. Many additions have been proposed to improve the iron liquor, and are used in several places; but, if the iron liquor be of good quality, and such as is easily obtained in trade, no addition that I know of will improve it in the slightest degree. It is customary in some establishments to prepare the cloth with chlorate of potash before printing. I presume the printers who use the process find some advantage in it or they would not make the expenditure; but as work quite as excellent can be produced without this preparation, either there is some difference in the conditions not generally known, or else the use of such prepares is simply a loss.

The thickening matter varies according to styles, etc.: flour is a good deal used; an artificial gum thickening at from 4 to 5 lbs. per gallon, and known as "purple gum," is also used very extensively; other gum substitutes and calcined farina are also in use. It is impossible to say what thickening is best without being actually on the spot, and though it is so important a part no written directions can be expected to meet particular cases.

The proportions of iron liquor to thickening must also depend in some degree upon the kind of thickening. One part of iron liquor at 28° to four parts water, thickened with flour, gives the black; one part iron liquor to eight parts of five-pound gum water gives about the darkest purple of lilac in use; one part iron liquor to 40 parts gum water gives light shades; but mordants so weak as one to 60, and one to 100, are sometimes employed for covers. The most usual colors are made with from 14, 18, and 20 parts gum water to one part iron liquor.

After printing, the pieces are aged and dunged, and then entered into the dye. From causes already stated the madder is used in its entirety, because no extract or decoction of its coloring matter can be made. It is bulky and gelatinous in water, so that a considerable proportion of this fluid has to be employed. But it is important to use no more water than is absolutely necessary for the easy running of the pieces: if an excessive quantity of water be employed, a large portion of the coloring matter does not fix, and is either wholly lost or returns upon the spent madder. The amount of madder will be, for economical reasons, apportioned as exactly as possible to the demands of the design; but, if it were not so, an excess could not be used without injury to the dyed colors and to the whites. To the colors, because the brown or dun-colored matter, dissolving more readily than the alizarine, would fill up the mordants, and dispute the entrance of the true coloring matter;

and to the whites, because the free alizarine would partially fix upon them, and would be difficult to remove.

The alizarine dissolves only slowly, and in small quantity, in water, so that madder dyeing occupies a greater length of time than other styles. About two hours is required to obtain the best results; a less time than an hour and a half would waste madder; if continued longer than two hours, at the usual temperature, the colors are liable to be injured without any corresponding economy of madder. The mordants commence taking the color at a temperature of from 100° to 120°, but only slowly; at 140° the dyeing proceeds rapidly, and could be wholly accomplished, but would not exhaust the madder; at 160° the colors become nearly saturated, at 180° is as high as it is advisable to carry the heat for the best colors. By raising the temperature to ebullition, a little less madder will suffice, but neither colors nor whites are at their best. As the dyeing does not commence until a temperature of 100° is attained, the goods and madder may be entered in water at that heat; in thirty minutes the temperature may be raised, in a gradual and regular manner, to 140°, and to 180° in another hour, keeping the temperature at that point until finished. Irregular heating is prejudicial to madder dyeing; if the steam should be cut off, and the dye fall twenty or thirty degrees in the course of the round, it would injure the colors very much. In France there is a system of dyeing madder colors at twice, dividing the quantity of madder into two unequal portions, dyeing with the lesser quantity at a low temperature, and then in fresh water dyeing with the remainder, using more elevated temperatures. I have tested this plan with good qualities of French and Turkey madder, but found no advautage in color, while there is greatly increased risk of accident, besides labor and expense; it might be useful in poor qualities of madder. The exposure to the air during dyeing is by some authorities considered of essential benefit to the colors; my experiments do not support this view, I only found that the more the piece was exposed to the air the more difficult it became to clear the whites; for all that the air resists, the dyeing might take place in vacuum.

The costliness of madder, and the certainty that none of the dyeing processes extract all its coloring matter, have led to the trial of numerous additions to it in dyeing, with a view of making it go further. The substances added have been mostly of the mineral nature, and have been used often at random without any particular aim; at other times they were employed with some specific intention, as, for example, to neutralize some supposed acid in the madder, to counteract the effects of lime in the water, to prevent the fixation of the brown coloring matter of the root, to make the real coloring matter more soluble and so on. The advantages which are said to have been derived from the use of such additions to the dye-beck are more imaginary than actual. In the publications of the Industrial Society of Mulhouse may be found several essays upon this subject, some by anonymous authors, others with well-known and respectable names attached to them; but the results are for the most part quite at variance with what can be obtained in England, working upon good French madder or Turkey roots. The most careful experiments that I could make in the same direction failed to give me the same results; the madder which yielded such symptoms of being improved by certain additions could not have been like the madder supplied by respectable import houses in Manchester. The "Bulletin de la Société Industrielle de Mulhouse" is not easily accessible to English readers, but a full *résumé* of these papers may be found in Persoz, " Traité de l' impression des tissus," ii. 504–510. The first point is the addition of ground chalk in madder dyeing. No one familiar with the literature of dyeing and printing can be ignorant of the letter M. Hausmann wrote to M. Berthollet, detailing that he, having removed from Rouen to Logelbach, found that he could not dye up the same colors as those which he had obtained at Rouen; how at length, by analytical examination, he found that the water at Logelbach was too pure, that the Rouen water differed from it by containing a quantity of lime salt; and how he had the happy idea of adding carbonate of lime or ground chalk to his madder, when all the dyes came up well again, and produced colors as fast and brilliant as ever. Since the publication of this letter, French writers hold it absolutely necessary that chalk should be in the water, either naturally or artificially, to produce good fast colors, and they consider that lime becomes an essential part of the color as fixed upon the cloth. The statements concerning this necessity for lime are by far too general, extend over too long a series of years, and have attached to them too many of the names celebrated in the annals of French dyeing, to be without some foundation in fact; but I can say, without fear of contradiction, that the Turkey roots and French madders, used through a series of years in the neighborhood of Manchester, were never improved by the addition of chalk, even when distilled waters were used for dyeing. Lime in some form is present in most samples of madder, and in France it is known that some madders are not only not improved but injured by adding chalk to them. The madder called Paluds, from the district which it grows in, containing a large amount of chalk derived from the soil, requires no addition; the Avignon madder is said to require some chalk added, and the Dutch and Alsatian madders imperatively demand it in rather large quantities. Such is the general statement made by French writers. M. Persoz mentions a case in which he found a sample of Alsatian madder to dye up perfectly well without any addition of chalk; it had been kept for a considerable length of time in a bottle, and he assumes that it had undergone some change by lapse of time. An anonymous writer, quoted also by Persoz, puts down all the alkalies and alkaline earths and carbonates, including chalk, as injurious to madder dyeing. This is in contradiction to the current statements throughout his work, and especially to the statements of that eminent practical authority, M. Daniel Koechlin. The statements of M. Koechlin prove too much to sustain the lime theory, for he states that alkalies in general serve the same purpose, and prescribes the exact quantity of carbonate of potash and soda which are to replace the ground chalk. M. Persoz states that good madder colors contain a definite amount of lime, and looks upon it as an essential constituent of the finished color. I have analyzed madder colors, but have not found lime as a regular constituent, and generally only found such traces as were due to the cloth itself. I cannot, therefore, concur in this theory of lime being necessary in madder dyeing as a general rule. Chalk is used to some extent in madder dyeing in many places. I do not know if it is general in England. To the extent of one pound in a hundred of madder it does no harm, if it does no good; but some houses use at the rate of five per cent. of ground chalk, and even more for pinks, but this I think is quite unnecessary, and even hurtful. I have proved the presence of undecomposed carbonate of lime in spent madder when chalk has been used, at the same time the liquors have had an acid reaction to test paper.

Besides chalk, no other substance has been recommended for general use in madder dyeing, but exceptional cases are not wanting in which additions of various articles to the dye have been thought beneficial or necessary. Bran, for example, was long employed with madder in producing the old London pink. Size or glue has been and is yet used in some places, under the impression that it gives better results, and enables the madder to dye further. Small quantities of potash and soda, as also oxalic acid, and cream of tartar are used, and may be beneficial owing to peculiar waters. The use of galls, sumac, and other astringent substances do not come under consideration, as these must be looked upon as real coloring matters themselves, and as such adding to the result produced by

madder, which addition may produce a simple exaltation of the
color, or an entirely different shade, according to the mordant
in use. With regard to these substances which are used in
madder dyeing without ill effects, those who employ them
should be the best judges of their utility, and their opinion is
worth more than that of a person at a distance and unacquainted
with the special and, perhaps, exceptional conditions under
which they work. Nevertheless a general opinion may be expressed, founded upon very numerous experiments, made under
a variety of circumstances, that with a fair quality of water
and good madder no addition is necessary, no addition is beneficial, and those additions which do not injure the madder only
leave it where it was as regards its dyeing powers, and the
quality of the colors it produces. In the second volume of
Persoz's work, already cited, copied also in Muspratt's Practical
Chemistry, are a list of substances tried as additions to madder, and before them figures which pretend to indicate their
beneficial or injurious action in dyeing. These figures (taken
from authors who have written anonymous essays) are simply
delusory, and bear upon the face of them evidences of their
unreliable character. In the list will be found a statement,
among others, that the addition of $\frac{1}{80}$th of sulphate of potash
to madder causes an increase in tinctorial power equal to
twenty-five per cent., and, further down, that the addition of
the same quantity of sulphate of soda diminishes the tinctorial
power of the madder by twenty-one per cent., making a difference between the two of forty-six per cent., or equivalent to
half the quantity of madder used. This hardly needed the
test of experiment to be condemned. I need scarcely say that
no madder obtainable in Manchester was improved even one
per cent. by the addition of sulphate of potash; and I am certain, on the other hand, that pure sulphate of soda, used in the
quantity laid down, did not injure it in any perceptible degree.
The deterioration by sulphate of soda was more likely to be
true than the improvement by sulphate of potash; and, in a
repetition of the experiment, I provided an impure sulphate of
soda, containing sulphate of iron, just as it came from the
makers, and by using a quantity of this, something greater
than that prescribed, I obtained a deteriorated result, which
could be carried to a perfect suspension of all dyeing powers
in the madder, by increasing the proportions. But it is evident that in this case it was the iron salt, and not the soda salt,
which was injurious. Something of this nature might explain
the minus result, but the plus, twenty-five per cent. of the sulphate of potash, is inexplicable. Other substances which are
placed in the same list as injurious are for the most part really

so, but those which are marked as beneficial, and improving the results of dyeing, sometimes to a very considerable extent, may be placed in the same category as sulphate of potash. I have tried them almost every one, viewing the results practically and commercially in comparison with pure madder colors, and I have not found any improvement in any case, but generally a deterioration. There is hardly a chemical salt, in either practical or laboratory use, which I have not made trial of as an addition to the madder dye for lilacs, reds, and blacks, and there is not one which gave an improved result which was of the value of the salt employed, and nine out of ten acted prejudicially, altering the shades, robbing the madder, staining the whites, or stopping the dyeing altogether.

After the pieces have been well washed out of the dye they are boiled with soap. The use of soap for clearing madder colors appears to have originated with the Turkey red dyers: its object in prints is twofold—first, to clear the colors from a dingy red coloring matter which spoils their shade, and, secondly, to make the white parts of the design bright by removing any color which is attached to them. It is considered least injurious to enter the pieces into the soap solution at a boiling temperature, or at the highest temperature which it is intended to employ: two or three soapings are necessary for the best quality of colors. A good deal of coloring matter is lost by soaping; I have made many attempts to clear and brighten colors without removing a portion of the coloring matter, but with very little success. Some eminent French authorities consider that the fatty acid of the soap enters into combination with the oxide of the mordant, and contributes to its stability and beauty. There is reason to doubt this statement in the ordinary cases of madder dyeing; beyond the difficulty of imagining such a combination as taking place under the circumstances, there is the certainty, that in many analyses no fatty matter is detected, and that purples are produced, of great beauty and stability, from commercial alizarine without the use of soap or fatty matters. No detergent answers as well as soap to remove the injurious brown coloring matters. The alkalies, caustic or carbonated, the borates, phosphates, and silicates, are all injurious to the colors. A final clearing is given by padding in solution of bleaching powder and steaming, or in the beck. In French processes, a passage in weak sours, between the soapings, is often prescribed; this gives a weaker purple, with less of the red tinge, and would be preferred in some markets. Madder purples are distinguished from all others by the magnificent purple color they produce when treated, first by sulphuric acid, at sp. gr. 1.4, and after washing, plunging in lime water.

Madder Pink.—The mordant may be either red liquor or the alkaline aluminate of potash. I believe the very best pinks are obtained by means of the alkaline or ash pink mordant; but, nevertheless, acetate of alumina is the mordant most commonly used, and excellent pinks may be obtained by it. The advantage of the alkaline pink is, that it will stand a hard drying; but a red-liquor pink must be very carefully and softly dried on the machine. The difficulty of working the alkaline pink on the other hand consists in evenly fixing the mordant, and then cleansing it from the thickening. To effect this it is necessary to permit the pieces to hang in a moist place long enough to thoroughly soften the colors; stoves, with steam escaping in, or the ageing machine, are best adapted for this purpose. After a sufficient penetration of the mordant has taken place, it is fixed by passing through a hot solution of muriate of ammonia or sal ammoniac, and then well washed. Dung is frequently used along with the sal ammoniac; but, though it may serve to scour the thickening off the cloth, it is not really essential, and is better omitted if the thickening is of a soluble nature. Besides sal ammoniac, muriate or sulphate of zinc may be used as the fixing agent, but preference is usually given to the former. It appears, from my experience, that the alumina mordant thus deposited is more susceptible to deleterious influences than the mordant obtained from red liquor, and is much more easily injured. Thus, an inferiority in the water, which would not tell upon a red-liquor mordant, will greatly injure an alkaline mordant. If the smallest trace of iron, for example, be in the water, it is all fixed, with wonderful rapidity, upon the alumina, and spoils the colors far more than it would those from red liquor: if the water also should contain any vegetable matters, they are attracted by the alumina, and the color injured; and if the pieces are left several hours between fixing and dyeing, they are subject to irregularities. The care required is consequently very great, and many printers have not been successful in using the alkaline mordant.

Whether the alkaline or red-liquor mordant be employed, the dyeing and subsequent treatments are the same. The temperature must not be pushed so high as for purples; should not, in fact, pass 160° or 170°. The madder must be of good quality, and more finely ground than is necessary for purples on account of the low temperature: the time of dyeing is from two to three hours. After washing from the dye, the first step is to soap, and this must be done at quite a low temperature—

not higher than 140° F. and last for twenty minutes; washed
out from this, the colors are cut or reduced by passing the
pieces in warm water containing very acid oxymuriate of tin.
This is a most important operation in obtaining good pinks,
and requires some attention and tact upon the part of the dyer.
The object to be attained is to reduce the dull red color to a
bright orange red, and, as the colors are not always equally
affected, the proportions of the cutting agent are not always the
same, nor the time in which the cutting is effected. If the
pieces are kept too long in the acid liquor, the color becomes
impoverished and bare; if not long enough, they will not soap
down to a soft toned pink. It appears better to have the
cutting solution strong, and perform the operation quickly,
than, on the contrary, to have it weak, and prolong the time
of immersion. The pink color is brought up by one or two
subsequent soapings, which may be worked at the boil. The
hue of color is improved by stewing the pieces in a pan for an
hour or two with soap liquor.

Oxymuriate of tin is sometimes replaced by sulphuric or
nitric acid, as the cutting agent, but the result of general experience is in favor of the oxymuriate.

It appears to be common amongst the French printers to add
tin crystals to the final soapings and stewings of the pinks, and
they appear to consider that a tin soap is produced which takes
some part in the color. I would not recommend any such
addition to good soap; it may be useful to a bad or alkaline
soap, but how it could do anything but injure a good quality
of soap is not easily understood.

The chief essentials to success in madder pinks are plenty
of madder, a low temperature in dyeing, and plenty of good
soap.

Madder Red.—The ordinary madder red is dyed upon a
strong acetate of alumina mordant, generally with addition of
some tin crystals to give brightness to the color, and throw off
any iron that might come into contact with it. Blotch reds
are now nearly all confined to garancine styles, on account of
the greater certainty and less cost with which they can be obtained.

Turkey Red.—This beautiful and remarkable color differs
from a common madder red by containing a considerable proportion of some oily compound in its composition, the nature
of which is not at all understood by chemists. The methods
of obtaining this red are very complicated, and differ very much
in different establishments. The following process is one followed by a house producing very good work, and may serve
as a general illustration :—

The pieces are not bleached white as for printing only, being well bottomed by liming and bowking. The dry pieces are padded in a mixture of Gallipoli oil and pearl ash, containing about 200 lbs. oil, 40 lbs. pearl ash, and 100 gallons water. This quantity is about sufficient for 4000 yards of calico. The pieces are exposed to the air in summer, and to the heat of a stove in cold weather, for twenty-four hours; then padded again in a mixture of oil, ash, and water, and again dried and exposed; and so on for as many as eight different treatments for dark colors. The excess of oil, or that oil which has not changed its character by oxidation and alkali, is now removed by steeping, and the pieces well washed. This completes the oiling, the utility of which is unquestionable, but the principles upon which its efficacy depends are at present hidden.

The next process is the galling and aluming, which are sometimes separate treatments, but in this process go together: 60 lbs. of ground gall-nuts are dissolved in hot water, and 120 lbs. of alum, and 10 lbs. of sugar of lead added to the liquor, which is made up into 120 gallons. The pieces are padded in this liquor, dried, and aged three days, then fixed by passing in warm water containing ground chalk; being washed out of this, they are ready for dyeing. The dyeing is in madder, mixed with a little sumac and with blood. For dark colors, the pieces undergo another galling and aluming after dyeing, aged, fixed, and dyed a second time. They are now of a very heavy brownish-red color, and are brightened by two or three soapings, or a passage in acid.

In other processes sheep's dung and cow dung are mixed with the oil, and other minor modifications introduced.

Garancine is now largely employed in Turkey red dyeing, and the operations of clearing and brightening much shortened.

Many attempts have been made to shorten the processes in the preparation for Turkey red, but it does not appear with much success; or, if any considerable change has taken place, it is kept secret by the discoverers.

The use of the oil in Turkey reds is, as before stated, enveloped in obscurity. Some chemists consider that the oil forms a true mordant, and have gone to the length of asserting that alumina is not at all necessary. What grounds there exist for so strange a statement are, as far as I know, utterly insufficient for this conclusion, and the result of my own numerous experiments are quite opposed to it. That oil does form a species of mordant is certain, but it is of so weak a nature as never to dye up more than a simple stain. The probability is that the oil, or whatever the oil is changed into, forms an excellent basis for the alumina and coloring matter, besides which the semi-transparency communicated to the cloth by the oil is of value in increasing the lustre of the color, and adding to its stability.

The use of galls or sumac appears evidently to be due to the increased affinity they give the cloth for the aluminous mordant. It appears, from the statements of several good authorities, that these astringent substances may be dispensed with, provided that, instead of using alum with a little alkali or acetate of lead, the ordinary acetate of alumina be employed.

Madder Browns.—The browns worked with madder colors are all derived from catechu, and very similar to those given for Garancine, page 241, and under catechu, page 129. The following two receipts will consequently suffice to complete the illustration:—

Madder Brown.

70 gallons water,
350 lbs. catechu ; boil ten hours, and add
100 lbs. sal ammoniac,
9 gallons acetic acid; thicken with ground gum senegal. The color is made from the above standard by taking
2 gallons above,
1 pint acetic acid,
2 pints acetate of copper (page 45).

Medium Madder Brown.

1¼ lb. catechu,
3 oz. sal ammoniac,
1 pint water; boil and dissolve, and add
6 oz. nitrate of copper at 80°,
4 oz. acetate of copper,
1 quart gum water.

This color will resist light covers of purple. The following will resist heavier covers, but it does not work well even with composition doctors :—

Resist Madder Brown.

1 lb. catechu,
8 oz. sal ammoniac,
1 quart of lime juice at 8°,
5 oz. nitrate of copper at 80°,
3 oz. acetate of copper,
2 lbs. gum senegal.

Magenta.—This is the name given to a red color obtained from aniline. There are several methods of obtaining it, but the best product appears to be obtained by means of Medlock's patent, that is boiling the aniline with arsenic acid. It is supplied to consumers in the liquid state. The method of its application to silk and woollen dyeing is very simple. See ANILINE, page 64. The most usual method of applying it in calico and delaine printing is by means of lactarine for the light or pink shades, and mixed lactarine and tannic acid for the crimson shades. For various methods of fixing see ANILINE COLORS. It is a very fugitive color upon cotton.

Magnesia.—Magnesia has an alkaline reaction, but feebler than lime; it is very sparingly soluble in water; it is expensive, and but little used in calico printing. Such uses as have been found for it depend upon its power of neutralizing acids, and exerting a slight alkaline reaction. It has been used with archil in Broquette's patented method; in the mixing of the color from the modified archil colors of Guinon, and as forming a discharge for indigo colors in combination with red prussiate of potash. Caustic magnesia has a powerfully injurious action upon dyed colors. If pieces of various colors from madder, indigo, logwood, or garancine, be boiled in water containing magnesia in suspension, they are in a short time surprisingly deteriorated and permanently injured. The cause of this action is not clear; probably the magnesia may displace some of the mordant of the colors, and not being itself able to yield bright lakes with coloring matters spoils the shades. Magnesia added to dyewoods, or to the water used in dyeing, is extremely detrimental; one per cent. of the weight added to a madder dye will entirely stop the dyeing, and a less quantity is very injurious. Magnesian salts are present in many waters, and may be the cause of failures in dyeing. The neutral salts of magnesia are not injurious to dyeing, it is only calcined magnesia and the carbonate which act in the manner described; a water containing magnesia may be corrected by the addition of a minute quantity of oxalic acid or a larger quantity of sal ammoniac.

Sulphate of Magnesia, or Epsom Salts, is used in calico printing to fix the lead mordant on chrome orange styles where colors are worked in combination, which would be injured by sulphuric acid. Sulphate of soda being cheaper, is generally used instead of Epsom salts. The remaining salts of magnesia are of no interest in a practical point of view.

Mahogany Tree Bark.—This bark contains coloring matters which are capable of being communicated to mordanted cloth, but they are of so dull a nature, and present in so small a quantity, as to render this bark much inferior to other dyeing matters; it is consequently not in use among British dyers.

Mahogany Color.—A reddish-brown color. The color distinctively called mahogany or acajou is produced upon calico by mordanting in a mixture of red and iron liquors; ageing, dunging, and dyeing in a mixture of equal weights of madder and quercitron bark, with addition of bone size.

Maize Color.—A low toned yellow orange. The method of obtaining this color requires no particular description, being the same as for orange. On wool, cochineal and fustic are used.

Mallow, or *Mallows Color.*—A plant, same as the French *Mauve*, yielding a bluish purple flower, gives the name of this color. There are, however, different colored mallow flowers, some of which are reddish purple, hence the name is not very precise in signification. The ordinary "mallows red" is exactly the same as dark crimson.

Manganese.—The metal manganese is obtained with difficulty, and is little known. In the state of black oxide of manganese it is an abundant natural product. There are two principal oxides of manganese, but only one of them forms compounds with acids in the general way, and that is the protoxide, formed of single atoms of the metal manganese and oxygen. It is this oxide which exists in all the commercial salts of manganese, and which is produced when caustic alkali is added to solutions of them. It is white at first, but soon undergoes a change from the absorption of oxygen; it assumes a reddish color, which finally passes into brown, and in that state it is no longer protoxide, but either peroxide or a mixture of both oxides. The other oxide of manganese is the one found native, and which is extensively employed by chemical manufacturers in the preparation of bleaching powder; it is never employed in dyeing or printing, and does not call for any further special notice.

Sulphate of Manganese.—This substance can be prepared by heating the natural peroxide along with strong sulphuric acid. On the large scale this is done in reverberatory furnaces; on the small scale, it can be done in any vessel that will stand the heat and the acid. It requires a subsequent purification to free it from earthy matters and iron. It is not much used in dyeing or printing at present, although proved capable of several applications. It serves to prepare the other salts of manganese from, as the acetate, muriate, and nitrate. It can be used as the bronze liquor, for producing manganese browns, but it is generally the muriate of manganese which serves for this purpose; it receives an application, in some places, for preparing cloth for indigo dipping, by which darker colors are obtained in shorter time than without. It has been recently patented as a substitute for sulphate of copper or blue stone, for resisting the indigo vat; but, I am informed, there are great difficulties in the way of its application, and it is not in use at present.

Muriate of Manganese (Bronze Liquor).—This compound is a secondary product in the manufacture of bleaching powder It is produced in very large quantities, so much greater than there is any demand for that it becomes a difficulty with some manufacturers how to dispose of it; but if a recent plan for converting it into the peroxide is commercially successful, this will be no longer the case. It can be made tolerably pure from the sulphate of manganese, by means of muriate of lime. As usually sold it is in a liquid of a slightly pink color, not likely to contain impurities, nor require any other test than the hydrometer. It is used for obtaining the manganese brown or bronze, by impregnating the cloth with it at a certain strength, and then passing in lime or ash. Like the iron buff, it must be well exposed to the air, in order to raise the color, or else to some oxidizing agent, as the chloride of lime. The first action is that the lime throws down the protoxide upon the cloth, and the second, that this absorbs oxygen from the air, changes its color, until it has absorbed all the oxygen it can, when it is in the state of peroxide, or thereabouts.

When the oxides of manganese are heated with alkalies in a manner favorable to oxidation, they absorb more oxygen, and form compounds with them, called manganates and permanganates. These compounds have rich colors, which, however, are easily destroyed by any substance which can take oxygen from them. A piece of calico dipped in a clear solution of permanganate of potash soon decolorizes it, while it gets permanently impregnated with the oxide of manganese. Deep and full shades of brown may be thus obtained, and are actually so produced upon silk and woollen; but as these fabrics must be oxidized in the process, it would be interesting to know whether there is not a deterioration of their strength. The permanganate of potash has been employed to mark calico, but it will not resist acid treatments. Since the permanganate is reduced by all the vegetable thickenings, it can only be applied as a topical color by being thickened with pipeclay, or some similar non-oxidizable substance.

Mangrove Tree.—The bark of this tree is capable of yielding several of the saddened shades to cotton cloth mordanted in alum. It was formerly employed for dyeing in Manchester, but, presenting no very desirable results, it has gradually disappeared from the market as a regular dyestuff.

Mauve.—As seen above, this is a French word, equivalent to the English mallows, but, on account of the extraordinary popularity of the aniline color called mauve, the term has become Anglicized, and means a violet or purple color. The mauve, distinctively so called, is the product of the action of oxidizing agents upon salts of aniline, and is nearly all made under Perkins' patent, of August 26, 1856, that is, by means of bichromate of potash and sulphate of aniline. Its application in dyeing and printing are given under ANILINE COLORS, page 64. The color upon woollen and silk is sufficiently stable, but upon calico it is very fugitive however applied, neither resisting the action of light or detergent agents. Its consumption for calico printing has, in consequence, become considerably diminished.

The *mauve noire*, or purple mallows, contains in its flowers a coloring matter apparently similar in some respects to indigo, but whether capable of application or not is unknown. Considerable quantities of these flowers are consumed on the continent, and it is suspected that, besides their uses in medicine and in doctoring wine, they are employed in some branch of dyeing.

Mazarine Blue.—A deep purplish blue color upon stuffs is sometimes called mazarine; it is simply dark Prussian blue, sometimes topped with archil.

Mercerized Cloth.—The process called mercerizing is so named from the veteran calico printer, Mr. John Mercer, who patented several processes of treating cotton cloth, Oct. 24, 1850. The cloth was subjected to the action of caustic soda, at a strength of from 40° to 70° Tw.; or sulphuric acid, at a strength of 150° Tw.; or a solution of chloride of zinc, at 145° Tw., heated to 150° or 160°. These processes, which were expected to improve the cloth for receiving color, have not been successful. For an account of the action of caustic soda upon cotton, see page 219.

Mercury, *Quicksilver.*—This metal, whose physical properties are well known, takes no part in either dyeing or printing; it is only seen in the shape of mercury gauges attached to steam boilers, in thermometers, and generally as the weight with which hydrometers are loaded. It is about thirteen times heavier than water, tarnishes when exposed to the action of heat and moisture, and forms a black substance which is an oxide of the metal. Mercury has a tendency to amalgamate with most of the metals as soon as it is brought into contact with them; it penetrates and deprives them of all their ordinary properties, making them powdery and soft; therefore this metal, and all its salts, must be kept out of contact with copper, tin, lead, or silver vessels; iron is not acted upon by the metal, but is attacked by the soluble salts of mercury. The compounds which mercury forms with metals are called amalgams, and are so distinguished from the compounds of all the other metals amongst themselves which are alloys.

The only compounds of mercury which have been much used are the bichloride of mercury, or corrosive sublimate, and the acetate of mercury; the iodide of mercury has been a little employed.

Bichloride of Mercury is a heavy dense crystalline salt, requiring much water to dissolve it, of most disagreeable taste, and is a virulent poison. Its principal employment has been for the purplish-red color from murexide, the amaranth or Roman purple—and for use in this it is generally mixed with acetate of soda to convert it, wholly or partially into acetate. Its chemical action in this case is to combine with the murexide, to form a salt of mercury, the composition of which is not well known, and which possesses the beautiful color which distinguishes this amaranth. As a rule, any color of which mercury, or a salt of mercury, forms an essential constituent, will be a loose color, for, although some compounds of mercury may be used in oil painting without fading, it must be considered that they are not exposed to the action of the air—the oil forms a varnish over them and protects them, but colors on calico and other fibres are exposed in a peculiar manner to the atmospheric influences, and also to light, under the combined action of which the compounds of mercury are unstable.

Acetate of Mercury.—This salt can be produced in a pure state by dissolving the red oxide of mercury in acetic acid. It crystallizes in pearly scales. For practical uses the acetate is made by adding acetate of soda to a solution of the bichloride.

Vermilion.—This is a compound of mercury and sulphur; its brilliant color is familiar, but it does not suit fibrous material; its great density also is an objection. Like red lead, this color can only be obtained in the dry way by means of heat; it cannot be precipitated or thrown down by mixing compounds of mercury and sulphur.

Iodide of Mercury is, perhaps, even a more brilliant color than vermilion, but it is very unstable, and changes even when kept in a corked bottle. It has been applied to calico, but it could never be any good. The process of fixing consisted in adding solution of iodide of potassium to bichloride or nitrate

23

of mercury, until the precipitate at first formed was redissolved; this was thickened and printed, and then passed in weak solution of the mercury salt to raise the color.

Metallic Colors.—Up to this time the application of metals, either in leaf or in powder, to textile fabrics has not met with any wide success; but there is reason to believe that if there were any cheap and regular methods of obtaining metallic effects there would be a demand for such a style. The chief methods which have been employed to fix metals are given under GOLD and HYPOSULPHITES.

Methylated Spirits.—This is a mixture of crude distilled spirit with a small quantity of naphtha, which the government permits to be sold at a merely nominal duty for trade purposes. It answers nearly all the uses for which spirits of wine were formerly consumed, and has been much employed in connection with dyeing and printing, as a solvent of the new coloring matter from aniline.

Mixed Fabrics, Dyeing Of.—Of late years this has become a distinct branch of dyeing, and is very much required, principally for mixed woollen and cotton goods. There are two kinds of dyeing, called double and single dyeing; in the first, or double dyeing, the woollen threads have a different color from the cotton threads, in the second both are to be of the same shade, or as nearly as possible. I give some brief hints of the methods employed in obtaining these results.

Double Dyeing. The Light Wool Blue and the Cotton Pink.—Dye the wool first with sulphate of indigo, in a bath made sour with vitriol, working at about 140°, when the shade is produced wash out and dye up the cotton a safflower pink. (See SAFFLOWER.)

The Wool Crimson and the Cotton Blue.—Treat the piece as if all wool for crimson, the cotton will not take the mordant, and, consequently, will not dye up; wash out clear, and dye the cotton Prussian blue, by first mordanting cold in a mixture of nitrate of iron, muriate of tin, and a little tartaric acid, rinse out well, and dye in prussiate of potash sharpened with vitriol. If the crimson is found to be dulled by the iron, a passage in weak spirits of salts, with some crystals of tin added, will revive it.

The Wool Yellow and the Cotton Blue.—Treat the piece as if all wool for yellow, the cotton will not dye; wash, and dye the blue as in the preceding case.

The Wool Orange and the Cotton Blue.—Treat the pieces as if all wool for cotton, *i. e.*, mordanting in tin and tartar, and dyeing in cochineal and fustic; the cotton is thereby only faintly tinged. Dye a Prussian blue on the cotton as before.

The Wool Green and the Cotton Pink.—The wool is dyed green by aluming and dyeing in fustic or sulphate of indigo, or by a mixture of picric acid and sulphate of indigo. The cotton is dyed pink by safflower.

The Wool Purple and the Cotton Blue.—The purple color is produced by working the wool for forty minutes in archil; the blue is obtained as before.

The Wool Chocolate and the Cotton Yellow at one Operation.—The piece is dyed with archil and turmeric, the bath being kept feebly alkaline. The result is that the archil goes to the wool, giving a lilac, and the turmeric to the cotton, giving a yellow. The pieces are then raised, and passed in a feebly acid bath with sulphate of indigo; the blue going to the wool converts it into chocolate, while the cotton is not affected.

The Wool Gray and the Cotton Pink.—Treat the piece with alum and tartar, and dye up the gray as for all wool with cochineal and sulphate of indigo; afterwards dye the cotton in safflower for the pink.

The Wool Orange and the Cotton Purple.—The cloth is mordanted in tin and tartar, and the color dyed with cochineal and fustic, at nearly the boiling point. To dye the cotton purple or lilac, a bath is prepared with clear water and the smallest quantity of logwood liquor that will effect the dyeing; the piece being cooled, is worked in this quite cold for fifteen or twenty minutes. If the process has been well carried on, there will be sufficient tin adhering to the cotton to enable it to dye up a lilac, while the low temperature at which it is worked prevents the wool from being affected. If, however, the color does not come up deep enough, it must be washed out of the logwood, winced in weak bichloride of tin for ten minutes or a quarter of an hour, washed, and again entered into the logwood.

The Wool Lilac and the Cotton Crimson.—Dye the wool blue with sulphate of indigo, in an acidulated bath, until dark enough; then drain and work in bichloride of tin at 4° for twenty-five minutes, and turn over into peachwood liquor, mixed with bichloride of tin quite clear, and work in or pass through a jigger until the required shade is obtained. If the cloth dyed blue be previously passed in sumac for half an hour before going into the tin, more of this metal will be fixed, and darker and more purplish crimson produced.

The Wool Green and the Cotton Chocolate.—Treat the piece as all wool for the green part (p. 256), then work in sumac liquor for half an hour and pass in a mixture of logwood, cochineal, and bichloride of tin, proportioned according to the

shade of chocolate required. If a chestnut or a yellow chocolate be required, some fustic liquor may be added.

The Wool Black and the Cotton Crimson.—The cloth is boiled
for half an hour in bichromate of potash, acidulated with sulphuric acid, say 6 oz. bichromate and 4 oz. sulphuric acid to
40 yards of delaine; but this will of course depend upon the
weight of the cloth.   Wash out, then dye in logwood, to which
a minute quantity of sulphuric acid has been added—not more
than just suffices to take the purple out of the logwood—then
work at the boil for about an hour, lift, and wash.   To brighten
the color, and to clear the cotton, work for a few minutes in
clearing liquor, made from bleaching powder and crystals of
soda, then wash and dye the cotton crimson by the method
given previously.

The Wool Royal Blue and the Cotton Pink.—The wool is
dyed Prussian blue by one of the processes given page 99,
preferably to the process of working in a nearly boiling mixture of yellow prussiate, acid, and tin salt.   The pink is given
by working in bichloride of tin, and then in peachwood; or
the safflower pink may be employed.

By these processes, and others which will readily suggest
themselves, almost any two colors may be dyed upon delaine or
fabric in which the warp or weft are of different fibres.   Only
wool and cotton have been mentioned, because they are practically the only ones which usually come under the dyers'
hands.   Silk very rarely occurs, but it can be done in the
same manner, but less successfully than wool.

Single Color upon a Mixed Fabric.—Sometimes the color can
be dyed by one process, and sometimes two different processes
have to be used, as will be seen in the following examples:—

Gray, by One Operation.—Work hot in decoction of logwood for half an hour, then lift and add copperas to the bath,
when dissolved work in for another half hour.   If the grays
are required reddish, some peachwood must be added; if yellowish, some fustic.

Royal Blue, by One Operation.—Mordant in a rather strong
mixture of nitrate of iron and crystals of tin, with a little tartaric acid; pass the cloth through several times, and leave
some hours before washing out.   Prepare a mixture of two
parts red and three parts yellow prussiate, five parts sal
ammoniac, and three parts oil of vitriol, dissolved in as little
water as it is possible to run the cloth in; heat up to 90°, and
work the cloth very well, gradually raising the heat up to the
boil in an hour; then lift, and add four parts of oil of vitriol
and a small quantity of crystals of tin, enter again, and work
at the boil until the color is well developed.   The high tem-

perature is necessary to get the wool well dyed, but it is unfavorable to the cotton, wich will be frequently found less deep than the wool; this can be remedied by afterwards treating the cloth as if for dyeing blue on cotton, quite cold.

Green, by Two Operations.—The cotton is dyed first a yellow, with turmeric kept slightly alkaline, then passed in nitrate of iron and tin crystals as for dyeing blue, washed and raised in yellow prussiate. The wool is dyed by mordanting in alum and tartar, adding at the same time fustic and sulphate of indigo; the cloth is worked hot until the shade on the wool is similar to that of the cotton.

Pink, by Two Operations.—The wool is mordanted in tin and tartar, and dyed up with cochineal to the shade. The pink on the cotton is obtained from safflower.

Light Blue, by Two Operations.—Dye the cotton first by mordanting in nitrate of iron and tin crystals, and raising in yellow prussiate; then dye the wool by working the piece warm in sulphate of indigo, acidulated with oil of vitriol.

Dark Chocolate.—The cloth is boiled in alum for about an hour, then worked cold for another hour in bichloride of tin at about 7°, and then dipped in a strong clear solution of turmeric, made by a boiling solution of carbonate of soda; the cotton takes up the coloring matter of the turmeric rapidly; when it is saturated the piece is rinsed and again worked in the bichloride of tin, and left to drain. The dyeing is completed by preparing a bath of peachwood, with a small quantity of logwood, in which is dissolved a quantity of alum and tartar; the piece is entered at 140°, and the heat increased to the boil, and taken out when the cotton and wool are perceived to have the same shade. If the colors are not equal other ingredients must be added to remedy this defect. If the wool, for example, is redder than the cotton, a little extract of indigo added soon changes it; if, on the other hand, the cotton is redder than the wool, the addition of some more logwood will equalize the shade; if the wool is too purplish, a little archil corrects it; if the cotton is too purplish, more of the red wood must be added. The dyeing of this color presents great practical difficulties and requires much experience.

Chestnut on both Wool and Cotton.—Proceed as in last case, as far as giving turmeric ground, then pass in a strong catechu bath, and raise in bichromate of potash. As the cotton takes up more color than the wool by this treatment, the shade is equalized by turning the cloth, for half an hour, in a boiling bath of turmeric and archil. The shades can be corrected, if at variance, by the same means as for chocolate.

Black on both Wool and Cotton.—The wool is first mordanted by boiling the cloth in bichromate of potash acidulated with sulphuric acid, afterwards the cloth passed, for half an hour, in a decoction of sumac, at 150° F., lifted and drained; then passed in nitrate of iron, at 6° Tw., for twenty minutes, and rinsed. The dyeing is completed in logwood and fustic, with the addition of a little tartar.

Another Method.—Steep all night in decoction of sumac, and work for an hour in a mixture of green copperas, blue copperas, and tartar, wash, and dye in a decoction of logwood hot, raising with copperas.

In all these processes much care and address is required to produce good and regular work; there is not much choice of materials, and most of the colors are fugitive. Although the wool and cotton may be very nearly the same in shade they do not long remain so in wear, the cotton fading much more rapidly than the wool under the same treatment or exposure. (See also DELAINE.)

Mordants.—A mordant is a substance which can exert an affinity for the fibrous material to which it is applied, and which possesses at the same time an attraction for coloring matters. It is necessary that it should possess these double properties, or it cannot be considered as a mordant. There are substances which combine with fibrous materials without showing any affinity for colors, and there are others which have an affinity for colors but which cannot contract any adhesion to the fibre. Neither of these can be called mordants. Mordants are not very numerous, and may be divided into mordants proper and mordants of a dubious nature; the first class consisting of those metallic salts whose oxides seem to effect an intimate chemical union with the fibre and coloring matter, and the second consisting of a number of substances which, possessing some affinities for coloring matter, do not appear to combine with the cloth in a chemical but rather to adhere in a mechanical manner. Belonging to the first are the mineral mordants, salts of iron, zinc, alumina, tin, and copper; to the second belong vegetable and animal matters, as galls, oils, albumen, caseine, and two or three others similar in nature. In wool, silk, and other animal fabrics, there naturally exists an affinity for coloring matters, but not to an extent sufficient to yield general good results.

From the above definition of a mordant, it is evident that it is an intermediary agent, uniting two substances which of themselves have no special affinity. Its powers extend on both sides, and must be equally effective on each; on the one to hold firmly to the cloth, on the other to retain the coloring matter in a close state of combination. There are very few colors which can combine with either vegetable or animal tissues without the aid of a mordant, and, of several which can do so, the majority are much improved, both in brilliancy and fastness by the presence of a mordant. The coloring matters of madder, logwood, fustic, etc., are only able to give a feeble stain to unmordanted calico, but, by the assistance of mordants, they communicate fast, full, and brilliant colors; on woollen, also, though several of the coloring matters can give a shade to the cloth, without a mordant it is poor and feeble in most cases, short of lustre, and possessing a very inferior degree of stability. On calico there are a few coloring matters whose peculiar natures allow them to contract an adhesion to the cloth without any mordant, and it cannot be said that the presence of a mordant adds in any way to the depth or stability of the colors which they can produce. These coloring matters are indigo, safflower, and anotta. On woollen the number may be extended to a few more, as picric acid, archil, and aniline colors. With the exception of indigo, which enjoys peculiar properties, unlike any other coloring matter all these are loose, unstable colors. They cannot be combined with mordants under any of the usual conditions of such combinations, and are obliged to stand alone.

There are certain conditions necessary to a mordant which should be always borne in mind, for, unless they are fulfilled, the mordanting will altogether fail or be deficient to a greater or less degree. Solubility is the first essential in a mordant; unless the metallic oxide, which is to act as a mordant, can penetrate into the very heart of the fibres it will not be able to resist washings, and it can only find such entrance by being in a state of clear solution. A capability of becoming insoluble is the second essential in a mordant. It is apparent that if the solution of a mordant could carry it into the fibre, the same solubility would carry it back again, unless means were taken to fix it permanently upon the spot to which it had penetrated. This condition of becoming insoluble is effected in a variety of ways. The metallic oxide is combined with a volatile acid, like the acetic, which flies off and leaves it insoluble in the fibre;—this is the most common and generally employed method;—or it is combined with any other nonvolatile acid, and methods taken to remove the acid by chemical means, such as passing in alkaline baths, exposing to vapor of ammonia, and the like; or, as in the case of nitrate of iron, a salt is chosen which holds one part of its oxide in a comparatively loose state, and, when diluted with water, allows the feeble affinity of the fibre to take it from the acid. This is the case also with alum, when a portion of the acid has been removed by the addition of alkaline salts. The vegetable and animal mordants are different in their method of adhering to the fibre. Oil, as used in Turkey red dyeing, undergoes a change by exposure to the air, which renders it insoluble and

irremovable by ordinary agents. Albumen and similar substances possess a mechanical affinity for the cloth; they are
applied in the soluble state, and become insoluble by heat, or
some other agent, which coagulates them; they embrace the
fabric in a reticulation of fibres, and may be considered rather as
holding to the surface than as retained in the interior of the
fibre. This leads to the question, which may be naturally
asked, what is the nature of combination which takes place
between the mordant and the fibre? Unfortunately there
cannot be any direct positive answer given to this question; it
is involved in doubt and controversy, and the evidence on the
various views is so inconclusive that no satisfactory result has
been arrived at. The insoluble particles of the mordant may
be retained in a simply mechanical manner by the fibre, which
is supposed to be hollow, and to act the part of a trap. The
fibres may be porous, and the acid solutions, penetrating the
pores, leave their oxides in contact with the walls of the pores,
from which they cannot escape, either by reason of the narrowness of the pores or some supposed angularity of the
metallic oxides. The mordants may form a chemical combination with the matter of the fibre, and so hold on to each other
until separated by more powerful chemical agents, or destruction by time; or the fibre may possess some active power of
adhesion on its exterior walls, for metallic oxides, and they
may be held together by virtue of the power of contact. All
these theories are held, and the evidence which can be drawn
from chemical or microscopical observations do not actually
tell more on one side than on the other. In the absence of
any satisfactory scientific account, we may take the practical
supposition that the particles of mordants are held by the
fibres in a state of chemical freedom, capable of exerting all
their affinities, and drawing to themselves those substances for
which they have an attraction.

The affinity which the mordants have for the coloring matters
is undoubtedly of a chemical nature; a given quantity or
strength of mordant can combine with only a certain amount
of coloring matter to produce a certain shade, and, as is the
case in most chemical combinations, the color of the resulting
compound bears no necessary resemblance to that of the constituents. The shades of color which any dyewood can give differ
for each mordant, and for various strengths of the same mordant; the same coloring matter yielding shades which appear
quite opposite in their nature, as, for example, madder, which
with weak alumina mordants gives pink; with strong, red;
with weak iron mordants, lilac or violet; and with strong, black
colors; while a mixture of iron and alumina mordants yields

various shades of chocolates. And these colors are undoubtedly derived or derivable from one single coloring matter. The power of the mordant is therefore very great in influencing the shade.

A given mordant has not the same properties upon all kinds of fibre. The tin mordant, so extensively employed in woollen dyeing, and yielding fast colors, is a very feeble mordant on cotton, and never gives colors comparable for fastness with those of alumina or iron. On the contrary, iron mordants are very difficult to employ on woollen, for reasons which have been previously stated.

A mordant may be applied to a cloth in three ways, with regard to the coloring matter. (1) It may be applied before it; this is the usual case in calico printing, for dyed goods, and frequently in piece dyeing. (2) It may be applied after the coloring matter; this is perhaps the most usual case in piece dyeing, where the pieces are first passed through the extract of the coloring matter and then through the metallic mordant; or (3) it may be applied at the same time as the coloring matter; this is the case with many colors in dyeing, and with all steam and spirit colors in calico printing. The first case, viz., that of applying the mordant before the coloring matter, is a necessity in printing designs upon the fabric to be dyed; it is necessary also that it should be fixed upon the cloth in a very perfect manner before going into the dye; because a design requires that there should be at least two shades of color, one of which may be the white of the cloth; and it is easily seen that if the mordant were loose, or floating about in the dye-beck, it would attach itself indiscriminately to all parts, and destroy the design. This fixing of the mordant necessitates several processes unknown in piece dyeing. The second case, that of applying the coloring matter before the mordant, can only be used in self-colored fabrics, and has arisen from motives of convenience and economy. It usually happens that the mordant is much cheaper than the coloring matter in regard to the quantities which have to be used, and the coloring matter can be more completely used up and less wasted by employing an excess of mordant than in the contrary case. The method of using the mordant and coloring matter together is one of limited application, because in the generality of cases insoluble lakes are formed which are not well adapted for giving good and fast colors, but in several cases of dyeing such a mixture is used. In calico printing it is possible to use the coloring matter in a concentrated state, and in combination with acids and other matters, which keep it and the mordant in a state of solution, for a time at least, until the affinities of the cloth come into play, assisted by extraneous agents, such as steaming, passing in alkalies, and the like, which cause the formation of an insoluble compound of the mordant and coloring principle, which then adheres to the cloth.   For the chief mordants see ACETATES, ALUM, IRON, and TIN.

Morine and **Moreine.**—The names of pure colorable principles extracted from fustic.

Morinda Citrifolia, *Sooranjee.*—A species of East Indian madder, said to dye up very durable, but somewhat dull colors. It is used by the natives to produce a species of Turkey red dye.   According to Dr. Anderson, who made experiments with a substance supposed to be roots of morinda citrifolia, it gives no colors to ordinary mordanted cloth, but, curiously enough, dyes up full colors with cloth prepared for Turkey red dyeing, Bancroft, and others, who have examined a substance under the same name, found no difficulty in dyeing common mordanted cloth with it.

Mosses.—The *Chondrus crispus*, commonly called carrageen or Irish moss, yields a mucilage, which has been used as a substitute for size in finishing.   It has also been employed as a substitute for gum in block printing; its thickening powers are, however, inferior.

Some mosses contain coloring matter: the *bryum stellare* gives a colored juice—brown at first, which soon changes into green, and finally acquires a blue-green color.   The lichens, which are nearly similar to the mosses, yield several colors.

Mungeet, *Manjit, etc.*—This is a kind of madder imported from the East Indies.   It does not contain much coloring matter, and appears to be rather the reedy stem than the root of a plant, as our European madders are.   It is used for making low qualities of garancine; it can be employed, but with doubtful advantage, as a substitute for madder in some classes of work : it does not stain the whites so much as madder.

Murexide.—The red color obtained from this substance has created a great deal of interest amongst printers and dyers since its introduction, about five years ago.   For purity and brilliancy of shade it was not excelled by any other color, and, though expensive, it was easy and certain in its application. Its fugitive nature has for the present limited its employment very much, but it is to be hoped that something may yet be done which will make this magnificent color better able to stand the effects of light and air, when it cannot fail to be frequently and regularly employed in certain styles.

It was known, upwards of thirty years ago, that when the uric acid, obtained from guano, or other sources, was treated with nitric acid, it yielded a white crystalline substance, called

alloxan, which stained the fingers and nails red, and a method
is given, in Leuch's treatise upon coloring matters, by which
woollen cloth may be dyed purple by steeping it in solution of
alloxan, drying, and then passing a heated iron over it. Not
much more notice was taken of this matter until the year 1853,
when Sacc and Schlumberger revived the idea, and communicated a paper to the "Société Industrielle de Mulhouse,"
accompanied by specimens of color upon woollen cloth. There
was not much in their communication that was really an advance
upon the statement in Leuch, as far as regards the coloring of
woollen cloth by alloxan, but there were many useful suggestions and statements, the results of an advanced chemical
knowledge upon the subject of uric acid and its transformations.
The authors attributed the production of the color by the heated
iron to the transformation of the alloxan into murexide, or
purpuric acid; but the idea does not seem to have occurred to
them to make murexide itself and try it as a color, and they
state definitely that they cannot obtain any color on cotton
cloth to withstand cold water. The sanguine authors of this
communication hardly dared to look upon this subject as one
which came within the limits of practical application, but rather
as an interesting experiment which must be confined to the
laboratory on account of its difficulty and the costliness of the
materials used. The making of murexide itself, and the possible
application of it as a coloring matter, would even at that time
be considered as impracticable; for even as a laboratory product it was a rarity, and beyond the supposed difficulty of its
preparation was the belief that its solubility and the absence of
any colored combinations of it with metals would prevent its
being of any use as a coloring matter. Some time afterwards
a French chemical manufacturer secured a patent (English
patent, dated Feb. 3, 1857), and sent abundance of murexide
into the market, along with a process of applying it on calico.
The first murexide sent into the market was a reddish purple
powder, dissolving in water with a fine purple color, leaving a
little residue undissolved. Improvements were soon made in
its manufacture, and there has been sent into the market murexide in crystals almost chemically pure, and with that green
metallic reflection peculiar to this body and the wings of certain
insects. There are two principal methods by which it can be
prepared, but the details would be too long and inapplicable
here, because it is a pure manufacturing process, and not as
such specially connected with printing or dyeing.

The composition of murexide is not known with any degree
of certainty; it may be looked upon as an acid body, or containing an acid, called from its color the purpuric acid; and

the purplish-red color it produces with oxide of mercury, or a salt of mercury, may be called the purpurate of mercury. The method of obtaining the color from murexide can be modified in several ways; but it is usual to make a strong solution of nitrate of lead, from three to five pounds per gallon of water, thicken, and while at blood heat dissolve it in the required quantity of the murexide, from four to eight ounces per gallon, stirring it well. This color is printed, aged a short time, exposed to the vapors of ammonia for a few minutes, passed through cold water, and then into a raising pit containing bichloride of mercury, or corrosive sublimate, at the rate of about two ounces to a gallon of water, and usually mixed with a small quantity of acetate of soda and acetic acid. The color may be considered as finished, or it may be again exposed to ammoniacal vapors to give it a modified shade, or passed in dilute acetic acid. The exposure to ammonia in the first instance is sometimes dispensed with, but I do not think so good or full a shade can be obtained without it. By using acetate of zinc instead of the bichloride of mercury an agreeable shade of yellow is produced, but not so much esteemed as the red. This color stands washing well in water, and is not injured by weak soaping; it retains its brilliancy for at least two years, when kept from air and light; but, unfortunately, a short exposure to a good light is enough to spoil it, and it must be classed amongst the fugitive colors. It is quite possible that some modification of this coloring matter may be found to give more permanent colors, but it is contrary to all analogy to expect that a mercurial mordant will ever do it. All the salts of mercury are subject to decompositions in a more remarkable degree than the other salts of metals used in printing and dyeing, and even among metals of the same chemical class they are distinguishable by their inferior stability. This is the case with mineral acids, and is more remarkable still with organic acids. It would be a matter for surprise if a compound of mercury, with a highly organized body like murexide, should not be alterable under the agencies of heat, light, and air. The murexide red undergoes some molecular change on the cloth not accompanied by a change of color, but connected with some change of form which renders it less adherent to the fabric. If a murexide colored dress be worn a few times under favorable circumstances, it will not differ much in shade from a piece of it which has been exposed to the same atmospheric influences but kept quiet. If the dress and the piece kept be subjected to washing together in cold water, the dress will be found to lose color to a perceptible extent, the water becoming colored at the same time; the piece not worn does not suffer to nearly the same degree. The same thing happens again, to a still greater extent, the worn dress losing much more than the unworn patch, until at a third washing the dress will have lost all its brightness and become yellow. This seems to be owing to the movement of the fibres one upon the other, inducing a change in the form of the mercurial pigment; for it may be observed that the parts most subject to rough movement, as the lower flounces which touch the floor, are most and soonest injured. It is probable that the precipitated purpurate of mercury is crystalline in form, and not amorphous; at any rate its behavior suggests something of that nature, and the taste of mercury which can be perceived upon masticating a little calico colored with it, seems also to point out a degree of solubility in the compound, or else a power on the part of the saliva to decompose the colored salt and appreciate the mercury. The red color is easily injured or destroyed by heat, it will not sustain the action of steam, and though a little soap does not much injure it, and may in fact be employed when cold to produce a modified shade, a boiling in soap is almost destructive to it. No attempts to fasten this color have been successful thus far, and though represented by some houses as a fast color, and sold as such, it is only in the name; there is actually no difference in the stability of the color thus guaranteed and that ordinarily produced.

Woollen does not take color well from murexide; it can be dyed with it, but it does not receive full and deep shades. The best color from the uric acid products for wool is derived directly from alloxan, which is a colorless body in itself; but being dissolved in water and applied to woollen cloth, and the material hung up in a room with ammoniacal vapors circulating in the atmosphere, it takes a deep reddish-purple or amaranth color. This change can be effected more rapidly by heating the woollen, but not so regularly. The color thus produced resists washing in water and weak soap, and is not particularly acted upon by atmospheric agency. The comparative expense of the alloxan, and the possibility of obtaining shades nearly similar with cheaper substances, will prevent this color from being used to any great extent.

Silk receives a fine red color from murexide. It may be communicated by passing the silk first in a bath of the bichloride of mercury, and then adding the murexide to the liquor; or, having two liquors, the mercury and murexide, separate, and passing from one to the other until the required shade has been obtained. The kind of red thus produced is not one in demand, and I believe is never made now. It is subject to the same influences and changes as the color upon calico, fading and washing out like it after an exposure of a few days to the air and light.

The orange yellow color which can be produced from murexide, by means of the salts of zinc being used instead of mercury, is a pleasant color, but not remarkable. A similar and hardly distinguishable shade can be obtained in several ways preferable on the score of economy and stability of color. It is not used.

Myrabolans.—An East Indian product, which contains a kind of tannin matter. It has been slightly used by the dyers of this country, as a substitute for galls and sumac.

N.

Nankeen Color.—This name is usually given to the shade of buff obtained from iron salts. In piece dyeing the cloth is run through copperas liquor, and then through lime water. For lighter shades nitrate of iron is preferable. The color may be softened both in appearance and to the feel, by finally working in warm soap suds. Anotta gives rather yellow nankeen shades, which may be combined with the chrome yellow. As nankeen is compounded of yellow and red, it may be produced by employing red and yellow dyeing materials in conjunction.

Naphthaline.—This is the name of a solid greasy substance which is obtained in large quantities from coal tar. It enters into an infinite number of combinations with the chemical elements, some of which approach so near in composition to natural coloring matters that strong hopes have been entertained of being able to produce them from it. Up to the present time, however, it is not known that any available color has been obtained from this source.

Nenuphar, *Nymphoea Alba*, *White Water Lily.*—This plant contains an astringent matter, which enables it to answer the same purposes as galls and sumac. It is widely used in the eastern parts of the continent in garancine dyeing, but I am not aware of its having been applied in England.

Nicaragua Wood.—One of the red woods similar to peachwood and sapan wood; the same apparently as Santa Martha wood. Its quality is variable, sometimes not worth one sixth of the price of Brazil wood, sometimes worth as much as one half.

Nickel.—A comparatively rare metal, distinguished by yielding salts of an apple green color. It has been tried both as a substantive color and as a mordant, but did not give any promising results. With dyewoods generally it gives the same kind of colors as iron.

Cobalt is a metal usually associated with nickel, and nearly resembling it in chemical character. It acts as a mordant, but does not yield any specially interesting colors.

Nitric Acid, *Aquafortis.*—Nitric acid is a compound of nitrogen and oxygen ; it is entirely a manufactured article, never being found free in nature; it is all obtained from either saltpetre, which is a nitrate of potash, or from nitrate of soda, the Chilian saltpetre. It is a strong acid, and possesses powerful oxidizing properties. It is not used in bleaching, and very little in either dyeing or calico printing, but it is employed in considerable quantities by the manufacturing chemists who make drugs for dyers and printers. It is used for making nitrate of iron, nitrate of lead, oxymuriate of tin, and several of those solutions of tin known as dyer's spirits ; it is used for a few colors in calico printing, and sometimes to cut madder pinks, that is, to reduce the red to a softer shade. It is occasionally employed to give silk a fast yellow color, by passing it through tolerably strong acid, and washing directly. It has the property of tinging all or most animal substances of a yellow color; it is used for etching rollers and deepening engravings, and in several other trifling cases. The strength of nitric acid is usually determined by the hydrometer, which is a good enough test between honest people, but it is possible to make it seem strong by putting vitriol in, and one or two other substances. Commercial nitric acid is occasionally contaminated with large quantities of spirits of salts or muriatic acid. It is the most expensive of the acids used in large quantities, and there are inducements to mix cheaper acids or salts with it. The test for muriatic acid is nitrate of silver ; for sulphuric acid, or vitriol, nitrate of baryta, diluting the aquafortis to be tested with pure water, so as to allow the tests to show properly. If any solid substance is present it is detected upon boiling down a little of the acid to dryness, in a proper vessel. If the acid be unadulterated its strength will be in proportion to the figure it stands at upon the Twaddle. The very strongest nitric nitric acid that can be made contains some water, so that commercial nitric acid is a mixture of a certain quantity of real supposed dry nitric acid and water, and by consulting the following table the quantity of real acid in commercial aquafortis, of any given strength, can be ascertained. The table is adapted from Dr. Ure's determinations.

*Table showing the Quantity of Anhydrous Nitric Acid in One
Hundred Parts of Liquid Acid, at various Densities.*

Twaddle.	Dry acid in 100 parts.	Twaddle.	Dry acid in 100 parts.	Twaddle.	Dry acid in 100 parts.	Twaddle.	Dry acid in 100 parts.
100	79.70	65	44.50	46	31.20	28	19.90
96	73.32	62	42.20	44	30.00	26	18.50
92	68.54	60	40.60	42	28.80	24	17.20
88	64.50	58	39.10	40	27.60	22	15.90
84	59.77	56	37.80	38	26.40	20	14.40
80	56.58	54	36.60	36	25.00	15	11.00
76	52.60	52	35.10	34	23.90	10	7.50
72	49.41	50	34.00	32	22.50	5	3.50
68	47.00	48	32.60	30	21.40		

The best method of ascertaining the actual value of nitric
acid is to determine the amount of other acids present, and
deduct them from the gross acidity of the sample under examination. The hydro-chloric acid is determined as chloride of
silver, and the sulphuric acid as sulphate of baryta. The
acidity of the whole may be tested by ascertaining how much
pure carbonate of soda a given quantity can neutralize. Pure
nitric acid for laboratory or special purposes is obtained by
precipitating the hydro-chloric acid from ordinary acid by
nitrate of silver, and then distilling. If there is no other impurity but hydro-chloric acid present, it may be expelled by
keeping the acid at near its boiling point, on a sand bath, for
several hours.

Nitric Oxide.—It is a gas, colorless, mixes with air or oxygen, and forms red or orange colored vapors, such as are produced in making nitrate of iron, or several kinds of dyer's
spirits. Its affinity for oxygen is very great; it is best prepared by acting upon nitric acid, with copper turnings; the
copper withdraws oxygen from the nitric acid to form nitrate
of oxide of copper, and the gas is evolved.

No applications; but is, I think, capable of receiving some.
It is a carrier of oxygen, and may, perhaps, be usefully employed as such; it is not applicable as an oxidizer for colors,
on account of the formation of the nitrous acid and its apparent
conversion into nitric acid. Acting under the impression that
the oxymuriate of tin, used for cutting or altering the shade of
madder reds, produced its effects by some nitrous acid, or hy-
ponitrous, contained in it, I tried the effect of the bi-oxide of
nitrogen mixed with air upon strips of madder red previously
damped. The action of the gas was prompt and energetic,
the whole color was changed to a yellowish shade; upon soaping, it devolved into a fine pink. If the gas was too strong,

or the fents left in too long, the change to yellow became permanent, and soaping only developed a cinnamon shade, which appeared to be quite as stable as other madder colors, resisting the action of soap, acids, and chloride of lime. This experiment only proves that the nitrous acid generated by the mixture of the gas with the air could produce the same effects as the oxymuriate.

Nitrous Acid.—This acid is formed when the bi-oxide mixes with air, and is the ruddy colored gas which is produced in several cases of the action of nitric acid upon metals, when the operation is carried on in contact with the air. To procure it in its pure state, nitrate of lead is heated in a retort, when it distils over. It has a suffocating effect when inhaled, supports combustion, and is readily decomposed. No applications at present: the remarks under the bi-oxide are applicable here as well.

Nitro-Cuminic Acid.—This acid is interesting on account of the colors which it yields. According to M. Jules Persoz, calico dipped in it made hot, and exposed to the sun, acquires a scarlet red color, which, by various treatments, is converted into a pure and bright pink.

Nitrogenous Matters.—This term, frequently found in treatises upon dyeing, indicates a class of organic substances which contain nitrogen. Most animal matters, such as flesh, urine, dung, milk, etc., are nitrogenous substances. The use of animal substances in cotton dyeing is thought to be connected in some way with the communication of a nitrogenized principle to the fibre, but the efforts of chemists to investigate this point have not yet succeeded in penetrating the obscurity which surrounds it.

Nona.—An East India root, apparently of the same kind as madder, being used to obtain the same colors. It is not imported, or, at least, is not known in England under this name.

O.

Oak Bark.—This bark has been slightly used in dyeing to produce shades of drab, gray, etc. Its value lies altogether in the astringent matter which it contains, the quality and quantity of which is not so suitable for the dyer as for the tanner.

Oils and Fatty Matters.—Oils of various kinds are used in printing and dyeing to a considerable extent, but in only one or two cases as directly connected with the production of colors; but these oils are so important in many respects as

24

subsidiary agents, that a knowledge of the properties of the principal of them is necessary. Oils are divided into two classes, namely, fat oils and essential oils; a fat oil being what is popularly understood under the name of oil, its distinguishing character being to give a greasy spot on paper which does not disappear by warming; the essential oils resemble turpentine, most of them have well defined smells, they are thin and volatile, and a stain made by a drop upon paper disappears when strongly warmed. They are none of them employed in printing or dyeing, except the oil of turpentine. The fat oils (and by this is meant to include those solid fats like tallow, which do not take an oily consistency in this climate) are again divided into two classes, called rancid oils and drying oils, from the manner in which they behave when exposed for long periods to the action of the air. An oil which dries up or forms a skin over the surface, or that shows any inclination to thicken into a resinous or gummy substance upon long exposure to the air, is called a drying oil; linseed oil is the best common example of this class of oils. When, on the other hand, an oil shows no inclination to skin over or dry up, but sometimes to become more fluid, and at the same time gives off a sour smell, and has that appearance and taste known as rancidity, it is classed among the rancid oils; some of these oils grow firmer on standing, but they never go tough or form skins. The animal fats, as tallow and sperm oil, are good instances of this class. The term "rancid oil" is not a good one, for some fats of this class do not go sensibly rancid by very prolonged exposure to the air, and probably never would become rancid; it actually means an oil which does not go thick, in contradistinction to one which does. The drying up of an oil on the one hand, and its rancidity on the other, are owing to the same chemical cause, which is the absorption of the oxygen from the air, and the production of some essential modifications in the internal composition of the oil. Exposure to the air is essential to this change, for in quite close vessels the drying oils remain thin and the rancid oils sweet; but as, practically, all common vessels contain and admit of a circulation of air within them, so these effects of drying and rancidity do make their appearance even in corked up phials, but of course only when so kept for great lengths of time; so much the more do their characters appear in oil casks and in oil reservoirs; but they are in their greatest state of activity when the oils are exposed to the air in thin films, as on paper soaked in them, or painted on wood, or as lubricants on machinery. The drying oils are suitable for painting and varnish making, and quite unfit for machinery, while, on the contrary, the rancid oils are adapted for all lubricating purposes, and quite unfit for painting, never becoming dry. The chief fatty matters in use on a print or dye works are as follows:—

Tallow.—This is the melted fat of animals, and one of the most valuable of all fats; it undergoes very little change on exposure to the air, and is solid at the temperature of this country. It is used as a grease for machinery, either alone or in combination with other fats or substances; principally, however, on account of its not being fluid, it is employed in warm places, or warm bearings, or on large wheels. It never goes acid when good. The best kind of soap is made from tallow, but on account of the comparative dearness of this fat it is seldom used alone in soap making, but mixed with other oils of less value. The soap sold as white curd soap should be a pure tallow soap.

Sperm Oil.—This is an animal oil, and the finest and best of all oils for lubricating machinery; it is very thin, and seems to resist the action of the air for any length of time. It is the most expensive of all the common oils, and is only used for particular purposes in lubricating and for burning.

Olive Oil (Gallipoli Oil).—This is perhaps the most important of all the vegetable oils, from the quantity which is produced in Europe, and the many useful purposes to which it is applicable. It is extracted from the fruit of the olive, the commercial quality being about the third pressing of the fruit, after two pressings have taken out the finer qualities used for domestic purposes. Besides the pure oil, it contains a quantity of vegetable matter of an albuminous nature, which makes it thicker and more ropy in appearance. This oil should be sweet to the taste and clear to the eye, of a dull yellow color, and an odor free from rancidity. It is often adulterated with cheaper oils; and it is a matter of the greatest difficulty to ascertain if this is the case by chemical means, still more so to judge of the nature of the adulterating oil and its proportion to the rest. Practical testing and comparing of qualities on the large scale are the only reliable means by which the value of a sample can be estimated; long experience enables a person to judge at once by the taste and smell what kind of an article is offered to him.

Gallipoli oil is equally fitted for lubricating machinery and making soap. It is not often used for this latter purpose in England, but the best known soaps of the continent, the southern countries especially, are all made from this oil. Gallipoli oil is employed by the Turkey red dyers as a preparation for the dyeing; for their use it must possess some peculiar properties not necessary to its goodness for general purposes.

It must mix up thoroughly with a weak solution of pearl ash, forming a milky fluid, which must not break or throw up any oily globules for a period of at least twenty-four hours. Not all the oil in commerce possesses this property; there are means of making it suitable when it is not so, but I believe that the bulk of common Gallipoli oil answers without any preparation. This property of forming what is called an emulsion is possessed by several vegetable oils, but by none in so perfect a manner as by this oil. Pure olive oil does not form a perfect emulsion, and it is supposed that the ordinary oil owes its qualities in this respect to the impurities it contains. An oil which will not mix with weak alkaline solutions by itself, will do so very readily if beaten up with the yolk of an egg, which seems to prove that there is something of an albuminous nature in the regular oil. It is quite plain, therefore, that an oil very good for one purpose, would be either inferior or bad for another use. Gallipoli oil is in regular use in color shops for mixing with colors, especially paste colors, to give them smoothness and enable them to work better; in gum colors it is used to prevent frothing.

Linseed Oil.—Linseed oil is of two kinds, the raw natural oil and the boiled oil. The boiled oil is only used in painting, and for a few other purposes where it is required to dry up into a kind of varnish. The raw oil is of a drying nature, but not so strongly marked in this respect as the boiled—the boiling being for no other purpose than to heighten its properties as a dryer, and to give it a little more consistency. Linseed oil is used in colors, the same as Gallipoli. It is usually lower in price, and there is an economy in employing it; for keeping down froth I believe it is rather better than Gallipoli. Linseed oil does not make good soap. Drying linseed oil is the basis of all paints, and also of most of the attempts made to produce oil colors for printing on calico. The chief difficulty in making an oil color fit to print on the machine seems to lie in depriving the oil of its flowing character or its greasiness; that property which causes it to spread beyond the limits of the design, and give an unpresentable appearance to the whole cloth. At the same time, nothing must be done which will injure the transparency of the vehicle, nor interfere with the hue of the colors to be employed. Boiled oil is inadmissible on account of its color, especially for all light or bright colored pigments. By repeatedly boiling oil in conjunction with earthy matters, it can be made into a kind of tough varnish, capable of being thinned down with turpentine, and when printed on calico not spreading beyond its proper limits. But this injures the color of the oil for all except the darkest colors, and the use of much turpentine is open to objection. Liebig published a method of converting linseed oil into a very drying oil without injuring its color. The process consists in shaking the raw oil up in a bottle with a basic acetate of lead and powdered litharge, along with water, until it had been acted upon and taken up a quantity of the lead in solution. It could then be washed by shaking with water, and the lead contained in the oil removed by agitating with weak oil of vitriol. By this process a drying oil can be obtained, which is all that can be desired, as far as color is concerned, but is deficient in other respects; it is as thin as the original oil, is quite greasy, and though it dries up pretty fast, it will run very much. The published processes take no account of an alteration of the properties of the oil caused by removing the lead which is held in solution by it. The presence of the lead is objectionable because it changes color on the cloth, becoming brown, probably owing to the sulphurous vapors in the air, probably from absorbing some oxygen, and becoming changed into the higher oxide; but its removal destroys the drying property of the oil. Without the lead it is little or no better than raw oil. To obtain a product having greater consistency, I boiled some of the oil made by Liebig's process without removing the lead; it became colored immediately and went worse as the heat was increased, until it was quite unserviceable. Oil from which the lead had been removed by acid behaved much as raw oil would do under the same circumstances, becoming colored to an objectionable degree as it boiled and became thicker. All the efforts I made to get a consistent colorless oil, which would print without spreading and dry up quickly, were of no avail. Some results I obtained of a promising nature by attempts to thicken the oil with various substances, as amber, rosin, and the like, and I was convinced that, sooner or later, the difficulties which seemed insuperable would be overcome, and oil printing become a regular and valued branch of calico printing. It resulted from my experiments that fast colors of great brilliancy and beauty could be obtained, and I did obtain them, but by processes too delicate and too costly to be practicable. I was of opinion that the cloth, in its state of looseness of texture, was not well adapted to receive oil colors, and that it would have to be prepared—the interstices being filled up with some thickening substance, and the whole fabric made smooth and absorbent—not for a permanent state, but for the application of the color, and while it was being fixed, to be washed off afterwards.

Other drying oils are *poppy oil* and *nut oil;* they are more expensive than linseed oil, but not too much so to be capable of being used in calico printing if their properties were particularly valuable. They are used by articles as drying oils.

Palm Oil.—This valuable product comes from the African coast; it has received its name from the tree whose fruit yields it. It is called an oil because it is fluid in the tropical climate where it is extracted. It is not used to any considerable extent by dyers or printers unless they make their own soap, and for this purpose it answers better than any other fatty substance. It is easily saponified, and suits very well all the needs of the printer and dyer of calicoes and woollens. It has a strong yellow color, of which it can be deprived by many processes, and brought into a white state, when of course soap made from it is white, while the soap made from unbleached oil is yellow.

Cheaper fish oils, as *cod oil* and *whale oil*, are not much in favor on account of their odor, which is very persistent, adhering to cloth through all kinds of processes, and making itself sensible to the smell in a disagreeable manner. They are employed in making soft soap, whence the odor of that substance is derived.

Spermaceti.—This is an animal fatty matter, obtained from a species of whale. It is not exactly an oil, although possessing many properties of oil; it mixes with oils and turpentine. It is sparingly used in color mixing for some colors, especially for a black which has a logwood basis. It gives brilliancy to colors, and is rather more manageable than other fatty bodies when mixed in colors.

Bees' Wax, which was formerly used as a resist, is now seldom employed. On specimens of calico printing which come from India the wax resist may be seen yet. It acts as a resist, and effectually so, in all cases where the liquors are not hot or strongly alkaline; it is a resist of the mechanical sort.

Action of Sulphuric Acid upon Oils.—When strong oil of vitriol is mixed with a fat oil, and left in contact for some hours, it causes a change to take place in the oil, which can then be dissolved in water. This method of treating oils, for applying them to dyeing, has been patented by more than one party, the patents of the dates June 22, 1846, to Mercer and Greenwood; and to the same parties on March 15, 1852, may be consulted upon this matter. The object of the inventors has been mostly to shorten the Turkey red process, but what degree of success they have met with I do not know.

Oil as a Mordant.—The use of oils as assisting cloth to absorb and retain coloring matters is of very ancient origin. It is used chiefly in Turkey red dyeing, and appears to be an introduction from the East, improved as to the manner of its application by the science of our own time and country. It plays a part simi-

lar to that ascribed to the astringent principles, capable indeed
of acting as a mordant by itself, but producing thus only dull
and feeble colors; but combining with other mordants, elevating their affinities and communicating stability to the colors,
produced by them. It seems necessary that the oil should
undergo some chemical change upon the cloth before it is fitted
for use as an assistant mordant; this change is most probably
an oxidation. The oil which has been exposed upon the cloth
for some hours or days is evidently altered in its properties in
some way or other; it is not removable by the same agents
which act upon quite fresh oil ; that though the oil most proper
for this purpose never passes to the resinous state of the drying
oils, it loses a great deal of its greasiness, and may be supposed
either to have partially dried up or to have formed some sapona-
ceous combination with the mordant, by which its oily nature
is disguised. We have no satisfactory information as to the
rationale of the Turkey red process; it is a series of empirical
treatments perfected by practice and long years of experience,
which science is as unable to improve as to explain. The Industrial Society of Mulhouse offer a prize, and have done for
many years, to any person who can give a satisfactory scientific
explanation and justification of the Turkey red processes, but
the prize remains unclaimed or unawarded—a sufficient proof
of the little knowledge possessed by chemists upon the matter.
What can be said is, that the oil enables the cloth to take the
mordant more readily and in greater quantity. If one end of
a strip of calico be dipped in the emulsion of oil and ash used
by the Turkey red dyers, and placed for a few hours in a warm
place, the fatty matter will be found to have contracted a permanent adhesion to the cloth, and if put into a madder dye it
will be stained of a dull red, not possible to clear off by the
usual clearing processes. If this strip, one end of which has
not been in the oil, be boiled in a weak solution of alum for
some minutes before dyeing in madder, a more striking difference is seen. The oiled part dyes up a red of no great depth,
but in great contrast with the other end, which remains colorless. These two experiments indicate clearly, firstly :—that
oil itself can attract coloring matters; and secondly, that it
possesses some power of withdrawing the alumina from alum
which is not possessed by the fibre. This is nearly all that is
clearly known of the properties and uses of oil ; besides assisting in the attraction of the coloring matter, it communicates to
it a permanence when fixed. This can be readily understood
upon general principles as attributable to the presence of the
oily matter, covering and shielding the changeable coloring
principle from the action of the destructive natural agents to

which it is exposed; an effect which is partially supplied by
the use of soap in other cases. An illustration of the affinity
of oil for calico and colors occurred to me in an unexpected
and puzzling manner, and its recital may be of some use perhaps to others. A number of pieces having been printed in
wrong mordants for madder dyeing were discharged, after about
a week's age, and before dyeing. The colors were for black,
chocolate, and purple; the discharge was simply a warm souring and slight chemicking to take the yellow off the cloth.
The pieces were printed again with various others some weeks
afterwards, and dyed as usual, but they showed the old pattern
through the back of the piece, and though not strong, it was
sufficient to spoil the color on the face and job the pieces. At
first I suspected the pieces had been originally printed in red
with crystals of tin in the red liquor, and it is well known that
the tin cannot be removed without a bowking or strong spirits
of salts treatment, and pieces for discharging containing that
color were kept separate. By testing the unprinted tab ends
I was certain there was no iron left on the cloth, but neither
could tin be detected. As tin is not easily found by chemical
tests when in very small quantities, the non-detection of it did
not prevent me still attributing the reappearance of the pattern
to it until, upon examination, it was found that those patterns
had never been printed in red, and they were traced to the precise lot spoken of as being printed in black, chocolate, and
purple, and it was evidently the black outline which showed
through the other colors to their injury. It was some time
before I fixed upon the oil, which was used in the color, as the
cause of the mischief, but I proved it very satisfactorily by
various experiments. Only about a half noggin was used to a
gallon of color, yet it had fixed upon the cloth, and had gone
through the hot sours, washing and chemicking, to re-appear
as a mordant several weeks afterwards. It was easy to prevent
the repetition of this accident when once its cause was discovered. It is probable that the chemicking, and the time
which elapsed between the discharging and subsequent dyeing,
enabled the oil to become oxidized and quite fast. If the pieces
had been reprinted, and dyed soon after discharging, I think
the color, attracted by the oil, would have disappeared in the
soaping. Not all oil produces this effect, it is only an exceptional case I believe, for evidently the acidity of the mordant
is not favorable to the oil forming an intimate union with the
fibre; it is, on the contrary, in the presence of alkalies that it
combines most effectually with the material of the cloth.

Cloth is oiled, in Europe, by means of an emulsion made
with pearl ash; but in the East, where the natives still produce

the finest and fastest reds known, it is applied in many various ways; but it appears that the best informed of them adopt a plan similar to the method in use with us, that is, diluting the fatty matter by some liquid which brings it, if not into solution, at least into a state of very fine suspension and division. Many of them, however, plunge the material to be dyed into the neat oil, and work it there, wringing it out and exposing it to the air. Many kinds of oil are used; in Europe it is generally an inferior quality of olive oil, but in Asia, fish-oil, lard, and other fatty matters are successfully employed.

I would notice here an impression held by some chemists that oil alone if properly modified would form a sufficient mordant for madder-red. I can find no ground for this idea, except in a very briefly reported and unconfirmed experiment of a pupil of M. Persoz, to the effect that prepared Turkey red cloth when treated by acetone, yields an oily matter, which, being transferred to a fresh untreated cloth, mordants it in an effective manner; and a further statement that M. Chevreul had tested a certain Turkey red color, which contained *very little* alumina. I do not share in the opinion of the sufficiency of the oleaginous body as an actual mordant, because the experiments quoted are not sufficiently clear, and because I have made many experiments without obtaining anything beyond a mere stain in the madder dye. It was stated in 1846 that M. Chevreul, who has made himself famous by his successful labors upon fats and oils, was trying to elucidate this matter; but, from the absence of any published account of his experiments, we may presume that he has not solved the question, which still remains a difficulty in dyeing chemistry.

Drying oils have been used as vehicles for pigments, but they are difficult of application and require special arrangements and skill to make them any good; for this reason, oil colors are scarcely ever found on calico. In some very superior qualities of continental work they may be seen producing very striking effects; but it is evident that they are applied by block upon carefully prepared grounds, and must, including the labor, be very expensive. I believe that in the course of time oil will be extensively used as a vehicle for colors in calico printing; but several improvements or alterations in the machinery will be necessary, and some more complete and effective means of taking the quality of greasiness out of drying oils discovered.

Olive Colors.—The color of the olive may be defined as a dark dull green, such a color as would be optically produced by mixing a pure dark green with black or brown, or what comes to the same thing, mixing some red with it; or again, as

in some practical receipts, mixing purple and green colors
together.   Thus an olive color for delaine is mixed by taking:—

> 2 gallons dark green color,
> 1 gallon dark purple (see DAHLIA).

In France, and in some parts of Great Britain, olive color
means a kind of brown.

Orange Colors.—Orange is a mixture of yellow and red,
and with the exception of the chrome oranges (see page 144),
these colors, in bath dyeing and printing, are produced by the
combination of red and yellow parts, as in the following
examples:—

Orange for Delaine.

> 1 gallon water,
> 2 gallons bark liquor at 18°,
> 5 lbs. starch; boil, and add
> 1 pint cochineal liquor at 8°,
> 2¼ lbs. crystals of tin.

This is rather a red-orange, a smaller amount of cochineal
would be better; for a bright yellow-orange, the cochineal
liquor may be left out altogether.

Orange for all Wool.

> 6 quarts bark liquor at 18°,
> 3 quarts cochineal liquor at 3°;
> 5 lbs. gum,
> 8 oz. oxalic acid,
> 1 lb. bichloride of tin.

In dyeing, orange colors are likewise obtained by combining
yellow and red elements; on woollen, the cloth is mordanted
in bichloride of tin and tartar, and then dyed in a mixture of
cochineal and fustic, proportioned according to the shades
required.

Upon silk, shades which may be called orange are obtained
directly from anotta, this dye stuff being dissolved in soft soap,
and the silk worked in until the right shade is obtained.   An
orange color from anotta for printing on silk is composed as
follows:—

Anotta Orange for Silk.

> 6 gallons water,
> 10 lbs. pearl ash,
> 4 lbs. anotta, boil down to one half:
> 1 gallon of the above and
> 1 gallon gum water.

Orpiment, *Yellow Sulphuret of Arsenic.*—Orpiment is a compound of sulphur and metallic arsenic; it has a good yellow color, and has been used to some extent as a coloring matter. It was applied to cloth by dissolving it in ammonia; padding the cloth in this clear liquor, and then hanging up till the ammonia evaporated and left the orpiment fixed to the fibre. Its principal use in calico printing is as a reducing agent; for, when mixed with caustic soda, or potash, it has a strong affinity for oxygen, and will take it from many substances—among others from indigo. It is used in this way in preparing the blue long known as pencil-blue, and which, in later days, was tried to be applied as "gas-blue," but without success. Orpiment is a component in many of the receipts for China blue, and it appears to fulfil some useful part in it, although China blue can be obtained without it.

Orpiment is poisonous, but not to the same extent as white arsenic, or arsenic acid.

Oxalic Acid.—Oxalic acid exists naturally in the juice of some plants in a state of combination with potash, and for many years there was no other method known by which it could be obtained than from the plant. At length it was found that when nitric acid acted upon sugar and some other vegetable matters of the same or similar composition, it produced oxalic acid, and until very lately all oxalic acid was thus produced. Messrs. Roberts, Dale, and another have recently patented quite a new method of making it from sawdust, and instead of an acid using alkalies. Oxalic acid is sold in a crystalline state, the crystals are small and soft; it is very acid to the taste, and does not dissolve to any great extent in cold water. It is a powerful and energetic acid, and, as proved by Dr. Calvert, has a very destructive action upon fibrous substances when heated to a high temperature with them, but at the ordinary temperatures of drying and steaming the pure acid has no injurious action, and may be safely used without fear. It is not very much employed in either printing or dyeing: it serves for a discharge in some cases; used to a small extent in several steam colors to form oxalate of alumina; occasionally employed as a mordant, and to form an acid oxalate of soda; used as a half resist on steam work.

Oxalate of potash is usually found in print-work, where it enters into the composition of some colors, and may be viewed as a milder form of oxalic acid.

The degree of purity of oxalic acid can be practically ascertained by heating a small quantity in a hollow metal cup: if pure it will pass away in vapor, and leave no residue of any kind; if it only leaves a film, it is not to be accounted bad;

but if it leaves a considerable quantity, which does not disappear at red heat, it is an indication of some impurity or adulteration.

In receipts containing oxalic acid it will be observed that the directions are uniformly to add it to the other ingredients when they are cold. This is on account of the action of oxalic acid upon the thickening, which it breaks up and makes watery if mixed with it when very hot. But at a temperature of blood heat there need be no fear of the oxalic acid thinning the color, and it is better to stir it in then than wait until the color is quite cold, because oxalic acid is only sparingly soluble in the cold, and it requires a great deal of stirring to get it dissolved.

Oxygen.—This is the active element in common air; in older chemical works it is sometimes called vital air.

Oxygen possesses active and powerful affinities which are assisted by heat. It combines with the metals, depriving them of all their metallic properties, making them into powders of an earthy appearance: these are *oxides*. It combines with the non-metals, producing acids. It is a producer and destroyer of color. It changes indigo-white into blue, and indigo blue it destroys, changing it into some colorless substance. The most powerful actions of oxygen do not take place with the pure gas; it is in the *status nascendi*, the nascent state, the moment of its being liberated from its compounds, it exerts its most remarkable oxidizing actions. Such is the case of bleaching or discharging with acids and bichromate of potash, with alkalies, and the red prussiate, with the peroxide of lead and other matters. Either the gas is in some physically different state at the moment of its liberation, or it has a chemical activity unknown to its free state; the latter seems most probable from the discovery of the body called ozone.

Oxygen, existing in the air, and being in fact the only active element in it, is the origin of most of the changes which take place in a spontaneous manner in nature. The gradual destruction and disappearance of organic matter can be all traced to the action of oxygen: the carbon is converted into carbonic acid, the hydrogen becomes water, and the nitrogen passes eventually into nitric acid. There is no element in nature with which oxygen cannot enter into combination, and so alter its appearance and properties to a most remarkable degree. Fluorine is said to be an exception, not forming a compound with oxygen, but too little is known of this element to justify the statement. The action of oxygen upon other matters is more or less energetic as the temperature is higher or lower; it seems probable that at sufficiently low temperatures it has no action, while at high temperatures it produces the most surprising effects. The fading of colors as well as bleaching, which is only a case of color fading, may be attributable in most cases to the action of oxygen, light assisting; colors upon fabrics cannot be preserved from the action of oxygen in any way except that of covering them with a varnish. Vegetable colors remain good and bright for centuries, when protected with oil or varnish, while the same would fade in a short time if deposited as an ordinary dye.

Ozone.—Ozone is the name given to a body whose actual existence and composition remain in question. It is perhaps a molecular modification of oxygen, it is not quite certain whether it contains hydrogen or not. Air and oxygen can be ozonized by electricity or by phosphorus, and the oxygen is then found to have an amount of chemical activity which it never possesses alone; it bleaches, it liberates iodine from iodide of potassium, oxidizes sulphurous into sulphuric acid, and performs other oxidizing actions all consistent with the supposition that it is pure oxygen, but with the certainty that it is in a very different state from common oxygen gas.

The subject of ozone is highly interesting in a practical point of view, for if ever oxygen is to be applied to the performance of those reactions which are indirectly attributable to it, such as bleaching or elevating colors, it must be through the medium of this ozone or some similar body. Common oxygen is to ozonized oxygen what a rod of iron is to a sharp sword, both of the same substance but in different states of activity and of very different powers. There is reason to hope that if ever the oxygen of the air can be ozonized in a practical manner, chemists will be able to effect those oxidations directly which are now accomplished in circuitous and expensive manners.

P.

Pastel.—This plant, formerly most extensively employed for blue dyeing, is the same or similar to woad; its botanical name is *isatis tinctoria*, and its coloring matter appears to be chemically identical with indigo.

Peachwood.—This wood is one of the red woods similar in all its characters to Brazil wood, although held to be poorer in coloring matter.

Pearl Ash.—A common name for a partly purified variety of carbonate of potash. (See POTASH.)

Phosphorus.—This interesting element has not yet received any application in dyeing or printing. In its ordinary state it is dangerous to handle on account of its easy inflammability;

but there is a modified state in which it is much less combustible, that is, the amorphous condition into which it is brought
by long continued heat or the action of iodine. I tried many
experiments with the amorphous phosphorus, but did not
succeed in making any useful application of it; it has powerful
reducing properties, can readily bring indigo into the white
state, and permit it to be fixed upon calico ; but it was difficult
to manage, alkalies seemed to bring it back to the active condition, and the unoxidized phosphorus adhered to the cloth with
pertinacity, and gradually seemed to burn the indigo with
which it was in contact. Vapor of phosphorus has been used
to produce a metallic dye upon some fibrous matters, by previously steeping them in solutions of silver, lead, or copper.
Phosphorus is easily soluble in bisulphuret of carbon, and can
be reduced to a fine state of division by melting it in urine,
and keeping it well agitated as it solidifies. Phosphorus forms
several acids with oxygen, the chief of which is phosphoric acid,
naturally existing in bones and other substances. It has not
received any applications in its free state, although attempts
have been made to use it for a discharge; it is a mild, non-cor-
rosive, but yet strong acid; it differs from all the previously
mentioned acids by being fixed in fire, not distilling or rising
in vapor. It forms salts with the oxides, which are as yet but
little used. The phosphate of soda has been slightly used in
calico printing; it acts the part of a mild alkali, partially neutralizing acids and acid salts. It has been used in dung substitute, as a solvent for lactarine, and in color mixing for printing
upon sulphate or citrate of iron mordants, when, by its alkaline
nature, it caused the precipitation of more iron than would
otherwise have been fixed, giving rise to double shades.

Picric Acid.—This is only lately introduced as a dyeing
material for silks and woollens : it has no affinity for cotton.
It is made in various ways, but always through the agency of
nitric acid upon some organic matter. The cheapest source
appears to be one of the oils separable from coal tar, called
carbolic acid, but many other substances can yield it. It is a
yellow crystalline powder, of an intensely bitter taste, not acid
to the tongue. It is very combustible, and the compounds
which it forms with potash and other bases burn like gun-
powder. It dissolves in warm water, communicating a fine
yellow color to it, and dyes wool and silk of a beautiful canary
yellow without any mordant being required. It has been
largely used in Lyons for silk dyeing; upon woollen its color
is too weak and transparent. It is a powerful coloring matter; one part giving a yellow tinge to more than one hundred
times its weight of wool or silk. Silk dyed with picric acid

can be detected by masticating it, when the peculiar bitter taste of this acid can be perceived. It does not work well with other colors, overpowering them and destroying them. Besides the name of picric acid it is known in chemistry as *carbazotic acid* and *nitropicric acid.*

Pigment Colors.—This name has been given to those colors which are in the state of powder, and insoluble in the vehicle by which they are applied to the fabric. The principal colors of this class in use are the ultramarine blue, zinc white, carbon gray, and one or two other mixed shades. From the fact of these colors being insoluble in water, it is evident they cannot obtain an entrance into the pores of the fibre, and that they are never more than superficially attached to the goods upon which they are printed. Hence arises the necessity of employing some thickening or vehicle which will fasten the colored powder upon the cloth; and if the color is to be fast in water it is further necessary that the thickening should not be dissolved by water. Suppose ultramarine blue is to be applied to calico, if thickened with gum or starch it can be printed, and when dry it will adhere to the cloth in a more or less perfect manner; but if the calico so printed was dipped in water the thickening would dissolve, and the ultramarine blue, having no power of itself to adhere to the fibre, would float away in the water, leaving only a few particles entangled in the threads. The most useful materials which are used for fixing this class of colors, namely, albumen and lactarine, have been treated of. It is owing to the fact that these substances undergo a change by steaming, which makes them insoluble in water, that they differ from the ordinary thickenings in not permitting the escape of the pigment when treated by water. Many trials have been made to fix pigment colors by means of varnish, solutions of the gum resins in volatile fluids, or by drying oils, but up to this time there is really no practical method of using these materials. It is true that pigment colors can be and are so applied, but the difficulties are very considerable, and the styles consequently limited in production.

The application of pigment colors in a perfect manner to calico printing is one of the most important objects which can be aimed at by an inventor. The methods at present are so defective, the vehicles so expensive, and even so uncertain in the fastness they communicate to the colors, that nearly everything remains to be done in this direction. It is, perhaps, too much to expect that any powder or substance applied merely upon the fibre should have the same degree of fastness as coloring matters which appear to be seated in the very interior of the fibre, and it is to be feared that any species of protecting varnish would have an elasticity less than that of the fibre, and would, consequently, crack by the ordinary wear of the material. But the really surprising manner in which albumen fastens a harsh gritty powder like ultramarine, gives encouragement to the hope that even more suitable vehicles can be procured. The advantages which pigment colors have in bloom and freshness, and the opportunities they present to an extended scope of design, are so considerable that I have no doubt they will before long receive the attention they deserve.

Pink Color, *Rose Color.*—Pink is a diluted crimson, and seems to differ from red of the same tone by the addition of a faint amount of blue or violet. The chief pink colors in calico printing are derived from madder and cochineal, and the methods of obtaining them are given in the articles upon these coloring matters, but some additional receipts will be found here. In dyeing, but not in calico printing, the pink from safflower is extensively used, and will be described under that head. The remaining pink colors, not before mentioned, are as follows:—

Brazil Wood Pink—Silk.

2 quarts Brazil wood liquor (sapan or
    peachwood) at 6°.
$1\frac{1}{2}$ lb. ground gum,
$2\frac{1}{2}$ oz. oxymuriate of tin.

Sapan Wood Pink—Steam.

1 gallon sapan wood liquor at 3°,
1 lb. pink salt,
8 oz. sal ammoniac,
1 oz. oxalic acid,
1 oz. sulphate of copper,
1 gallon thick gum water.

The pink salt, now very seldom employed, is a double chloride of tin and ammonia. It is necessary to remark that all these wood pinks are of a low class, and very much inferior to the cochineal pink, both in beauty and permanency.

Spirit Pink—Standard.

2 quarts sapan wood liquor at 14°,
4 oz. sal ammoniac,
2 quarts gum water,
1 pint oxymuriate of tin at 120°,

to be reduced with gum water according to shade; not to be steamed, but washed off after three days' hanging in a cool place.

Another Spirit Pink.

1 gallon sapan wood liquor at 8°,
1½ lb. starch; boil, and cool to 100°.
¾ pint oxymuriate of tin at 120°,
¼ pint acetate of copper.

It will be observed that in the pinks from the sapan wood there is usually some copper salt, this is for the purpose of oxidating the coloring matter. In many cases chlorate of potash may be advantageously substituted, as in the following receipt:—

Pink for Calico—Steam.

2½ gallons sapan wood at 8°,
½ gallon cochineal liquor at 8°,
1 quart nitrate of alumina,
1½ lb. alum,
1 oz. oxalic acid,
4 oz. chlorate of potash.

These ingredients mixed together warm, and then added to

6 gallons gum water.

A greater or smaller quantity of gum water may be used according to the shade required.

Common Cochineal Pink.

1 gallon cochineal liquor at 6°,
heat to 170°, and dissolve in it
6 oz. alum,
3 oz. cream of tartar,
½ oz. oxalic acid.

This standard, reduced with two parts of gum water to one part standard, will give the most commonly required shade. The best cochineal pinks are from the ammoniacal cochineal (p. 164), the following receipts will illustrate their composition. The preceding cochineal pink is no more than a light red, while a good pink has a delicate hue entirely different, and can only be obtained from the ammoniacal cochineal.

Pink for all Wool.

1 gallon water,
8 oz. solid ammoniacal cochineal,
8 oz. ground cochineal; boil to 3 quarts,
1 gallon gum water,
3 oz. oxalic acid,
6 oz. bichloride of tin.

25

On woollen, the pink from ammoniacal cochineal alone would be too blue, therefore a quantity of ordinary cochineal is added. Instead of making the decoction as above, liquors of corresponding strength could be employed.

Pink for Silk.

1 gallon ammoniacal cochineal,
6 oz. binoxalate of potash,
3 oz. oxymuriate of tin,
1 gallon gum water.

Another Pink for Silk.

1 gallon ammoniacal cochineal at 6°,
4 oz. alum,
½ oz. oxalic acid,
3 lbs. ground gum.

Cochineal Pink for Delaine.

1 gallon ammoniacal cochineal at 10°,
2 oz. cream of tartar,
8 oz. alum,
4 lbs. ground gum.

See the article on cochineal for the precautions necessary to be employed in using the ammoniacal cochineal.

The pink colors obtained from the aniline products are applied by means of lactarine and tannic acid, as before described.

Pink Colors by Dyeing.—Silk is dyed pink by mordanting in bichloride of tin, and dyeing in decoction of ammoniacal cochineal. Common shades are obtained by using peachwood instead of cochineal.

Wool is dyed in precisely the same manner as for crimson, mordanted in a mixture of oxymuriate of tin, tartar, and alum, and dyed up the required shade in cochineal liquor, or for the best colors in ammoniacal cochineal.

Common pinks on cotton cloth are merely weak reds, the safflower pink is the one generally employed for fancy shades.

Pink Salts.—A name given to the double chloride of tin and ammonia. It was formerly employed instead of the other salts of tin, in the wood pinks; but it is now very seldom met with in commerce, and is replaced by using both muriate of tin and sal ammoniac in the colors.

Pipeclay.—This substance, or the very similar, china clay, is employed in calico printing, as a constituent, in certain resists. It is a resist of the mechanical sort, as distinguished from chemical resists, which act by producing chemical changes in the mordants and colors. The clay, being tenacious, covers the fibre and receives the superimposed mordant or color which does not consequently reach the fibre; upon washing, the clay is detached, carrying with it the mordant or color. It is employed in resists for indigo styles, and frequently in fancy styles to reserve colors from the action of the cover or ground color.

Plum Color, *Plum Spirits.*—The term plum, is used amongst dyers to describe a reddish purple color, not unlike the dahlia shade of the printers. It is obtained from logwood, as a coloring matter, and tin as a mordant, and is generally obtained from a mixture called a *plum tub*, made by mixing a decoction of logwood with a solution of tin, called *plum spirits.* As there are several plum shades, so there are several ways of mixing a plum tub, and besides the method above given, alum and logwood are employed, and for a red plum, peachwood and tin salts are used. (See SPIRIT COLORS.)

Polygonum.—A species of plants of which the *P. tinctorium* has attracted much attention, because it produces a blue coloring matter similar or identical with indigo. The Chinese are said to obtain blue and green colors from species of the polygonum.

Pomegranate Bark.—This bark is used in dyeing to a small extent; it yields colors analogous to those obtained from quercitron bark, and has been chiefly used for shades of drab and gray.

Potash.—This alkali is the most powerful of all the bases known to chemists, and was formerly much employed in bleaching, but on account of the cheaper price at which soda, is now obtainable potash is seldom used. The chief compounds of potash interesting to the dyer and printer are caustic potash and carbonate of potash, in its impure forms of American potash or pearl ash. Other salts of potash, the chemical action of which are more referable to their acid than to the base, are not treated of here, but may be found under appropriate headings—as OXALATE, TARTRATE, PRUSSIATE, etc.

Caustic Potash.—Caustic potash is usually sold as a liquid which is a variable mixture of real potash and water; when the liquid is boiled down sufficiently a mass is obtained, becoming solid when cool, which is still a mixture of dry potash and water. The solid caustic potash of the druggists' shops contains usually about 20 per cent. of water. Liquid caustic potash is prepared from commercial carbonate of potash, sold under the names of potash, American, Canadian, or Russian potash, or, when refined, as pearl ash. The difference between

caustic potash and carbonate of potash is, that the former is
free from carbonic acid, with which the latter is combined.
The operation of making ordinary potashes caustic consists in
abstracting the carbonic acid from them; this is done by means
of quick lime, in the following manner: The commercial
potash is dissolved in water until the solution marks about 20°
Tw.; the solution heated up to the boiling point, in an iron
boiler, quick lime is added by degrees until about one-half of
the weight of the potash has been added; the boiling is continued, with uninterrupted stirring, until a portion of clear
liquor, taken from the boiler and mixed with dilute muriatic
acid, gives no effervescence; the heat is then withdrawn, the
clear liquor syphoned off, and boiled down in another pan to
any required degree of concentration. The bottoms consist of
carbonate of lime which, retaining some of the caustic potash.
are usually washed once or twice with water before being
thrown away. Caustic soda is made in precisely the same manner, substituting soda ash for potash.

Caustic potash and soda are more energetic in their actions
than their carbonates; this can be attributed to their being free
from the neutralizing power of the carbonic acid. The common
idea that the lime communicates some caustic principle to the
potash or soda is erroneous; the lime communicates nothing,
but on the contrary, removes something, which is carbonic acid.

The caustic alkalies have many applications in the arts which
are referred to throughout this work, but caustic potash is not
nearly so much employed as caustic soda. It has, however,
advantages in the preparation of the alkaline pink mordant in
making soft soaps, in dissolving anotta, and generally in all
cases where it has to go upon the fibre, because it produces the
deliquescent, while soda gives the efflorescent, carbonate. The
following table will give an idea of the relative proportion of
alkali and water in liquid caustic potash of various strengths.
For exact determination of the value of caustic solutions the
chemical test given in the article of ALKALIMETRY must be
employed:—

*Table showing the amount of Dry Potash in one hundred parts
of Liquid Caustic at various densities.*

Degree Twaddle.	Potash.	Degree Twaddle.	Potash.
66	28	40	19
63	27	34	16$\frac{1}{2}$
60	26	29	14
56	25	24	12
53	24	19	10
50	22$\frac{1}{2}$	14	7
45	21	10	5

Carbonate of Potash.—Pure carbonate of potash is a white crystalline salt, commonly known as salt of tartar. Pearl ash and American potash consist of carbonate of potash mixed with impurities, that is, other salts of little or no value. There are no external characteristics by which the value of commercial potashes can be estimated in a satisfactory manner. Carbonate of potash is not largely employed in printing or dyeing. Pearl ash is used in Turkey red dyeing for the emulsion of oil; it is used as a solvent for anotta and safflower, and in a few cases of bleaching and scouring.

Bisulphate of Potash.—This is a very acid salt, and is employed in a low class steam work as a substitute for tartaric acid. It contains two atoms of sulphuric acid to one of potash; one atom of sulphuric acid is in so feeble a state of combination that it acts nearly as strong as vitriol itself, and consequently there is always a risk of corroding the fibre if the least excess be employed.

Nitrate of Potash, or *Saltpetre.*—This salt is very seldom used in printing; in some cases where it is prescribed it may have an oxidizing action, but usually its beneficial effect, if it have any, may be traced to its hygroscopic character.

Preparing.—The series of operations technically called preparing are employed for the purpose of giving to the fabric a uniform coat of the higher oxide of tin, in order that it may receive the class of colors known as steam colors. By whatever sequence of treatments this is effected, and there are many different ways of performing it, the object is simply to get a sufficient quantity of tin into the cloth to serve as a kind of basis for the colors afterwards applied. The utility of this preparation is unquestionable, for the colors which are obtained from a well prepared cloth are incomparably superior to the same colors printed on an unprepared or badly prepared cloth. The nature of the action of the deposited tin upon the colors does not seem difficult to explain; in the first place it is a mordant, and so combines with a portion of the coloring matter, but this is probably the least useful action of the tin; in the second place, it serves to saturate the necessary acidity of the majority of steam colors, or to act upon the salts contained in them, favoring the production of basic compounds, which are the real foundation of all colored lakes. To illustrate this explanation, let a color be made of weak cochineal liquor, without any addition but the thickening, and printed upon the prepared cloth and steamed; the result, upon washing, will be merely a dull stain, because the tin which is upon the cloth cannot, in its existing state, combine with the coloring matter of the cochineal beyond a very small extent. Now make the same cochineal color slightly acid with oxalic acid, print upon the same prepared cloth, and also upon unprepared cloth, steam and wash off. The prepared cloth will this time show the 'pattern in a light pink, well defined though faint; the unprepared cloth will hardly show a trace of color. These two trials prove that the acid has acted upon the tin in such a manner as to allow it to combine with the coloring matter of the cochineal, has in fact, brought it out of the cloth into atomic contact with the color, forming an insoluble lake, which remains adhering to the cloth. The unprepared cloth having taken no color shows that the acid alone does not fix the cochineal. To pursue the subject further, prepare, say a weak cochineal red with crystals of tin, cochineal liquor, and gum water, and print a prepared and unprepared fent with the color, steam and wash off; it will be found that the prepared cloth has a fuller, deeper, and richer color than the unprepared cloth. The theoretical explanation is as follows: The tin crystals during steaming give up oxide of tin to the cochineal, and liberate the muriatic previously combined with the oxide of tin; after this has proceeded a certain distance the amount of muriatic acid liberated becomes so great as to put a stop to the further decomposition in the case of unprepared cloth, and the formation of the colored lake is at an end; but in the case of the prepared cloth the muriatic acid falls upon the tin of the prepare, and forms muriate of tin with it, and brings it into contact with the color, at once permitting a further decomposition, and at the same time a further and more deep seated deposition of the colored compound of tin and cochineal. Not only does the tin in the cloth act as a mordant itself, but it so acts upon the tin in the color as to greatly increase its mordanting powers. In the case of alum mordants a very similar explanation may be given; the acid of the alum is partly saturated by the oxide of tin, sulphate of tin produced, and a basic alum, both of which form lakes with the coloring matter. A still closer and more extended consideration would serve to show how the same prepare is not equally suited to all styles, nor answers the same under different conditions, as to steaming, kind of cloth, etc.; but here it would be necessary to go into minute details, which are different for every works, and for almost every color. Nor would it be possible, even on these grounds, to explain the preference which is here given to one method of preparation, and there to a very different one; nor give a satisfactory reason how it is one manager can prepare at one operation with stannate, and another, doing the same class of work, must combine stannate and oxymuriate in a complicated process. Yet such is the state of affairs, and each one believes that he has the only good and real method of preparing for his particular styles, and considers success a sufficient justification for a number of unnecessary if not hurtful treatments of the cloth.

Preparation by Alkaline Stannate.—In the preparing salts, or stannate of soda, the tin is held in solution by the soda, which, if neutralized by an acid, loses its hold on the tin and parts with it. The process of preparing is based upon this fact: the cloth is padded in a clear solution of the stannate until thoroughly saturated, it is then passed into weak vitriol sours and washed. The process may be repeated two or three times; it is very simple, requires very few precautions, and is, in my opinion, capable of preparing any kind of cloth for any kind of work; but herein, I am aware, a great many will not agree with me. The only points likely to be missed in preparing with stannate, are, having too much liquor in the cloth and letting the acid get too weak. In the first case, the tin is not soundly deposited, probably, because the stannate is not promptly decomposed by the acid, a good deal of tin gets into the sours, which is a bad sign. In the second case, there is a danger of all the stannate washing off. Nothing would appear simpler than keeping the sours well up, and yet I know that a great deal of bad and uneven work is due to nothing else but deficient souring; in one case, where there was a great deal of trouble on account of irregular work, I found the supposed sours quite alkaline! Since the alkalinity of the stannate varies, and the quantity of it taken up by the cloth is irregular, this point must be well looked to. The sours should always be very sour, and the cloth not hurried through too quickly.

The strength of stannate to be used depends upon its degree of purity, that is, the percentage of tin which it contains. For a quality containing about 20 per cent. of tin, a strength of 24° Tw., is about as high as can be safely or profitably used; two treatments of this strength are sufficient for the darkest styles. The sours should be about 6° Tw. The order of operations may be put down as follows:—

Preparing for Dark Steams—Calico or Delaine.

1. Pass in stannate at 24° Tw.
2. Leave wet two hours, or more.
3. Sour with vitriol sours at 6°.
4. Wash and whizz.
5. Pad in stannate at 24°, a second time.
6. Leave two hours, or more.
7. Sour with vitriol sours at 6°.
8. Wash and dry.

For light shades, the same operations, but only half the strength of stannate. For delaines, the same treatment, up to No. 7, when the cloth must pass into a rather strong solution of bleaching powder, and afterwards into sours.

I give some other processes which have come within my observation, and which answer, more or less perfectly, the requirements of the various styles.

Delaine Prepare for Blue.—Make a mixture of equal parts of bichloride of tin at 120° Tw. and muriate of tin at 120°, and set the preparing box or cistern with this mixture until it stands at 10° Tw. Pass the delaines four separate times through this solution, and then through sours at 4° Tw., and then through chemic and sours.

Double Prepare for Delaine—Chocolate and Dark Greens.—Pad four times in stannate of soda at 12° Tw., and let rest on the rolls for four hours, at least, and twenty-four hours, at most; sour at 5° Tw. twice over, and pass through clear water and squeeze open. Next pass four times in the mixture of muriate and bichloride at 10°, same as above, and sour immediately in vitriol sours at 4°, and then through chemic and sour, wash and dry.

Preparations for All Wool:—

For Dark Blue.

50 gallons water,
30 lbs. crystals of tin,
1 lb. sulphuric acid;

pass through this solution twice, and leave it in for some hours; then wash out very well, and wince in weak chemic for twenty minutes; finish with a weak sour, wash very well, and dry cool.

For Lilac, Chocolate, and Wood Colors.

50 gallons water,
· 12 lbs. crystals of tin,
8 lbs. sulphuric acid;

pass through twice, leave for some hours, and wash very well. For green, pale blue, and reds, the same prepare may be used, but must be finished in chemic.

Another Delaine Prepare for Blue.

6 gallons bichloride of tin,
10 lbs. crystals of tin, }
1 gallon water; }

mix, and then add caustic soda at 20° until the precipitate at first produced is re-dissolved; an excess of soda is to be avoided.

Reduce to 15° Tw., pass the cloth in twice, and leave on the rolls for three hours, then pass for ten minutes in an acid mixture composed as follows :—

> 100 gallons water,
> 4 lbs. muriate of ammonia,
> 14 lbs. sulphuric acid ;

then was off, and chemic as before.

Although I am of opinion that the stannate would answer all purposes of prepare for calico and delaines, it is a more general opinion that the wool will not take tin enough from the alkaline solution, and the usual practice is to give a first passage in the stannate for the cotton, and a final passage in an acid solution for the purpose of filling the wool. Although prepared cloth remains good for a considerable time, it is not advisable to prepare beyond the immediate prospects of printing, for there have been known cases in which delaines seem to have undergone some change by long standing in the prepared state; freshly prepared cloth is the best. The cloth should be dried soft and cool; if dried on the tins, it should be taken off damp and hung up to finish the drying. If circumstances permit, the delaines should be well whizzed and allowed to dry spontaneously, especially if they are for wood colors, as chocolate, lilac, etc. By a dry heat, it appears as if the oxide of tin was put into some allotropic condition, in which its affinity for coloring matters is very much reduced, and this seems to be the case more frequently with the stannate than with the acid prepares.

Privet Berries.—The ripe berries of the *ligustrum vulgare*, similar, or identical with the privet hedges of England, are capable of communicating green colors to cotton, with an aluminous mordant, and dark colors to alumed wool. They are said to be employed on a small scale, in Italy, for dyeing silk of a bluish-green color.

Proteine.—This name was given in trade to some kind of a nitrogenous substance, intended as a substitute for albumen and lactarine in fixing pigment colors; it was only partially successful, and has, I believe, ceased to be offered.

Prussiate of Potash.— *Yellow Prussiate, Ferrocyanide of Potassium.*—This useful salt is made by fusing animal substances, as guano, hoofs of cattle, etc., with potash and iron. Its chemical components are carbon, nitrogen, oxygen, potassium, and iron; the iron which it contains is in a peculiar state of combination, and does not show like iron in its usual state in salts; but the action of acids upon it is to decompose it and make this iron apparent, turning it then into Prussian blue by combining it with another part of the same salt. Yellow prussiate, as generally sold, is in a nearly pure state. If a simple examination of it leaves its quality doubtful, there is no other way of trying its goodness than either regular chemical analysis, or making colors from it in comparison with prussiate of a known quality. It is used in dyeing for making various shades of blue, by passing the goods alternately in some iron bath and then into the prussiate made slightly acid; in printing no iron can be used, and the blue is made from the prussiate's own iron. It is not used except as a producer of blue, although many of the metals give particular colored precipitates with it, none of them have been found of any practical value except the Prussian blue, which is a prussiate of iron. It is used to make the Prussian and Chinese blues used in finishing and spirit colors: from it red prussiate is made; and it serves to prepare prussiate of tin, or tin pulp, for steam blues.

Red Prussiate of Potash, Chloro-prussiate.—This salt is made from the yellow prussiate, by passing chlorine gas over or through it. Its properties are somewhat different from those of yellow prussiate, though of the same general class; it is essily distinguished, by not giving any blue color with a per-salt of iron, which the yellow prussiate does. Mr. Mercer found that a mixture of red prussiate and caustic potash possesses peculiar powers of oxidation, the most interesting of which, to the calico-printer, is its discharging effect upon indigo blue. If a piece of dip blue be soaked in solution of red prussiate and dried, and then dipped in moderately strong caustic, it will be entirely bleached; the same thing happens if it be dipped at once into a mixture of red prussiate and alkali. This interesting reaction has not been taken much advantage of on account of some practical difficulties and the comparative expensiveness of the process. The discharging of the indigo blue is owing to its oxidation, and in so far resembles the regular process in which bichromate of potash is used. The difficulty in applying this discharge to calico lay in its acting too quickly, and its being difficult to thicken the mixture of caustic and red prussiate; all the ordinary thickenings were more or less oxidized and destroyed; at the same time the mixture lost its discharging power. It has been proposed to use calcined magnesia instead of potash to mix with the red prussiate. No action takes place until the cloth is steamed or otherwise heated, then the blue is discharged. I have made trial of this plan, but I do not think that it is likely to be employed, for it is even more expensive than using potash and quite as uncertain.

Red prussiate is used by dyers for obtaining peculiar shades of blue. If cotton cloth be passed in the usual way through nitrate of iron and rinsed, it will make no blue when put into red prussiate, but if it is afterwards passed into muriate of tin liquor it strikes a blue directly, which has a good shade. The shades of blue produced by the yellow and red prussiates are not precisely the same, one being preferred in one place and another in another, according to the peculiar demands of trade; by mixing the two it would be possible to modify the reflection and hue of the color produced. Red prussiate is not much used in printing; it is found in some receipts for dark steam blues, and for a few other colors, as myrtles and chocolates.

A liquid is sold as a substitute for red prussiate under the names of chloro-prussiate liquor, red prussiate liquor, etc. It is the mother color from which red prussiate crystals have been obtained, and contains all the chloride of potassium produced in making the yellow prussiate into red, besides any impurities which may be formed in the process; it is not a safe substitute, nor is it always a cheap one compared with the pure crystals.

The chemical name for red prussiate is ferridcyanide of potassium. The best criterion of its purity is the size, color, and clearness of the crystals; the iron test may be applied to see if it contains any yellow prussiate unchanged. It should lose no weight upon drying, and should dissolve completely in water without residue.

Prussian Blue.—This is a prussiate of iron, and it may be either a ferrocyanide or a ferridcyanide of iron, as it is made from yellow or red prussiate. It is insoluble in water, destroyed by caustic, which forms one of the alkaline prussiates from it, and leaves the red oxide of iron; some varieties dissolve in oxalic acid, and others do not. Muriate of tin and oxymuriate bring it into a kind of solution, and in this state it is applied as a spirit color. Dissolved in oxalic acid it forms a good blue liquor for finishing, correcting the yellowish tone of garancine whites, when these are not well cleared. Not much is known of the actual composition of the various Prussian blues.

Prussiate of Tin, or tin pulp, is used only as an ingredient in the making of steam blues on calico and delaines. It is made by mixing proper quantities of muriate of tin and yellow prussiate; the more water employed in mixing them, the finer will the pulp be and the better to work. If practicable, the prussiate should be dissolved separately in one half the water, and the muriate mixed with the other half, and both then poured together into the vessel in which the pulp is to settle. Another pulp is made from the bichloride or perchloride of tin, instead of the muriate; but I am not sure that there is any advantage in using it instead of the preceding: some very fine blues have been produced by using a mixture of the two pulps.

The following receipts for making tin pulp for steam blues have been in use in different places :—

Ordinary White Tin Pulp.

	No. 1.	No. 2.	No. 3.
Prussiate of potash,	4 lbs.	8 lbs.	9 lbs.
Muriate of tin at 120°,	2 qts.	5 qts.	6 qts.
Water,	6 gal.	10 gal.	10 gal.
Yield of pulp,	2 gal.	4 gal.	6 gal.

The prussiate is first dissolved in one half of the water, the muriate of tin then mixed with the remainder of the water, the two solutions are then mixed together and well stirred to break up the precipitate into a fine pulp, thrown upon a filter, or else washed by decantation, and then drained down to the given bulk.

Blue Tin Pulp.

9 lbs. prussiate of potash,
5 gallons hot water,
3 quarts bichloride of tin at 100°,
5 gallons cold water.

The two solutions are mixed, stirred up well with about a dozen gallons of water, and then drained down to six gallons.

In mixing dark steam blues it frequently happens that large quantities of prussic acid are developed from the hot colors, and sometimes men working over the pans or mugs are stupefied or sickened by the smell. There should always be good ventilation, but in default of that some ammonia liquor sprinkled about will relieve the place a good deal; or, inspiring its vapor, not too strong, is perhaps the best restorative for men affected by these exhalations, combined in severe cases with dashing cold water on the face.

Puce Color.—A color resembling that of the flea (from the French *puce*, a flea) a kind of chocolate with a purplish hue. The color, which is known in calico printing as puce, is the brown or chocolate, which may be obtained from the lead salts by fixing the oxide in lime and raising it to the state of peroxide by means of warm chloride of lime. (See BROWN and CHOCOLATE for the shades of this class.)

Purple Colors.—Purple is compounded of red and blue; in the common idea of what is purple, the red predominates over the blue; but there are, of course, a vast number of hues and shades of purple not to be defined by language. The terms red purple, or blue purple, serve to indicate the color which predominates. For the bluer and lower tones of purple, the words violet or lilac are more generally employed, while, by common consent, purple is confined to express deep and full shades of color.

For the purple from madder, see page 330; for the most usual steam purple obtained from logwood, see DAHLIA, page 190.

Common Steam Purple—Calico.

3 gallons logwood liquor at 10°,
3 gallons red liquor at 18°, }
12 oz. crystals of soda, }
1½ lb. oxalic acid; dissolve, and add
18 lbs. ground gum senegal.

The purple yielded by logwood and red liquor, or alum, is rather red, and never very deep; in order to strengthen it, and make it more blue, the prussiates are combined as in the following receipt:—

Dark Purple—Calico.

3 quarts logwood liquor at 16°,
3 quarts red liquor at 20°,
3 oz. crystals of soda,
6 oz. red prussiate of potash,
6 oz. oxalic acid.
5 lbs. gum senegal.

The purple just now in vogue for delaines and calico is the aniline mauve, thickened with lactarine, and containg tannic acid, it does not call for any receipts. I give one or two of the old purples for delaines.

Purple for Delaines—Standard.

1 gallon red liquor at 18°,
6 lbs. ground logwood;
　　　steep hot for several hours, and add
2 oz. binoxalate of potash,
2 oz. oxalic acid;

leave for twenty-four hours, then strain, and use the clear as a standard; a blue liquor for mixing with this standard is made as follows:—

Blue Liquor for Purple.

3 pints red liquor at 18°,
1 pint neutral extract of indigo,

well stirred until dissolved. The color is prepared by mixing the standard with this blue part.

Purple Color.

2 quarts standard,
½ pint blue liquor,
2 lbs. British gum.

The blue liquor in this receipt is mainly for the benefit of the wool, which, under the same conditions, takes a redder tint than the cotton from logwood. The dark purple given, page 190, is stronger and bluer than this one, since it contains prussiate of potash.

The purple colors upon all wool are obtained by combining the red of cochineal with the blue of sulphate of indigo. The receipt for lilac, p. 313, may serve as a model for purple on wool.

The purple for printing on silk is somewhat different, and nearly approaches that for calico. Here is a receipt:—

Purple for Silk.

2 quarts logwood liquor at 3°,
2 quarts peachwood liquor at 3°,
2 lbs. alum,
1½ lb. sugar of lead;

warm and stir well, leave for several hours, then take

1 quart clear liquor,
1 quart gum water,
3 oz. oxymuriate of tin,
2 oz. nitric acid at 20°.

The nitric acid might be advantageously replaced by nitrate of alumina and oxalic acid.

The aniline purples are also successfully printed on silk.

Purple Colors by Dyeing.—For the ecclesiastical purples on wool a fast color is required. which is obtained by first dyeing a blue in the indigo vat, and then dyeing a cochineal or lac scarlet upon the top. The color, though not very brilliant, is very durable. The common class of purples upon wool are from logwood and extract of indigo, mordanted in alum, tin, and tartar. The following process may be taken as an illustration:—

125 lbs. wool (merino),
18 lbs. white tartar,
12 lbs. sulphate of alumina,
4 lbs. oxymuriate of tin;

dissolve the salts in the water, enter the stuff, and work for three hours at the boil; take out, and leave for a day before washing.   Prepare a boiler with 2 lbs. of white tartar, 8 lbs. of sulphate of alumina, and as much logwood liquor and extract of indigo as is required to produce the color intended.   Old vats dye best; a purple vat will last a month, and the older it is the better it dyes.   Purples are also obtained from cochineal for the red part, and sulphate of indigo for the blue; the cloth is mordanted in tartar, alum, and oxymuriate of tin, and dyed in a mixture of ammoniacal cochineal and sulphate of indigo. Purples are also obtained from archil as the red part, and extract of indigo for the blue, leaving out the tin in mordanting. Purple shades are also dyed with aniline purple, made blue, if required, by a passage in an acid bath of sulphate of indigo.

Purple colors upon cotton are derived from logwood with a tin basis; the goods are first grounded in sumac, by leaving them in a hot decoction of it for several hours, then wrung out, and passed in oxymuriate of tin at 2° for half an hour, then dyed up in logwood, and raised with oxymuriate.   This class of colors are loose and fugitive, soon fading upon wear and exposure.

Silk is dyed purple for common styles by mordanting in tin, and then dyeing in a mixture of logwood liquor and sulphate of indigo.   Aniline purple colors are extensively employed upon silk, and beautiful shades are obtained from archil and cudbear.

Purple, French.—A preparation of archil, by Guinon, Marnas & Co., of Lyons, introduced into trade a few years ago. By some changes made in the method of extracting the archil from the lichens, an improved product is obtained, which yields much faster colors than any previously known to dyers from the same material.

Purple Heart, *Purple Wood of Guiana.*—The wood of the *copaiba pubiflora*, very hard and dense, principally used for making ramrods, contains also a colorable principle which assumes a purple color when exposed to light.   It is capable of communicating colors to fabrics, but I am not aware of its being employed for dyeing.

Pyroxilized Cotton.—Cotton which has been treated with concentrated nitric acid, or a mixture of strong nitric and sulphuric acid.   (See page 214.)

Pyroligneous Acid.—Crude and impure acetic acid when obtained from the distillation of wood.   (See ACETIC ACID, page 49.)

Q.

Quercitron Bark.—This dyeing matter, as its name indicates, is the inner bark of a tree ; the *Quercus tinctoria,* or black oak, growing in several parts of America. It was introduced into England, at the close of the last century, by Dr. Bancroft, well known for his treatise on " Permanent Colors." It soon came into use, as being cheaper and stronger than the yellow coloring matters then known in the trade. It is sold in the ground state, has a yellow color, a bitter taste, and a peculiar smell. It dyes up good yellows upon wool and cotton—on the first with a tin mordant, and upon the second with an alumina mordant. The coloring matter is very soluble in water, and is much used for steam colors, under the name of bark liquor ; its principal use in this respect being for compound shades, from dark chocolate down to the lightest drabs, grays, etc. Its yellow, either by itself or with blue in forming a green, is not liked so much as that which can be obtained from other sources. With iron mordants it gives shades of gray, olive, and black, not good in themselves, but which combine well, and modify the shades produced by other dyewoods. The ground bark is liable to adulteration with mineral matters and worthless vegetable substances, the manner of detecting which are the same for all coloring matters, and have been before alluded to. Bark is much used in garancine dyeing. When quercitron bark is mixed with sulphuric acid and water, then steamed, as in the making of garancine, a product is obtained which has somewhat higher powers of dyeing than the original bark. This method of treating bark has been followed in some places, but is of no real advantage, and is now quite abandoned.

An extract of bark is sold under the name of Flavine (see page 226): other preparations which contain the coloring matter in a concentrated state are also found in trade.

R.

Realgar, *Red Arsenic.*—This is one of the sulphides of arsenic, similar in its composition to orpiment. Its only use in printing or dyeing is as a reducing agent for bringing indigo into solution.

Red Colors.—The chief red colors are derived from cochineal, lac dye, madder, and garancine, and the methods for obtaining them have been given when treating of those substances. A less important but still largely used class of red colors are obtained from the red woods so called, of which Brazil wood is the chief type; the barwood red dye is described, page 73, and I give here a few receipts showing the methods of applying the coloring matters of sapan wood, peach-wood, etc.

Steam Red, for Calico.

3 gallons sapan wood liquor at 10°,
2 lbs. alum,
1 quart bark liquor at 18°,
1 quart red liquor at 20°,
10 oz. crystals muriate of copper.

The oxidizing agent here is the muriate of copper, which may be replaced by a mixture of nitrate of copper and sal ammoniac, or else by chlorate of potash, which is more usual in England, as in the following receipt:—

Steam Red, for Calico.

6 quarts sapan liquor at 8°,
1 quart bark liquor at 12°,
1 quart nitrate of alumina,
1 lb. alum,
3 lbs. starch,
2 oz. chlorate of potash.

The nitrate of alumina must also act in this color as an oxidizing agent; if it be omitted, and a corresponding amount of alum used, the chlorate may be increased to six ounces. Attention has been several times drawn to the fact that the wood reds are not good without chlorate of potash or some oxidizing agent; it was formerly the custom to pass the reds in chrome, but this was bad for the other colors. The use of chlorate of potash for wood reds is due to a Lancashire color mixer, and has been of great service in the lower class of chintz styles.

The wood reds are inadmissible upon delaines, being so much inferior to cochineal, but are sometimes used upon silk. (See PINK.)

In dyeing common reds upon calico the cloth is well mordanted by steeping in hot sumac liquor for several hours, and then worked in oxymuriate of tin (red spirits) for a sufficient length of time to enable it to take up as much tin as possible, then washed and dyed in a mixture of peachwood and fustic, taking about 3 lbs. of the red to 1 lb. of the yellow wood, and finally raising by adding a quantity of the red spirits to the dye. The use of fustic in dyeing, and bark in printing, is in order that the yellow they communicate may brighten the otherwise crimson of the red woods into a scarlet shade. If, consequently, a crimson red be aimed at, the yellow wood must be left out, and if a bluer crimson is required a small proportion of logwood may be added. The wood red upon silk is dyed the same as calico, but the sumac treatment may be omitted.

Inferior reds upon wool may be obtained by mordanting in a mixture of equal parts of alum and bichromate, and then dyeing up in peachwood and raising with alum.

Red Liquor. (See ACETATE OF ALUMINA.)

Red Woods.—The woods known by this name are those which give a red or crimson color to yarn or cloth mordanted with either tin or alumina. The chief varieties in use are sapan-wood, Brazil wood, peachwood, Lima wood, and barwood. Brazil wood may be taken as the type of the red woods, which differ from one another rather in the quantity than the quality of the coloring matter contained in them.

Resists, *Reserves.*—A resist in calico printing is some composition applied to parts of a fabric in order to prevent the deposition of color or mordant upon those parts. The parts thus protected may be the uncolored fibre alone, which it is intended to keep white, or it may be some colored part which is required to be preserved unaltered while the remainder of the cloth is covered with some colored design. Resist compositions intended for this latter purpose are usually called pastes, and color so preserved is said to be "pasted."

Resists may be either chemical or mechanical, or both combined, according to the nature of the color or mordant to be resisted, and the manner in which it has to be applied.

Resists for Iron and Red Liquor Mordants.—The best and only safe resist for these mordants is lime juice; if of a fair quality, and employed at the proper strength, it leaves nothing to desire. Good lime juice, at 15° Tw., thickened with calcined farina, will resist all the iron and almina mordants in use for madder work; at 30° Tw. it is strong enough as a resist for garancine styles.

Muriate of tin is used in conjunction with red liquor, in order to give the latter a power of resisting weak iron mordants. It would not answer as white resist, because some of the oxide of tin would become attached to the cloth, and dye up a red stain.

Oxalic acid, tartaric acid, and the acid sulphate of potash will act as resists, but are very inferior to lime juice or citric acid. These substances will resist for a day or two, and may be used when only a short period elapses between the printing and dyeing; but after a few hours the oxalate, tartrate, or sulphate of iron, or alumina begins to suffer decomposition, and deposit oxide upon the fibre, which attracts sufficient of the coloring matter to make the whites bad. Citric acid does not act so, for I have kept lime juice resists five years between printing and dunging, and the whites were as perfectly clear then as three days after printing. (See CITRIC ACID, page 148.)

Resist for Catechu Colors.—The best resist for this purpose appears to be citrate of soda thickened with calcined farina. It may be made by taking lime juice at 30°, and adding caustic soda at 58° until the lime juice is quite neutralized, or made a little alkaline. This resist will also throw off light chocolates and purples, but is rather defective in cutting power, and will not resist dark paste chocolates.

Citrate of soda, or potash, forms also a good neutral resist, but is seldom employed without being strengthened with pipeclay, or some of the other mechanically resisting substances.

Reserves, or Protecting Resists.—These are mostly applied by the block, and are in great part mechanical, as will be seen in the illustrations.

Resist Paste for Chintz.

6 gallons water,
15 lbs. arsenate of potash neutral,
40 lbs. pipeclay,
30 lbs. calcined farina.

If the arsenate of potash (the arsenate of soda will also answer very well) be acid it must be neutralized with caustic. The arsenate of potash has some resisting powers of itself, but its use in combination with pipeclay is owing to its drying up in a dense and somewhat gummy mass which gives solidity to the pipeclay, and makes it more capable of withstanding the effects of friction.

Another neutral paste is made from citrate of soda mixed with pipeclay and calcined farina; it is well adapted for styles which are to be covered with iron buff or chrome green.

Indigo Resists.—These have been given in the article on indigo so far as regards the white and chrome yellow resists. I append here two or three resists for a style to be dipped in indigo and dyed in madder, the active agent in which is a compound of copper with fat, obtained by mixing nitrate of copper and soda together.

Copper Soap Resist.

10 gallons water,
50 lbs. gum Senegal,
100 lbs. pipeclay,

well incorporated together; then add a hot solution composed of

>50 lbs. of soap,
>5 gallons water;

and then mix very gradually, and with constant stirring,

>50 lbs. nitrate of copper at 80°.

This mixture, which should be perfectly smooth and homogeneous, constitutes the resist, and it can be combined with mordants which will fix upon the fabric, while the fatty salt resists the indigo.

Resist Red for Indigo Dipping.

$1\frac{1}{2}$ gallon red liquor at 20°,
2 lbs. starch,
6 oz. crystals of tin,
3 quarts of copper soap resist.

A chocolate and a black can be prepared in the same manner. The pieces are dipped as usual, cleaned as perfectly as possible from the copper, and dyed in madder.

Resists for special colors, not included in this article, will be found under the head of the colors, or by reference to the index.

Rosin.—Common rosin has become an important substance in bleaching cotton goods; its effects are remarkable, and could scarcely have been anticipated. It is the residue of the heating of the juice which flows from certain trees—the pine species in particular. It varies considerably in quality, and consequently in price; some kinds contain a sufficient quantity of turpentine, when arrived in this country, to make it worth while to distil them over again, to extract what the unskilfulness of the foreign distiller has left in. The quality of rosin can be generally estimated by a simple inspection of it. It should be clear and transparent. Its color does not matter much for bleaching purposes, if it is otherwise good, but its transparency seems to be an important matter; if milky-looking it usually contains water, and is weak. It should not be dirty—containing bits of chips, gravel, sand, etc., for of course these deteriorate its value, however good the pure portion may be. Practical dealers lay some stress upon the absence of small specks upon the rosin, which can be observed, if present, upon holding small pieces to the light. When held in the flame of a gaslight or candle, the rosin ought to melt without spitting or sputtering.

Rosin Soap, *Prepared Rosin for Bleachers.*—Rosin combines with potash or soda to make a kind of soap; but it is not a real soap, according to the chemical definition of that substance; but the alkali causes the rosin to become soluble in water, and that solution possesses properties of the same nature as soap. The prepared rosin is mostly sent to the bleachers ready made; but I understand some bleachers add the pounded rosin to the soda in the kier, and boil it until dissolved, and then enter the pieces. The results of many experiments I have made upon this point seem to show a decided advantage in having the combination made separately, chiefly because they can be better made, and with less loss of time than dissolving it in the kiers. It is made in two or three different ways: Firstly, by boiling crushed rosin with soda ash and water; secondly, by boiling the rosin with weak caustic; and thirdly, by boiling it with strong caustic soda. Whichever way it is made it ought to dissolve well in water, without leaving any sediment. It is very variable in its quality, containing an irregular amount of water, and sometimes a deficient quantity of alkali. Its appearance cannot always be depended upon as an indication of its quality, and nothing but chemical analysis will point out if there is the due proportion of rosin and soda with regard to water. It is not difficult to make; some bleachers make their own, and more might do so with advantage, both in a pecuniary point of view, and as having a more regular and reliable product. In making the prepared rosin by means of soda ash, the ash must be weak to begin with, and the pounded rosin added by degrees, as it dissolves in the boiling liquor. The quantity which ash can dissolve depends upon its strength, which is variable, but a little practice soon shows when it has dissolved as much as it can properly; when this point is arrived at a little more ash is added, and the boiling continued until all the soap separates from the liquor, and rises to the top as a pasty mass. When it has cooled a little it is scooped off and is fit for use—the bottom liquor being mixed with water for the beginning of another operation. In making it from weak caustic the same method is followed; if it does not rise to the surface some soda ash or common salt put in it will compel it to do so. The method of making it from strong caustic is only applicable to small quantities. The following proportions and methods may be adopted: Six gallons of caustic soda, containing $1\frac{1}{4}$ per cent. of real alkali (marking, if pretty pure, $44°$ Tw.), is heated in an iron boiler until it gets near the boil, then 112 pounds of crushed rosin are added by degrees, with constant stirring, and the heat continued until perfect combination has taken place, which can be ascertained by taking a little of the stuff out and observing if it is all of one consistency, containing no visible particles of rosin, and when put into hot water, dissolving with only a little milkiness. This process gives very good results when the proportion of alkali is right, and the boiling has been continued long enough; it requires some care in the making, principally to prevent the rosin burning at the bottom of the pan. Larger quantities than 112 pounds of rosin can be prepared at once, but the difficulties increase, and the danger of burning is greater; but, as the whole process does not take more than an hour and a half, it is quite easy, even with a small pan, to keep a large bleaching works supplied with prepared rosin. If the prepared rosin is short of soda, or if the combination between the rosin and soda has not taken place in a perfect manner, the rosin is liable to be thrown out upon the pieces during the bowking in such manner that it will not wash off. The sour fixes it still more, and it remains upon the cloth to its injury, especially when the cloth is intended for dyeing. If the pieces are dryed over tins the rosin will sometimes collect on the skrimp rail in considerable quantities. I have known several ounces of a mixture of rosin and cotton fibre thus collected, and seen pieces of sound calico torn across by the half-melted rosin on the bar, holding them against the pulling of the tins. This can always be avoided by increasing the amount of soda to the rosin, and attending to the heating of them together. The action of the prepared rosin in bleaching is that of a soap, dissolving the natural and added resinous matters contained in the fibre, and so loosening the dirt which these kept as it were fastened to the cloth.

Gum Thus.—This resinous body is the exudation of a different tree from that which yields common rosin. It is employed by some bleachers who conceive that it gives better results than ordinary pine rosin. For bleaching for calico printing, I believe that it is no better than the cheaper common rosin, if this latter is well prepared; but gum thus seems more readily and perfectly soluble in carbonated alkalies, which will be an advantage in some methods of bleaching. Such qualities of gum thus that I have had experience of were much softer than rosin and contained volatile oil, and, besides, were greatly contaminated with leaves, twigs, dirt, etc.

Roseaniline, *Roseine.*—This is the name given to the base which constitutes the Magenta color as manufactured by Simpson, Maule, and Nicholson, upon Medlock's patent. As sent into the market, it is in the state of acetate of aniline, dissolved in some mestruum, the nature of which is not generally known.

Rosolic Acid.—An acid body obtained from coal-tar, so called, because it gives pink-colored salts with some bases. Hopes were entertained of isolating the coloring matter, and applying it as a tinctorial substance; but recent investigations have shown that there is no prospect of any useful color being derived from it.

Ruby Color.—A color of a deep rose red; the only dyed color, distinctively so called, that is obtained upon silk by means of cudbear. To produce it the cudbear is boiled in water and strained, and the silk, without mordant or preparation, worked in the clear liquor until the required shade is obtained. The addition of ammonia at the end of the dyeing, or working the silk separately in ammoniacal water, gives a bluish hue to the ruby; on the other hand, very weak sours, or tin salts, convert into a reddish hue.

By combining the fustic and logwood with the cudbear, and raising in tin, a variety of hues and shades can be produced.

S.

Safflower.—This substance, called also *carthamus*, is the flower of a plant growing in the north of Africa, and in some other warm climates. It contains two colors—a yellow which is loose and valueless, and a red which is very beautiful. The yellow dissolves in cold water, but the red is insoluble; and as the yellow would injure the red if the whole safflower were to be used together in dyeing, it is always washed in cold water to remove the former coloring matter. The method of washing consists in putting the safflower loosely into bags, and leaving these in a slight stream of pure water until, upon pressure, it is found that the water runs away nearly colorless. Another way is to put several bags into a large vat or cistern with water, and trampling the bags with the feet; but this seems to cause a loss of color. Where pure running water is not available, the best process is to wash the safflower upon a fine straining cloth by pouring water upon it until it ceases to pass through yellow. Care must be taken that no portions of the safflower escape washing, or only inferior shades of color will be produced. The red coloring matter is soluble in weak alkalies, and crystals of soda or pearl ash are usually employed to dissolve it; heat is not necessary and is even injurious, the operations succeeding best at the natural temperature. When the solution of the coloring matter is made, it is set free again by neutralizing the alkali with some acidulous body; lemon-juice, citric acid, tartaric acid, and vitriol are used by different dyers; it does not seem that it is of any importance which acid is used, notwithstanding all the minute directions which have formerly been given in this way, and the great importance which seemed to be attached to it. No time should be lost after the addition of the acid in placing the articles to be dyed, for it appears that the affinity of the colored particles for the fibre diminishes in proportion to the time of their precipitation or liberation from the alkali. The stuff to be dyed is then worked in the liquid until the color is exhausted or the desired shade obtained. It yields shades of red from the finest light pink to a flame-colored poppy red. Its shades can be heightened and assisted by other colors for the darker shades; as for example by anotta, turmeric, etc.; but these are not admissible for the lighter and brighter shades.

A method of obtaining finer colors is sometimes used; the red coloring matter is taken up by finely-carded cotton wool from the dye tub; this is gently washed in clear water, and then the color taken from it by crystals of soda, and the stuffs dyed in it after the addition, of lime juice. It is said to give better results, but it is questionable whether it is worth the trouble. Inferior qualities of safflower might require some such treatment to give good colors, but good qualities do not.

Safflower Colors.—Although safflower yields the most delicate shades of color which the art of the dyer can produce, it has the disadvantage of being one of the most unstable of all coloring matters; being excessively susceptible to the action of light and alkalies, it can neither stand exposure nor washing. Its chief consumption is in silk dyeing, and in light fancy articles of cotton not intended to be washed.

Few dyers now wash or extract the coloring matter of the safflower themselves, preferring to purchase an extract ready made which contains the coloring matter in a concentrated state and by means of which they are enabled to obtain more regular results. The process of dyeing is simple in the extreme; a sufficient quantity of the extract is added to as small a quantity of water as the goods can be worked in and well mixed up. The goods being ready for the dye, a small quantity of acid is mixed with the liquor, citric acid seems to be generally preferred, but weak vitriol, acetic or tartaric acid can be used; the goods are entered cold and worked about until all the coloring matter is absorbed, or until the desired shade is obtained. As a finish, the goods are passed through water made very slightly acid with tartaric or citric acid. Some dyers work the goods in the safflower before adding acid, lifting them out and adding the acid, and then entering and working again.

About four ounces of safflower will dye a pound of cotton cloth light pink, eight ounces will dye a full rose pink, and from 12 oz. to 1 lb. will die it a full crimson. In order to take up this quantity the cotton must be several times dyed in fresh solutions of the coloring matter.

No mordanting is used for safflower colors since none of the mordants known either increase the affinity of the cloth for the coloring madder or give it any greater stability.

In fancy cotton dyeing safflower is used to top many dyed colors; combined with Prussian blue it gives lavender or lilac shades, with anotta it yields scarlets.

The quantity of safflower required to dye silk is nearly the same as for cotton, and the process and colors precisely similar.

There is one peculiarity about the dyeing with safflower which is worthy of consideration; in all other cases the dyer is generally careful to have his dye stuffs in a perfectly soluble state for dyeing, but here the exceptional method of precipitating the coloring matter from its solution before the goods are entered for dyeing is practised. It is curious to speculate how the colored particles can be so rapidly and so completely attracted by the fibre as is found to be the case in practice.

Saffron.—Saffron cannot be said to be a regular coloring matter, being mostly used for other purposes, but it contains an intensely yellow principle, not soluble in water, but soluble in spirits, oils, etc. It is used to color confectionery and varnishes, and is sparingly used in topping thread dyed yellow with chrome and lead. It is not a fast color, but used in this last way it communicates a very desirable shade to thread for fancy articles.

The pure yellow coloring matter of saffron has not been separated in a sufficiently pure state to be analyzed. It has received the name of *Polychroite*.

Sal Ammoniac, *Muriate of Ammonia.*—This is the only salt of ammonia in regular use for the purposes of the calico printer. It is found in two different states—in large lumps of a fibrous structure, and in small crystals. The former variety has been sublimed, the latter has been crystallized from an aqueous solution. Preference is given to the sublimed sal ammoniac as being the purest. I have found samples of the crystallized sal ammoniac very pure, but often also contaminated with metallic salts, as chloride of zinc and chloride of lead. The chloride of lead is objectionable, but the chloride of zinc would not be injurious in small quantities. With the exception of a trace of iron, and that in an insoluble state, I have not found any impurity in good qualities of sublimed sal ammoniac.

The uses of sal ammoniac in color mixing seem to be referable to its power of forming double salts with metals, and thus in some unknown way regulating their action upon the coloring matters. It is especially useful in conjunction with copper salts, when the nature of the color requires their presence: it has been satisfactorily proved by experiment that without sal ammoniac a much larger quantity of copper would be required to produce the same effect, even if any practical quantity could effect the results attained by a combination of the two salts.

Salmon Color.—This color may be defined as a kind of buff, somewhat redder and warmer than the common iron buff. The color yielded to cotton and silk by anotta is frequently called a salmon color; to obtain the shade by means of the compound colors it is only necessary to mix the red and yellow colors in the proper proportions.

Sapan Wood.—One of the red woods, the concentrated decoction of which is largely used in calico printing, as the most economical red part for compound colors. In chemical and other properties it is identical with Brazil wood.

Santa Martha Wood.—One of the red woods, which appears to be the same as the wood more frequently called peach-wood, similar or identical in its properties to Brazil wood.

Santal Wood, *Sandal Wood, Red Saunders Wood.*—This wood is very similar to barwood, it contains a red coloring matter which is very little soluble in water. The wood is extremely hard, and can only be employed in bulk when ground excessively fine; it is said to communicate a harsh feeling to wool dyed with it, probably on account of hard resinous matter present in the wood. With aluminous and tin mordants santal wood dyes brownish-red colors of considerable depth. It appears to be very largely used in France as a constituent in dyeing woollen of a dark blue color, for soldiers' uniform. This color is obtained as follows:—

French National Blue.—The wool is first dyed a strong blue in the indigo vat, and then (for every 100 lbs. wool) boiled for one hour with a mixture of

30 lbs. sandal wood,
1½ lb. logwood,
2½ lbs. archil,
1½ lb. gall nuts.

At the expiration of an hour, 2½ lbs. green copperas are added, and the wool further worked until the color is deemed to be sufficiently raised. The use of sandal wood in this case is only to fill up the blue, and give it a brown or bronze shade.

Saw Wort.—This plant, the *serratula tinctoria* of Linnæus, according to Bancroft, affords a good substitute for weld, dyeing up a bright lemon yellow color, of considerable durability.

Scarlet Color.—Scarlet may be considered as red with the addition of a little yellow. The processes for producing this color are given in the articles on COCHINEAL and RED COLORS.

Shaded Styles.—By this term I intimate a style of work produced by printing colors which fall upon one another, and at the point of contact produce a shade different from the body of the piece. There are several methods by which this may be accomplished ; one or two illustrations will suffice.

Two Shades of Red for Madder.—A red liquor of such strength as will produce the lightest shade is taken and mixed with nitrate of alumina, equivalent in strength to four pounds of alum per gallon, thickened, and printed in say a stripe, a pad, or a Bengal pin, and dried up. The nitrate of alumina cannot act as a mordant, and the color, if now dunged and dyed, would only produce a shade corresponding with the strength of red liquor ; but if some color which will precipitate the alumina of the nitrate of alumina be printed upon the first pattern, it will then cause a deposition of alumina in those places, which will, consequently, dye up a dark red. Acetate of soda, properly thickened, is suitable for this purpose ; wherever it falls upon the nitrate of alumina it effects a double decomposition, producing acetate of alumina and nitrate of soda ; the acetate of alumina, by ageing, deposits its alumina, and forms a dark red mordant.

Two Shades of Purple—Madder.

1½ gallon of 16 lilac,

that is,' gum water, which contains one-sixteenth of strong iron liquor ; dissolve in it

1½ lb. green copperas.

Print and dry. Then for the precipitating color take ╎

1 gallon water,
2½ lbs. acetate of soda,
5 lbs. gum.

And print, age, dung, and dye, as usual. The chemical action here is the same as in the preceding case, acetate of iron is produced, the metal of which is taken up by the cloth which cannot take it up from the sulphate.

Instead of acetate of soda, a solution of neutral arsenate of soda may be employed, standing at about 14° Tw.

Another method is carried out as follows :—

Chocolate for Shades—Madder.

3 gallons red liquor at 18°,
5 gallons own liquor at 8°,
20 lbs. flour ; boil, and add
1 gallon lime juice at 40°.

Purple for Shades—Madder.

8 gallons water,
1 gallon iron liquor at 24°,
13½ lbs. flour ; boil, and add
7 pints lime juice.

In these receipts it will be observed that the mordants are, when thickened, strong mordants and calculated to yield dark shades ; but the addition of citric acid converts a quantity of the iron into citrate, which cannot fix, and the shades produced are only such as would be due to the iron still left as acetate ; but by printing a composition of arsenate or phosphate of soda across the aceto-citrate mordant the iron is precipitated, and forms a mordant for dark colors.

Darkening Color for Shades.

1 gallon gum water,
8 lbs. crystals phosphate of soda.

There is a good deal of uncertainty and irregularity in this style of work unless closely attended to. The lime juice method is the best and most regular, but a good quality of lime juice must be employed ; if it contains any notable quantity of saccharine, or extractive matter, there is no saying how much, or how little iron it will deposit, or how it will answer the phosphate or arseniate precipitating color.

Silicon, *Silica.*—This element is the basis of sand and many minerals ; it is little known in its free state. Its combination with oxygen is silicic acid, which exists pure in rock crystal, and a few other natural minerals. It has no applications as an acid, being insoluble in water in its ordinary state. Silicic acid combines with bases giving silicates, all of which are, in their neutral state, insoluble in water. The alkaline silicates of potash or soda are soluble in water. Common glass is a neutral or acid silicate of potash or soda, containing an excess of silicic acid ; if an excess of the alkaline material be used a glass is equally produced, but it is soluble in water. Powdered flint or quartz is dissolved by a boiling concentrated solution of caustic soda, producing a silicate of soda ; such a solution was known to the older chemists as liquor of flints ;

a solution of silicate of soda is now extensively employed in the cleansing of printed goods instead of cow dung, and is known as one of the dung substitutes.

Silicate of soda is made on the large scale by fusing clean sand with soda ash until a glass is produced, which is broken into small lumps and boiled with water until a solution of the required strength is obtained. Silicic acid is one of the feeblest acids, consequently the addition of any of the ordinary acids to a solution of the silicate of soda effects the decomposition of the salt, the silicic acid being displaced in a gelatinous state; if the solution be concentrated it becomes semi-solid by the free silicic acid in suspension. Silicic acid, thus liberated from its compounds, may enter into a solution to a considerable extent in water and dilute acids; it is rendered insoluble by boiling or evaporation to dryness. Solution of silicate of soda is decomposed by the carbonic acid of the atmosphere; a strong and nearly neutral solution, when exposed in an open vessel to the air for a few days, becomes solid; this may form a test of the alkalinity of a sample. Silicate of soda for dung substitute should be as neutral as possible, but it is necessarily alkaline, and is not well fitted for cleansing any styles of work likely to be injured by alkalies. Silicate of soda has been used as agent for fixing ultramarine blue and other pigment colors; the color was ground up well with a solution of silicate at about 90° Tw., printed without any other thickening, dried soft and hung in a cool place, then fixed by passing in solution of muriate of ammonia or common salt. The results were not generally satisfactory, the color was difficult to print; uncertain as to its fastness, sometimes all coming off in the fixing liquor, and even when well fixed communicating an unpleasant harshness to the cloth. Grays for the printing machine are in some continental establishments passed through silicate of soda; they are said when thus prepared, to give a better impression, to last longer, and to absorb less of the color from the white piece. Silicate of soda has been mixed with soap and the compound highly recommended; but it possesses no detergent properties, and can only be looked upon as an adulteration. Silicate of soda has a very injurious action upon unsoaped madder work.

Silk.—Silk is the fine, strong, and apparently solid thread which several insects wind round themselves as a protection while undergoing their metamorphosis. It is originally fluid or semi-fluid, and exudes from an opening in the lip of the worm, but soon solidifies in the air. The color is usually of a golden yellow, but sometimes quite white; the thread often exceeds 1300 feet in length, and is consequently the longest

fibre known, the pure silk contained in this length will not
weigh more than a couple of grains, from which fact an idea
of its extreme fineness mny be formed. Its diameter is about
$\frac{1}{2000}$ of an inch, under the microscope it appears as a cylindri-
cal fibre, without any twists or evidences of structure. Silk
contains, naturally, a large quantity of gummy substance
attached to it which must be removed before it can be dyed ;
it is best removed by boiling the raw silk in soap and water ;
a loss of weight equivalent to one-fourth, or one-third of the
weight of the silk takes place in boiling off. For the behavior
of silk towards the various drugs used in dyeing, see FIBROUS
SUBSTANCES, pages 213 to 226.

  Soap.—Soap is a combination of a fatty matter and an
oxide, but the name is usually applied to the compounds of
fatty matters with potash and soda only. Soap is manufactured
by heating together some fat or oil with dilute caustic alkali,
until they coalesce into a homogeneous mass. There is no real
difficulty in making soap, and it might be frequently made by
the consumer to advantage; many printers make their own
soap, and where much madder work is done the economy is
conspicuous. The most valuable soap is made from tallow and
olive oil, but a cheaper and excellent soap for dyeing and
clearing purposes is made from palm oil ; the strong yellow
color of the unbleached palm-oil is not found to be any disadvantage in its use for madder work, or for bleaching woollen
or delaine goods. For use on print works, a soap may be
made by boiling in an iron pan palm oil and caustic soda at
16° Tw., in the proportion of about two and a half pounds of
caustic to one of oil. Combination takes place in two or three
hours ; upon cooling the mass sets as a soft yellow solid, which
is very suitable for dissolving rapidly in the becks on account
of the large amount of water it contains. This soap contains
all the glycerine of the oil, as also whatever impurities there
may exist in either the palm oil or the caustic soda. It answers
very well for all uses on a print works ; but its peculiar method
of manufacture requires a nice adjustment of the quantity of
oil and alkali, and is not likely to be successfully carried out
without some chemical knowledge.

  Soap is used in bleaching silk, some qualities of woollen,
and some fine kind of cotton goods. It is used in several
cases of dyeing, either to soften the water or to modify the
shade of color, especially in silk dyeing. It is much used in
madder styles, and very generally employed as a final operation to modify dyed colors and to clear white grounds. The
detergent action of soap depends upon its power of dissolving
or rendering fatty matters emulsive and removable by water.

Dirt may be defined as dust, with some oleaginous matter, which renders it adhesive; the grease or fat being acted upon, the dust is easily removed by water. For simply detergent purposes very alkaline soaps may be employed, such as the rosin soap used in bleaching, or common soft soap; but when the action is of a mixed nature, as upon dyed goods, the soap must be of a mild, neutral character. Soap may be bad or defective by being deficient in alkali, or by having it in excess. A good soap for finishing madder work should be slightly alkaline; it is more economical than a pure neutral soap, and gives as good or better results. If the soap be too alkaline, it acts harshly upon madder colors impoverishing them, especially the reds, while the lilacs and purples are turned to a disagreeable reddish hue. Pinks soaped with a too alkaline soap will be flat and dull, without bloom; catechu drabs and browns will be much deteriorated. Work thus injured may sometimes be improved by passing in water made sour with sulphuric acid, washing and re-soaping. If a soap is deficient in alkali it takes a greater quantity and longer time to effect the clearing of goods; frequently the whites are left dull, and a general aspect of flatness pervades over the colors. Such a soap is improved by adding eight or ten ounces of soda ash to the water before dissolving the soap in it. An alkaline soap, on the contrary, is improved by the addition of muriate or oxymuriate of tin in small quantity to the water. If soap made from a strong smelling oil be used in madder work, the finished pieces will retain the smell in a very persistent manner. I made a large quantity of soap from linseed oil, which answered very well, but was objectionable on account of the smell of paint which the pieces emitted when kept in the warehouse; cod oil and whale oil are objectionable on the same account. As a rule, any smell which is possessed by oil applied to the pieces is retained by them, and especially by delaines and woollens. An instance came under my notice of an oil introduced as a substitute for Gallipoli; I examined it and reported unfavorably of its qualities—it was a product of distillation, and contained many bad smelling oils. Its cheapness caused it to be used, however, and in a few days the delaines were complained of as having a disagreeable smell—they were returned from the warehouse as unsaleable. There was no clue to the cause of the smell; the pieces were washed, winced and dashed, without removing the odor, and the fault was variously laid upon the gum, the water, the steaming, and the drying. As soon as the pieces were submitted to me, I recognized the odor of one of the oils separated in the course of my analysis of the new oil. Upon inquiry it was found that it had been used in the color mixing; Gallipoli being again used, the cause of complaint disappeared.

Soft Soap.—Soft soap is mostly made from potash instead of soda; but there are many oils which give soft soaps with soda —the fish oils particularly. Soft soap is commonly a stronger and harsher soap than the hard soaps. Inferior oils are mostly consumed in making commercial soft soap, but very fine neutral soft soaps can be made if proper care is taken, and sufficiently pure materials employed. I analyzed a very good soft soap made from olive oil. It answered very well for delaine bleaching, but was expensive. There are some cases where a soft soap might be advantageous, if it were equally as good as a hard soap. Common soft soap is used in a few cases of color mixing; it serves to dissolve anotta, and to make some resists. It should not be used in soaping for any delicate colors. It may be employed in moderation for brightening indigo colors, as China blue. The quality of soft soap is thought to depend in some measure upon the existence of white particles diffused through the mass, producing the appearance called "figgy,"— but this is no real test of its quality; first rate soaps are sometimes quite uniform, and the figged character can be communicated to an article of an inferior nature, containing glue and other useless matters.

Substitutes for Soap.—Since the excise duty was taken off soap a vast impetus has been given to the trade, and a great number of patents have been taken out for adulterating soap with substances more or less hurtful to it. The result is the disappearance of soap, properly so called, from ordinary use, and the substitution of waxy compounds of glue, rosin, ground bones, gelatine, earthy matters, and similar substances, having but little soap mixed with them. Such cheap substitutes as the printer or dyer is likely to have offered to him will be of this nature, and likely to prove more beneficial to the maker than the consumer. So much damage can be done by the use of bad soap that the printer ought to insist upon being supplied with a soap which contains nothing but water, soda, and fat. Anything else will be only an addition of weight without value, and likely to be hurtful and useless. I have made many experiments upon substances similar to soap, with a view to substituting them for it, but with no notable success; it seems to be fatty matter which is wanted, and nothing else will do: even a small percentage of rosin appeared to be hurtful. I found that animal black could be made to clear the whites of madder work, but it left the colors very dull and poor, and even soaping afterwards would not restore them to a proper shade. All alkaline substances, as the carbonates of potash

and soda, ammonia, the alkaline borates, silicates, and phosphates, give bad and worthless results without soap. The only
way in which soap can be saved with advantage to the work is
in adding a little soda ash to the water in the beck if the water
is hard. Sometimes a beckful of water will destroy a pound
or more of soap in precipitating the lime; eight or ten ounces
of soda ash, added before the soap goes into the beck, will frequently save this.

Analysis of Soap.—The usual test for soap is to try it upon
madder work comparatively with a good quality. With care,
good results may be obtained upon small quantities of soap—say
half an ounce—but when a sufficient quantity of material is at
disposal the trial should be on the large scale. The following
method may be pursued in the analysis of soap—it gives sufficiently close results for technical purposes, and does not take
much time. In all genuine samples of soaps there will be
nothing but water, fatty matters, and alkali. The water can be
determined by taking 100 grains of the soap and heating it in
a porcelain dish, with a gentle heat, until all the water is dissipated; the heat may be pushed as high as 260° F., or until the
dry soap begins to exhale an odor of fatty matter; this will take
place about fifteen minutes, and give more exact results than a
water bath or oil bath, which would require as many hours.
For the determination of the alkali another hundred grains may
be dissolved in about three ounces of water, and an excess of
dilute sulphuric acid of a known strength, added to it; there
should be about twice as much acid added as would be necessary to wholly decompose the soap. The solution containing
the liberated fatty matter must be kept hot until all milkiness
has disappeared, and the oil floats clearly upon the top; then
a weighed quantity of pure beeswax (about fifty grains) is added
to the solution, and all kept hot until the wax is completely
incorporated with the fatty matter; the whole is then allowed
to cool. Upon cooling, the wax solidifies along with the fatty
matter, forming a well-cohering cake, which may be removed
from the liquid, washed with a little water, and the washings
added to the original solution, then carefully dried and weighed.
The excess of weight above that of the beeswax employed is
the quantity of fatty matter present. A gradual alkaline solution is then added to the first liquid until it is neutral; it is
found how much of the acid first used was neutralized, and the
amount of alkali calculated from that.

Soda and its Salts.—Caustic soda is prepared exactly in
the same manner as caustic potash, and possesses properties
which are very similar to it. Caustic soda is sold either solid
or in solution; the manufacture of solid caustic soda upon the

27

large scale is of very recent introduction. The product sent
out by some of the large firms appears perfectly well suited to
the wants of a dye or print works. Liquid caustic soda is sold
in carboys at a strength of about 60° Tw. The following table
shows the percentage amount of real dry caustic soda in 100
parts by weight of the liquid at different strengths of Tw.:—

Twaddle.	Dry Soda.	Twaddle.	Dry Soda.
66	23	40	13
63	22	34	11½
60	20½	29	10
56	19	24	8
53	17½	19	6½
50	16	14	5
45	14½	10	3½

The uses of caustic soda are very numerous, being cheaper than
potash ; it is used whenever a strong alkali is required, whether
for raising colors, scouring, soap making, neutralizing acids, or
dissolving coloring matters.

Soda Ash.—Soda ash, which is extensively used in bleaching,
is an impure carbonate of soda. Its whole value rests upon
the amount of pure carbonate it contains. It should be quite
soluble in water, of a good color when opened, and not inclined
to change by exposure to the air. Soda ash which goes yellow
usually contains some sulphuretted compounds. For bleaching,
an ash of a bluish tint is preferred. There is a difference of
opinion amongst bleachers as to whether a soda ash for bleaching should or should not contain soda in the caustic state. It
is quite certain that even pure caustic soda can be used safely
in bleaching, but at the same time it seems proved that caustic
soda does sometimes weaken cotton in a very remarkable manner ; what the conditions of this apparently contradictory
behavior of the fibre may be due to is not clearly known, but
it is evident that caution must be used in employing a caustic
kind of soda ash. Some samples of soda ash contain 12 to 15
per cent. caustic soda, and must assuredly behave very differently to soda ash which does not contain any. Soda ash is sold
at different prices according to the percentage of real soda
contained in it ; soda ash for bleachers and printers' purposes
should be 48 per cent. at least; if the strength is lower than
this, the amount of common salt and sulphate of soda is likely
to prove embarrassing in some operations.

Crystals of Soda.—Crystals of soda are made from soda ash,
and are a compound of dry carbonate of soda and water. In
round numbers, three pounds of crystals contain two pounds

of water and one pound of dry carbonate of soda. In most cases the carbonate of soda in the crystals is less contaminated with other salts than is the case in soda ash, and it should, consequently, be preferred in all delicate operations. In a warm dry situation the crystals lose water, becoming white externally ; they are not chemically changed or injured, as is vulgarly supposed, but really improved, because a given weight contains more of the alkali than previously. Carbonate of potash may be distinguished from carbonate of soda by the property which the former has of absorbing moisture from the air, becoming damp ; crystallized carbonate of soda has the opposite tendency, to give up its water to the atmosphere.

Bicarbonate of Soda.—This salt differs from simple carbonate of soda by containing twice as much carbonic acid. It is a milder alkali, and on that account receives a few applications in dyeing, where the more energetic carbonate might be hurt-ful. Genuine bicarbonate of soda is a nearly pure salt, and when made red hot, to expel its water equivalent of carbonic acid, leaves pure carbonate of soda used for analytical purposes.

Sulphate of Soda.—This salt, commonly known as Glauber's salts in the crystallized state, and as salt cake in the anhydrous condition, is employed in calico printing to fix lead mordants preparatory to dyeing them orange or yellow. The only impurity likely to be injurious in sulphate of soda is sulphate of iron, which I have known to spoil work. It can be detected by the usual tests for iron. By dissolving such impure sulphate of soda in hot water, and adding carbonate of soda, all the iron may be precipitated, and a pure sulphate obtained.

Nitrate of Soda is sometimes prescribed in color mixing; its utility appears to be owing to its possessing feeble deliquescent properties by which it is enabled to draw moisture from the air ; and so keep the color it is mixed with soft.

Common Salt, commonly known as chloride of sodium or muriate of soda, is also sometimes used in color mixing on account of its deliquescent properties.

Sorgho Red.—A new kind of red to which attention has been drawn ; extracted from *sorghum saccharatum,* or Chinese sugar-cane, said to dye good and durable colors upon wool and silk ; a tin mordant being used. From the nature of the reports available it does not seem likely to be of any importance in dyeing.

Spinel Mordant.—A name given to a combination of alumina and magnesia, which Wagner recommends as a mordant preferable to alumina alone. He appears to have derived the idea from the analysis of the Indian yellow (see EUXANTHIC

ACID), in which alumina and magnesia exist in single atoms, and because this is also the composition of the mineral called spinel, he gave this name to it. There are no accounts of this compound mordant having been practically employed.

Spirit Colors, *Couleurs d'Application.*—Several spirit color receipts have been given in the preceding pages; the name is derived from the use of the oxymuriate of tin, or tin spirits, in their composition. They are low-class colors and possess very little durability; on account of the excessive acidity of the colors they will not stand steaming, and the process of fixing simply consists in hanging the pieces for three or four days in a cool place, and washing out the excess of tin, and thickening by passing gently through cold water.

Stannate, *Stannic Acid.*—A term derived from *stannum*, the Latin name for tin. The only stannate in use, is the stannate of soda, extensively used as a prepare for steam colors. (See TIN and PREPARING.)

Starch.—Starch is a widely-diffused vegetable product; it exists in a vast number of plants, fruits, and trees, and seems to be one of the fundamental bodies of organic life. Its composition is very similar to that of sugar, being a compound of carbon with hydrogen and oxygen, in the proportions requisite to form water. It is extensively used in printing and finishing, but does not in either case exercise any actions of a purely chemical nature; as a thickening it is only a vehicle for conveying the color or the mordant to the fibre; as a finish it is only to give stiffness or fulness to the cloth. But its actions in many cases involve the play of chemical affinities, and should be minutely known. Pure wheaten starch, when closely examined under the microscope, is found to be composed of very small globules. In commerce it is found in a peculiar state of aggregation, incorrectly said to be crystallized; the quality of the starch is often judged and determined by the appearance of these columnar masses called crystals. No other starch but that from wheat takes the same form in drying. It is not prudent, however, to depend too much upon this as a test; for I believe the crystalline character can be communicated to other starches, and that it is not an essential character of wheaten starch, but rather an accidental one, due to a partial decomposition and breaking up of some of the globules, which communicate a gummy nature and adhesive character to the remainder, or to a residue of unremoved glutinous matters. Starch does not dissolve at all in pure water when cold, it mixes up, but then settles down, leaving the liquid clear; it dissolves in hot water, swelling out to a great extent; it begins to dissolve, or the particles to burst, at about 150° F., but color cannot be well thickened at this heat, it must be boiled to get a good result. Starch boiled with acids or acid liquors thickens at first but afterwards becomes thin, owing to the destruction of the starch and its conversion into sugar; colors should not, therefore, as a general rule, be boiled until they begin to grow thin again—although in special cases this is prescribed, and is an advantage, but it is usually unnecessary, and likely to injure the color.

A good wheaten starch is white and clear, has a sweet taste on the tongue, or at least an absence of bad taste, and, before dissolving in the mouth, shows an adhesiveness to the tongue; when mixed with water it should give a white milky fluid, without any particles of dirt floating on the top, and should settle down quickly, forming a solid hard mass at the bottom of the fluid. As a trial for its thickening powers a quantity may be boiled with water in the usual manner; two proportions should be taken, one thicker than is generally required, and another thinner—for instance, one trial at one pound to the gallon, and another at two pounds per gallon, and both boiled with the usual precautions. The manner in which it behaves on boiling, as well as its appearance when boiled, should be observed. A good starch will thicken gradually and evenly throughout, not in lumps; it will keep smooth all the time with only a moderate amount of stirring, and when boiled will be of a clear, transparent, gelatinous appearance—not milky and opaque, nor breaking off short when lifted with a stick. At two pounds per gallon it ought to be pretty stiff while hot, to pour out slowly, and for the most part adhere to the sides of a gallon mug, when this is inverted for a short time; at one pound per gallon it should flow smooth and oily, without appearance of water or breaks in it. When cold, the thick trial should be very stiff, and feel tough and solid in the hand; the skin should be of a tough leathery nature, and no water should be floating about—it will not be so clear as when hot, but still should be partially transparent; the thinner trial should be also of increased consistence, and not show any water; it should be smooth, and not containing lumps. There are besides these characters a great number of others, too minute to record, which are combined in forming the opinion as to the quality of a sample of starch. It is a practical question, and nothing but a number of trials, upon all kinds of starches, will enable any one to form a correct opinion upon this matter.

Starch is sometimes adulterated with mineral substances, as gypsum, sulphate of baryta, or mineral white, China clay, etc. The existence of these substances makes a starch boil rough and opaque: they can be discovered by burning some of the starch in a proper manner—if much earthy matter be left as a residue, it will be a sign of adulteration. It is sometimes understood that starch for finishing contains mineral matters, and a proportionable reduction in price is made, but oftener there is only one party cognizant of it; at any rate, a starch containing added mineral matter ought not to be used in mixing colors, however good it may be as a finishing starch. Inferior qualities of starch, under the names of seconds, slimes, and hair powder starch, are extensively used in the trade, and may be economically and easily employed in numerous cases; for it is not necessary, in making colors, that a starch as pure as is required for domestic purposes should be used; what is required is a good sound article, free from adulteration, not injured by acids or fermentation, and, if otherwise good, it does not matter whether it be in powder or in crystal, perfect white or a little grayish. Starch is sometimes injured by some of the gluten of the flour being left in it. Such a starch does not keep well, soon goes watery, or putrefies, emitting bad smells. By scattering a little of this kind of starch upon a red hot iron plate the gluten makes itself apparent, by giving off a disagreeable animal smell, like burning woollen, or leather, or the hoofs of horses. This kind of starch has never a good color, and, if in crystals, has a flinty hardness. Good starch does not contain more than ten or fifteen per cent. of water; the latter is the largest quantity it should lose in drying, at moderate temperatures.

There are other kinds of starchy substances in occasional use for printing and finishing which deserve a notice, as potato starch or farina, rice starch, and sago flour—which is not a flour at all, but nearly pure starch.

Sago Flour.—When this is purified from chips, leaves, dirt, and coloring matter, it can be advantageously used in finishing, as a partial substitute for the more expensive wheaten starch. It works up softer as a paste than farina, and I think could be used in color mixing with safety and economy. It serves to make gum from.

Rice Flour Starch.—This has been employed in finishing and thickening, but not to any considerable extent I believe. It is more used in finishing than as a thickener. It is employed in the manufacture of artificial gums.

Steam Colors.—Colors which are developed and fixed by steam. The method of fixing colors by steaming seems to have been arrived at by slow degrees, so that it is difficult to name a date at which it was introduced. The French admit that the first experiments in this direction were English, but claim for themselves the practical application of steam colors as a style.

Nothing now seems more natural than that printers should have tried to apply the dyers' ingredients in a more concentrated state to cloth, and then submitted the colors to heat. In fact, the experiment was often tried, but always, at first, with dry heat; the pieces were hung up in stoves heated by red-hot flues, or the printed pieces were passed over hot callenders, or even pressed with a kind of smoothing-iron made hot. The results, of course, were unsatisfactory, and it is easy to point out now that the absence of moisture and the inequality of the heat were sufficient causes of failure. Upon seeing the process of steaming colors for the first time, every one who has not had occasion to study the matter expresses surprise that the pieces are not wet and the colors do not run into one another. There is no doubt that if ever it occurred to the earlier printers to use steam as it is now used, they would be deterred from the experiment by the mistâken apprehension of the steam wetting the cloth and causing the running of the colors. As an illustration and to show that an ingenious experimentalist had some such idea, we may quote an experiment recorded by Dr. Bancroft; on his researches upon quercitron bark, shortly before the close of the last century, the idea of fixing the color by the heat of steam occurred to him, and he printed some calico with a mixture of tin salt, bark liquor, and gum, dried it, and having wrapped it up in soft paper so as to prevent marking off, he put it into a bag made of stout drill which he had carefully saturated with wax, so that no steam might get into it, and tied and waxed it up and then suspended it up in steam. He was partially successful, but of course he would have been more successful if he had not taken such excessive pains to keep the steam from touching his printed calico. It does not appear that steam colors were much worked until about the year 1830.

The steam has no other action than that which is due to its heat and the presence of an atmosphere more or less saturated with moisture; the special systems of steaming which are in use owe their adoption to differences in styles or manners of working which it is impossible to treat in a satisfactory manner, depending as these differences do upon very minute points of detail. In some print works it is thought necessary to hang the pieces twenty-four hours in a cool place before steaming; in others, the pieces, hard and dry, are taken direct from the printing machine to be steamed. In some places the steam is used dry, in others it is used very damp; evidently the nature of the thickening and the previous state of the cloth as well as the quality of the colors must here be taken into consideration. Again, one house steams half an hour, takes out, airs

and steams again for another half hour, while another establishment steams for an hour at once. In one works the column
alone can be satisfactorily used, in another the same class of
work is found to be best done in the ordinary kennel. The
conditions necessary to success in steaming are, consequently,
not definable without an·exact acquaintance with all the con-
necting circumstances. Some general principles, however, may
be laid down. There must be as much free moisture either in
the steam or in the cloth as will suffice at least to take the hardness out of the cloth; if the pieces come off the steam decidedly dry there will be irregularity, except in light work; there
must be sufficient steam to keep a good volume passing away;
this is necessary to remove all free acids vapors, which, if
allowed to collect in the chambers, may injure the lighter colors,
or the cloth itself. The damper the steam the sooner will the
steaming be done, but (not to speak of steam so wet as to cause
the colors to run), steam which is too damp causes the colors
to sink too far into the cloth, and takes the bloom off them.
The more quickly the steaming can be fairly accomplished the
better the result, and this in ratio with the dampness of the
steam, which should, consequently, be so regulated as to hit the
medium condition as nearly as possible.

Substantive Colors.—A term first employed by Bancroft
to indicate those colors produced by coloring matters which
fixed upon fabrics without the necessity of a mordant. Thus
indigo, turmeric, and safflower, are vegetable substantive colors,
and iron buff and lead puce are mineral substantive colors.

Sugar.—Sugar is only sparingly employed in printing; as
mentioned under chromium, it is used to reduce the chromic
acid in forming some chrome shades. But, though it is not
directly employed, it is often present in cases where it is
not suspected, and exercising such chemical actions as it
possesses. A kind of sugar, known as grape sugar, from
having the same composition as the natural sugar existing
in grapes, is easily formed from vegetable substances by the
action of acids; and it not unfrequently happens that
saccharine matters exist in colors, produced from the prolonged action of acid and heat upon the thickening. A species
of sugar is found in madder roots, and also in some varieties
of artificial gums, particularly the light-colored gums, made
by the action of acids. The presence of sugar is injurious to
the fixing of mordants; saccharine matter acts, like many
soluble organic substances, as if it suspended the chemical
properties of the salts with which it is in contact, especially in
metallic salts. I thickened some iron liquor with molasses, and
different mixtures of gum and molasses; the colors printed

well, but when hung up to age became sticky and damp, especially the one which had no other thickening but the molasses; when dunged and dyed, it was found that the molasses in the pure state had entirely prevented the fixation of the iron. The outline of the pattern could be just discerned in certain lights; the others were bad in proportion as they contained more or less molasses. The influence of saccharine matter in dyeing is not so marked; it does not produce any visible effect unless in excessive quantity.

The kind of sugar which is made by boiling weak sulphuric acid with potato starch, and which is known as glucose, or grape sugar, possesses powerful reducing properties, it has been long known that it would dissolve indigo in contact with alkali, as orpiment or protosalts of tin do, making a kind of pencil blue. A patent has been recently taken for a new method of fixing indigo by means of glucose, or grape sugar (see GLUCOSE); the indigo, finely powdered, being mixed with the alkali, and the glucose, printed and steamed. An application of the same glucose is mentioned under COPPER. It seems probable that this substance, at present almost unknown in printworks, may turn out useful in some cases, when practical men are more familiar with its properties, and probably serve some useful purpose in the direction of indigo colors.

Sulphur.—Sulphur is found in the market in three states, as flour of brimstone, which is the purest quality; roll sulphur, the second quality; and the crude sulphur, as it is imported. The only use to which sulphur is applied in the arts we are treating of is in the bleaching of woollen goods or delaines; for this purpose the crude sulphur answers sufficiently well. Sulphur does not dissolve in water, but it is dissolved by slacked lime and water, and by caustic potash and soda, with boiling, producing compounds called sulphides or sulphurets, which were formerly used in bleaching, but have been long in disuse. Sulphur enters into the composition of wool and some other animal fibres, producing special phenomena with some colors. In the author's opinion the greater affinity of the animal fabrics for certain colors is in some way connected with the presence of sulphur in them, if they are deprived of their sulphur they are usually incapable of receiving good colors. Various attempts made to incorporate sulphur artificially in vegetable fabrics have been unsuccessful, no improvement having been obtained. A good quality of sulphur is known by its burning freely, and not leaving much residue unburnt. The actual test for the quality of sulphur is to burn a certain quantity and observe the weight of incombustible matter remaining. Roll brimstone does not leave one per cent. of unburnt matters; crude brimstone leaves more, according to its degree of cleanness; that quality is to be chosen which leaves the least residue.

Sulphuric Acid, *or Oil of Vitriol.*—There are three varieties of sulphuric acid to be met with, the Nordhausen or fuming sulphuric acid; rectified sulphuric acid or ordinary oil of vitriol; and unrectified sulphuric acid, of a dark color, known as brown vitriol. The Nordhausen sulphuric acid is seldom seen in England, it is prepared by distilling calcined sulphate of iron; its manufacture is confined to a few places on the continent. It is the strongest form of sulphuric acid, fumes in damp air like muriatic acid, whence it is called fuming or smoking oil of vitriol. It is said to be better for making sulphate of indigo than any of the other acids, producing a blue of a rich purplish shade. Whether this is really the case or not is doubtful; a couple of samples of Nordhausen acid that I experimented with did not show any advantage over concentrated English vitriol.

The rectified sulphuric acid is extensively used in bleaching, and also in dyeing and printing; its uses are too familiar to require enumerating. When at its greatest strength it marks from 169° to 170° Tw., it should not be below 166°. It is extremely corrosive, disorganizing vegetable textures and organic compounds with great force. This it is supposed to accomplish by means of its great affinity for water, which causes it to take the elements oxygen and hydrogen from compounds containing them. There are some organic substances which resist the destructive action of this acid in a remarkable manner, amongst these are many of the pure coloring matters, especially alizarine, the coloring matter of madder. When sulphuric acid is diluted with water its corrosiveness is suspended; for example, cotton cloth may be kept in dilute acid for a long period without injury to its strength. But if a piece of calico were steeped in dilute acid, and then exposed to the air, the water would evaporate until the acid in the fibres became concentrated enough to disorganize them. Concentrated sulphuric acid withdraws moisture from the air, and should therefore be kept in covered vessels.

Brown vitriol, or unrectified sulphuric acid, is not so strong as the rectified, and is colored; there is no reason why it should otherwise differ from the stronger sulphuric acid, but in manufactories where both qualities are produced it will usually happen that it is inferior in purity. Brown vitriol is usually sold at 150° Tw., and in point of acidity is about 15 per cent. in value below rectified vitriol; it is not worth purchasing unless it is 20 per cent. cheaper than the latter. For all purposes where great purity or the highest degree of concentration is not an object, it can be used with advantage. Brown vitriol freezes at about the same temperature, as water, while the strongest vitriol requires a much lower temperature, and is hardly ever frozen in England. The freezing of brown vitriol is rather common and sometimes productive of serious accidents; it does not freeze wholly, but at the bottom of the carboy, as if it were a crystallization; upon the tilting up the carboy the solidified mass falls against the neck and may break it, scattering the still fluid acid around. If this solidification is detected, the carboys should be placed in a warm situation in warm water; it is not prudent to pour warm water into the carboys. I am informed that a very little water added to brown vitriol effectually prevents the solidification by cold; if the water present be more than is necessary to form the deuto-hydrate, no freezing takes place.

Impurities, Analysis, &c.—Sulphuric acid may contain several impurities detrimental to its use in the arts. It may contain lead and arsenic, which, though poisonous, are not likely to do any injury in its applications in printing or dyeing. It may contain some of the gas, bioxide of nitrogen, which has been known to injure colors. This gas, which is used in the manufacture of the acid, can be largely absorbed by it in the concentrated state; if there is much present it can be detected by mixing a pint or so of the acid in question with an equal bulk of water, when the peculiar smell of nitrous acid will be perceived. A more delicate test is to pour some of the suspected acid upon clean crystals of sulphate of iron; a pink coloration, deepening to claret and brown, will take place if any of the nitrous compound be present. Acid containing this compound should not be employed in garancine making, it destroys coloring matter; such an impure acid has been known to completely discharge a delicate cochineal pink on silk, which was being passed in it to brighten it. A solution of ammoniacal cochineal is proposed as a test for this nitrous sulphuric acid, but it is not as good as the sulphate of iron test. Some manufactures add the contents of their nitre pots to the brown vitriol; so that sulphate of soda may be looked for. In all ordinary cases the hydrometer is a sufficient test for the goodness of sulphuric acid. I subjoin an abridgment of Dr. Ure's table of the strength of sulphuric acid at different densities:—

Degree Twaddle.	Strong sulphuric acid, per cent.	Degree Twaddle.	Strong Sulphuric acid, per cent.
170°	100	60°	40
162	90	43	30
140	80	28	20
120	70	13	10
97	60	5	4
77½	50	3	2

For exact purposes the sulphuric acid would be estimated by the amount of alkali it would neutralize, or by the weight of sulphate of baryta produced by a given quantity of it. The compounds of sulphuric acid with the metals are called sulphates.

Sulphurous Acid.—When sulphur burns in the air it produces sulphurous acid gas. It is this acid gas which is the bleaching agent in all cases where sulphur is used for bleaching woollen, silk, or straw. This gas can be procured by other means than burning sulphur, but not so economically. When strong sulphuric acid is heated along with charcoal, it is decomposed, and gives off all its sulphur and sulphurous acid, the charcoal at the same time being oxidized. This gas is soluble in water to a considerable extent, and it is possible to bleach goods with such a solution. The bleaching action of this gas is not a real decolorizing effect, for the color which it appears to destroy can be revived by either neutralizing or expelling the sulphurous acid, which is not the case with chlorine gas, the active element in ordinary bleaching powder. The effect of sulphuring upon woollen goods is not simply that of whitening, it gives also lustre and brilliancy, and communicates an elasticity, accompanied by a degree of harshness, easily perceptible to the fingers. A large quantity of this gas enters into some intimate combination with the woollen fibres; it cannot be removed by washing in cold or warm water, it is not expelled by a temperature at 212° F., and may remain unchanged in the fibre for the space of six months at least. The sulphurous acid may be removed by alkalies and by sulphuric acid; it can be changed into sulphuric acid by the action of bleaching powder and acids, when it is readily washed out by water. The existence of this acid gas in delaines is unfavorable to the production of good colors; in woollen dyeing it is found also an obstruction, and usually it is only the finer goods, for dyeing in bright light shades, which are sulphured.

Sulphurous acid exists in the air of those towns where coal is burned, and the author believes he traced a frequently recurring accident in dyehouses to its presence. In dyehouses which are not well ventilated, there is a constant dropping of condensed water from the beams, roof, &c.; if these drops fall upon mordanted goods before dyeing, they cause the appearance of light spots in the pattern, due to the removal of a portion of the mordant. The drops have a lesser effect upon dyed goods, but still perceptible; it is upon light lilac grounds that the greatest effect is visible, frequently sufficient to render the piece so damaged unsaleable. This effect was usually attributed to acetic acid from the mordants arising with the steam and, becoming condensed, falling down again. I obtained a sufficient quantity of the droppings for analysis, and found hardly a trace of acetic acid, but instead sulphuric acid, to the amount of 11.8 grains of the mono-hydrated acid to a quart of the droppings. The extraordinary and unexpected nature of this result caused the analysis to be questioned, but it was confirmed by a second analysis, made about a month afterwards. This is, no doubt, an unusually bad case; the dyehouse was old, in the city, and a sulphuring stove was in constant work within fifty yards of it. There can be little doubt that the sulphuric acid resulted from the oxidation of sulphurous acid carried by the air; the constant moisture, elevated temperature, and porousness of the old beams, being conditions well calculated to facilitate the change. Dr. Angus Smith states that he has found sulphuric acid as well as sulphurous in the air of towns. Some of the acid may, therefore, have been condensed upon the beams as such. The remedy for such a case as this is ventilation and frequent washing of the roof and walls of the dyehouse. Sulphurous acid combines with bases forming sulphites; strong acids decompose these compounds, the sulphurous acid being evolved. Sulphites are antiseptic, and are used to preserve animal matters from putrefaction. A small quantity of sulphite of soda will preserve a solution of albumen and cochineal sweet much longer than they would naturally remain so.

Sulphuretted Hydrogen.—This body is frequently found in nature. It is the active principle of the medicinal sulphuretted waters, and is produced in most cases of putrefaction of animal matters, giving rise to offensive odors. It is sometimes found in steam boilers, and, rising from the steam, occasions accidents to steam colors. Colors containing lead or copper are blackened by this gas; chrome oranges having lead bases are blackened by steam which contains sulphuretted hydrogen; and the dark metallic reflection upon some colors containing copper is owing to the same cause. Fents dipped in acetate or nitrate of lead are frequently up in the steaming chambers to absorb this gas. The addition of sal ammoniac,

oil, and turpentine to colors appears to make them less easily
acted upon by it. It is generally in water containing sulphate
of lime and organic matter that this inconvenience is felt. It
is worst after the water has rested for some time, as, for example, all night; it is only a common precaution to blow off a
good quantity of steam each morning before using it for goods.
The water in the boiler should also be more frequently blown
off when it is in this state.

The "coppering" of steam colors above mentioned cannot,
in every case, be traced to the presence of sulphuretted hydrogen, because it has been known to take place when leaded
fents were not blackened. It is very probable that some organic vapors or gases of a strongly reducing nature may be
the cause. It has been suggested also that the metallic appearance may not be due to the reduced metal at all, nor to
its sulphide, but to a peculiar formation of the colored lake.
This idea is based upon the fact that most of the organic coloring matters do, in some well-known form or other, acquire a
quasi metallic reflection, and in the steaming some unknown
causes operate to throw them into this peculiar condition.

Sumac.—This coloring matter is the ground up leaves and
smaller branches of a shrub which grows in many parts of the
world. That which comes from Sicily is the most esteemed,
and brings the highest price, but several other countries produce an useable article. It has a greenish-yellow color, bitter
astringent taste, and, when good, a smell reminding of tea, or
sometimes of new hay. Its quality can be judged of by its
color to a considerable extent; it should be bright and clear;
some samples are dull and have a brown faded look. These
are nearly always inferior. The difference of shade can only
be discovered by an experienced eye, or when there is a good
sample to compare with.

Sumac has the same chemical properties as galls, containing
the same acids, tannic and gallic; but, in addition, it has a
certain amount of yellow coloring matter, which, though nearly
worthless in itself, modifies its effects upon mordanted cloth.
Sumac being much cheaper than galls is extensively used as a
substitute, and in dyeing it answers the purpose very well, but
in printing it as an extract its yellow color would interfere too
much with the effects of the tannin matter contained in it.
Sumac is used as an addition to the garancine dye. It is employed in the production of many shades of colors in dyeing,
as drabs, olives, grays, etc. There are no reliable determinations of the quantity of the various principles in sumac. Davy
has certainly given some results, but the method he pursued

was not likely to yield trustworthy numbers; his statement is
as follows:—

Sicilian sumac contains . . 16.2 per cent. tannin,
Malaga sumac contains . . 10.4 per cent. tannin,

while he gives gull nuts as containing 27.0 per cent. of tannin,
which is below the truth, even for very inferior qualities.

The chief consumption of sumac is probably in cotton dyeing,
where it is the preliminary treatment for nearly all the fancy
shades to steep the cotton for some hours in decoction of sumac.
The astringent matter of the sumac is thus firmly combined
with the cotton, which can now be easily mordanted with
either tin or alumina, which form the basis of the colors.
Sumac liquors have a strong tendency to become acid, which
must be guarded against in those cases where an iron or alumina
mordant is concerned, since the acidity is sometimes strong
enough to dissolve out weak iron mordants.

T

Tanner's Bark.—This is an astringent material, of variable
nature. It is mentioned as being used for the production of a
gray color. The cloth is mordanted in a mixture of equal
parts of iron and red liquors, aged, cleansed, and dyed with
about two pounds of tan per piece. When cleared by a slight
soaping, and a subsequent passage on very dilute acetic acid,
agreeable shades of gray are obtained.

Tannic Acid, *Tannin.*—The substance called tan is well
known for its uses in converting skin into leather. There are
a great many substances capable of tanning skin, but it is found
that they all possess an astringent or acid body, nearly the
same in properties and composition, and it is this body which
is called tannin, or tannic acid. However, the name is more
generally intended to represent the pure astringent principle
of gall nuts, which is now an article of commerce, and extensively employed by calico printers for fixing the aniline colors.
Tannic acid is prepared from gall nuts by digesting them with
aqueous ether, which dissolves out scarcely anything but the
pure tannic acid. Rather more than half the weight of good
galls is obtained. When dry, the tannic acid has a brownish-
yellow color, and an intensely astringent taste, but no perceptible acidity to the tongue. It is soluble in water, and gives
precipitates with nearly all the metals and coloring matters.
Its chief characters, with reference to fibrous matters and drugs,

will be found in the articles upon GALL NUTS, ASTRINGENT MATTERS, etc.

Tartaric Acid.—This acid is very extensively used in calico printing for the class of steam colors; it is not much used in dyeing in the free state, but in combination with potash, as tartrate of potash or cream of tartar, it is very largely employed in woollen dyeing.

Tartaric acid exists in the juice of the grape, which is the only practical source of it; it falls out of wine in the state of red tartar, impure cream of tartar, or bitartrate of potash; the pure acid is obtained by converting this salt into tartrate of lime, decomposing it with sulphuric acid, so as to combine all the lime with the sulphuric acid, and leave the tartaric acid free. It crystallizes in large clear crystals; they are very soluble in water, dissolving to a syrup; their taste is agreeably acid. As obtained from respectable houses tartaric acid is in nearly a pure state, but it is said to be occasionally adulterated with the bisulphate of potash. This adulteration may be discovered by taking some of the pounded crystals and heating them to low redness; if the tartaric acid is reasonably pure there will be nothing but a black coal left, which would burn away at an increased heat, and which has no taste; if there be any bisulphate of potash present it will show as a white ash, which will have the well known acid taste of the bisulphate, and which will not burn away at any heat, or with any length of time. A sample of acid can be judged of by simple inspection if it is in crystals, but if ground, only testing can tell what its quality is. Tartaric acid is at once a strong and a mild acid; it has powerful affinities, but it is not corrosive, and does not injure the finest fabric to which it is applied. Besides its acid properties, tartaric acid possesses properties of a singular nature, in masking or hiding the characters of metals with which it is mixed: for example, if caustic potash be mixed with the sulphate of copper dissolved in water, it will precipitate or throw out all the copper, but if tartaric acid be previously mixed with the sulphate of copper, potash will not produce any precipitate; many other metals behave with it in just the same manner. Tartaric acid is used principally in steam colors for blue and green; its effect in steam blue is to take potash from the yellow prussiate and leave its acid, the ferrocyanic, at liberty to react upon the other components and upon itself to produce the blue; the tartaric acid forms bitartrate of potash, which continues the action, being itself an acid salt. It is also much employed as a discharge on dipped blues and Turkey red.

Cream of Tartar.—Tartaric acid combines with potash to form two compounds, known respectively as the tartrate and bitartrate, the first containing only half as much acid with reference to the potash as the other. The simple tartrate is very little known in practice, it is a neutral and very soluble salt. The bitartrate is the form in which the wine deposits the tartar mixed with other substances, and with the coloring matter of the wine; in this state it is called either simply tartar, red tartar, or argols. It can be used in dyeing in this state, but it is usually purified by dissolving in water, separating the impurities and re-crystallizing, in which state it is called cream of tartar; it is in crystals of small size, hard and gritty to the teeth, and having a slightly acid taste. A further purification brings it into a white crystalline powder, and it is thus used in the finer styles of printing and dyeing. It is very little soluble in water, and cannot be made to mark more than two degrees Twaddle at natural temperatures. It dissolves much more if mixed with potash or soda, but then it loses some of its principal properties. In calico printing its uses are few, and mostly such as could be replaced by tartaric acid, but the cream of tartar is thought to act more mildly upon some delicate shades, as cochineal pink. In dyeing it is used for mordanting woollen goods in conjunction with alum and the salts of tin. It is not quite clear in what the chemical action of the cream of tartar consists in this case; in the first place, it seems to correct bad waters; if they are very hard, and contain much lime, it appears to prevent its falling on the cloth to its injury; if the water contains iron it forms combinations with it which keep it from injuring the stuffs to be mordanted, probably by holding the iron in some condition in which its active properties are for the time being suspended; it may take some part in the saturation of the strong acid of the alum or muriate of tin, permitting their bases to pass more easily into the fibre of the wool. As an addition to the actual dyeing, tartar acts, no doubt, in two ways: first, as a corrective with regard to lime and iron in the water, and second, as a mild acidifying agent; for wool does not take colors well unless the bath is slightly acid, and tartaric acid and cream of tartar are the best acids which can be used for the purpose.

Tartaric Acid Substitutes.—The cost of tartaric acid and the irregularity of its supply, have led many persons to seek for a substitute. Several have been patented, and others privately offered in the market. Arsenic acid has been proposed, but has not answered, and in low class printing the bisulphate of potash is frequently used instead of tartaric acid.

I have had occasion to analyze several of the substitutes offered for sale. The majority of them were of the most worthless character, consisting simply of sulphuric acid with some

28

common salt, with sometimes addition of oxalic acid, alum muriate of tin, and other similar substances.   There are, doubtless, many cases in which dyers use tartaric acid or tartar, when nearly any other acid would answer as well.   In such cases these pretended substitutes might pass, but for most of the uses of tartar these substitutes could not be employed with safety.   Some other substitutes contained a considerable proportion of tartaric acid in a crude form, and were worth the price asked for them.

Tea Color.—A dull green color, similar to that of dried tea leaves.   The chief color, known as tea green, is that obtained from salts of chromium, according to the methods given in page 146.   I give here one or two recipes, in which the same shade is obtained by different means:—

A tea color for raising in lime may be obtained by making a mixture of nitrate of lead and nitrate of copper, padding or printing the piece with the mixture, and raising in chrome.

Tea Drab Color—Calico or Delaine.

1 gallon bark liquor, at 3°,
1 quart copperas liquor, at 30° Tw.,
5 oz. nitrate of copper, at 80°,
5 oz. extract of indigo,
1½ gallons thick gum water.

The above and the following have not much green in their composition, and might perhaps be more correctly called olive drabs.

Tea Drab, for Wool.

1 gallon catechu liquor, at 20°,
½ gallon peachwood liquor, at 12°,
3 oz. extract of indigo,
6 oz. alum,
6 oz. oxalic acid,
2 quarts thick gum water,
1 pint cochineal, crimson color.

Thermometer, *Heat Glass.*—The degree of heat which is employed in the operations of color mixing and dyeing has so great an influence upon the results, that it is necessary to have some measure of temperature less uncertain than its effects upon the senses.   Some dyers and color mixers may be able to accomplish their work by ascertaining the temperature of fluids by the hand, or similar means; but those who aim at accuracy and regularity should be familiar with the thermometer or heat-glass.

The action of the thermometer in ordinary use is based upon the expansibility by heat of the mercury contained in the bulb; when the medium surrounding it is warmer than usual, the expansion of the mass of metal below forces a column of it up the narrow or capillary tube above; when the medium is colder than usual, the contraction of the bulk causes the entry of that in the tube into the bulb, and a consequent fall of the column; and as this expansion and contraction are constant for the same variations of heat, a good measure of actual temperature is obtained. On Fahrenheit's thermometer the scale runs from 32°, or the freezing point, to 212°, the boiling point, of water. The only precaution required in using the thermometer is not to plunge it too suddenly from a hot to a cold liquid; there is a risk of breaking the instrument, or of interrupting the continuance of the column in the hair tube; when this latter accident occurs, it may be remedied by gently tapping the instrument, or by tying a cord securely to the upper part of the instrument and whirling it round with velocity. The scales of thermometers used on the continent differ very much from that of Fahrenheit. The one chiefly used, called the Centigrade, has the freezing point marked 0°, and the boiling point 100°; another one, used in the northern and central countries of Europe, Reaumur's, has the freezing point at 0°, and the boiling point at 80°. To convert the degrees upon the scales of these thermometers into the corresponding ones upon Fahrenheit requires only simple multiplication; each degree of the Centigrade thermometer is equivalent to a degree and four-fifths of a degree of Fahrenheit's, and each degree of Reaumur's is equal to two degrees and a quarter of Fahrenheit's; by multiplying the degree of either of these thermometers by the proper numbers, and adding 32° to the product, the equivalent degree of Fahrenheit will be found. Receipts and processes frequently arriving in England from France and Germany, with the temperatures marked in degrees of one of these thermometers, a table is here given showing the correspondence between the three instruments for a sufficient number of degrees; the fractional parts of a degree are omitted in the Fahrenheit column:—

Table showing the Corresponding Degrees upon the Centigrade, Reaumur, and Fahrenheit Thermometers.

Cent.	Reau.	Fahr.	Cent.	Reau.	Fahr.
100	80	212	50	40.0	122
98	78.4	208	48	38.6	118
96	76.8	205	46	36.8	115
94	75.2	201	44	35.2	111
92	73.6	198	42	33.2	108
90	72.0	194	40	32.0	104
88	70.4	190	38	30.4	100
86	68.8	187	36	28.8	97
84	67.2	183	34	27.2	93
82	65.6	180	32	25.6	90
80	64.0	176	30	24.0	86
78	62.4	172	28	22.4	82
76	60.8	169	26	20.8	79
74	59.2	165	24	19.2	75
72	57.6	162	22	17.6	72
70	56.0	158	20	16.0	68
68	54.4	154	18	14.4	64
66	52.8	151	16	12.8	61
64	51.2	147	14	11.2	57
62	49.6	144	12	9.6	54
60	48.0	140	10	8.0	50
58	46.4	136	8	6.4	46
56	44.8	133	6	4.8	43
54	43.2	129	4	3.2	39
52	41.6	126	2	1.6	36

Thickenings.—In piece or yarn dyeing the mordant is applied in a simple fluid state, the object being only to impregnate every part of the fibre in an equal manner with the mordant or coloring matter. But in printing, it is required that the mordant shall be applied only to certain parts of the cloth, the remaining part being either left white or occupied by some other mordant or coloring matter. The capillary attraction of the fibres is such, that if a drop of mordant in its fluid state be applied to a piece of cloth, it spreads in a circular form far beyond the size of the drop placed on, but not in an equable manner; the spot upon which the drop was first placed holding the greater part of the fluid, and the surrounding portions less and less as they are further removed from the first boundaries of the drop. This inclination of liquids to spread beyond the limits of their first application is overcome by the addition of various matters to them called thickenings, such as gum, starch, etc. These substances act by themselves, setting up an attraction which disputes more or less successfully the capillary attraction of the fibre, and retains the applied mordant within the limits of the design.

This is the first use of thickening matters; a second is to permit the application of a larger quantity of mordant to the cloth than could be managed with thin liquors : this property of thickenings is sometimes taken advantage of in dyeing where no design is required.

The use of the thickening matters employed in calico printing is simply transitory ; while most of the other substances employed carry some traces of themselves on the finished product, the gum, starch, red flour employed as thickenings are only temporary in their application, and have to be all removed before the colors are finished. The great expense of the thickening matters, and the complete loss of all the raw material, should draw the attention of printers particularly to this point to see if it is not possible by mechanical means to dispense with the use and waste of such large quantities of substances which are for the most part derived from articles of human food. The Industrial Society of Mulhouse, always alive to the wants and necessities of printing, having offered a prize to any one introducing into the market thickening matters capable of replacing those now in use and not made from articles used as human food. I believe it is possible to go higher than this, and to ask for some means of doing without thickenings altogether. I am convinced that this is possible and practical, and that the skill and ingenuity of mechanical science will not be long turned in this direction without reward.

The art of thickening colors lies at the very root of calico printing ; upon it depends so much in the way of obtaining good results that it may be considered as the most important part of color mixing, and that a color mixer will be good, bad, or indifferent, as he intuitively perceives the importance of this branch of his art, and is successful in carrying it out. To give receipts and furnish the best materials is nothing without there exists at the same time an intelligence to comprehend the action of the various materials, and an ability to put them together in such a way that they can produce their full effect. Food receipts fail in the hands of unskilful color mixers, and the purest and most expensive drugs are only thrown away ; while an expert hand can produce good colors from inferior articles and at less price. The difference all lies in the putting together of the materials, and the properly blending them with the thickening matter most suitable to them and the particular styles they are intended for. It is not possible in a book to enter into all these matters with the minuteness they deserve, nor to communicate all the knowledge necessary to success ; it is so much a practical matter that after all has been said that can be upon the subject, a great deal more that is essential will be left unsaid, for no description or formulæ can explain what the finger and thumb can feel, or the accustomed eye perceive, in a made-up color. Such generalities upon thickenings as will lead to the understanding of particular cases will be given here, and other connected matters may be found in the section on gums.

Many mineral matters, such as pipeclay, white lead, gypsum, and the like, could prevent the spreading of a fluid mixed up with them; but all mineral matters of this nature are heavy, not soluble in the fluids, and consequently fall to the bottom, leaving the supernatant liquor thin, and of course producing irregular results. Such substances cannot be generally used by themselves; when in combination with some of the vegetable thickenings they are frequently employed to give density and toughness to the paste, and are thus particularly useful in resists. There may be colors of such an excessive chemical activity, that the vegetable substances employed in ordinary circumstances are acted upon to the mutual destruction of both the thickening and color. Such are chromic acid and the permanganate of potash; if the application of these is necessary or useful it will have to be through the medium of some mineral thickening like pipeclay.

The vegetable substances used for thickening colors may be divided into three classes : first, starches, flour, and thickenings insoluble in cold water ; secondly, gums proper, whether natural or artificial, which are soluble in cold water; and, thirdly, artificially made-up mixtures of the first and second classes, partly soluble and partly insoluble, partly gummy and partly pasty.

The thickenings of the first class require a much less weight per gallon to give the desired viscosity to a mordant than those of the other two classes. This is their first and most striking characteristic, and it will be well to consider here the influence which the weight or mass of thickening has upon a color without taking any of the other properties into consideration. Good starch thickens sufficiently well for most purposes at twenty ounces per gallon of water, flour at about the same or a little more, and gum tragacanth gives a consistent color with only one-half this quantity ; on the opposite side, calcined farina requires eight or nine pounds to be added to a gallon of water to bring it to a proper state of viscosity. Is it indifferent whether a color or a mordant be thickened with starch or calcined farina? It is not only not indifferent but a matter of the greatest importance. As a general rule, it may be stated that the smaller the quantity of thickening in a color the darker the shades produced. If iron liquor at six or seven degrees Tw.

be thickened with starch or flour, a good black may be obtained from it in the madder dye; if the same strength of iron liquor be thickened with calcined farina, it will not dye up a black but only a shade of purple. To obtain a black from calcined farina thickening would require the iron liquor to be at least double the strength given. The same rule holds in other colors; the depth of shade is always in some ratio to the amount of thickening matter. The mass of thickening impedes the access of the particles of the mordant or color to the fibrous substances intended to receive them. It is evident that if the color was made of an excessive degree of thickness, none of it would leave the thickening to go to the fibre; it is only because the thickenings contract upon drying that the fibre receives from them any of the color or mordant enveloped by them; and in proportion as this contraction is greater or less in relation to the original bulk of the matter when applied, so is the amount of color or mordant communicated to the fibre. Starch swells out when boiled into a bulky vesicular mass, which may be looked upon as a sponge holding the liquor with which it was boiled; the drying of the starch on the cloth is equivalent to the squeezing of the sponge, the liquor leaves it because it can be dried up into a small compass. Calcined farina, when dissolved, may be looked upon as a sponge also, but one of a denser kind, with less room for liquids and more difficult to squeeze dry because of its solidity. When it dries on the cloth a good share of the mordant or color never touches the fibre, being entangled by the intervening mass of the thickening with which it remains in contact until washed away in the cleansing or dunging.

Another and dependent effect may be noticed, that is, the penetrating power of the different thickenings. Printed with rollers of an equal depth of engraving, it will be found that the starch or flour thickened color has penetrated through the cloth, and shows plainly and strongly upon the reverse side of the piece, while the calcined farina thickened color has penetrated much less, and may not be at all percepible upon the back of the piece. The observation of this, and the knowledge that the coloring matter, which is visible on the wrong side of the piece, costs the printer money without adding to the effects produced, leads to questions concerning the economy of thickening matters in a comparative point of view. One pound of starch will cost less than five pounds of calcined farina or other gum; but there is a possibility of this difference being wholly or partially made up in the quantity of madder required. If the watery starch color permits the metallic mordants to penetrate to parts where they exhaust the

dyeing material without adding to the color, it becomes simply
a matter of calculation and trial to ascertain if this does not
overbalance the difference in the cost of the gum and starch.
In madder dyeing it is not only a question of economy, but of
goodness of color, not to let the mordant penetrate too deep
into the centre of the fibre; for the deposition of the coloring
matter only takes place in a perfect manner upon or near the
surface: it is there only that the mordant becomes saturated,
and its natural color overcome by the true lake formed.   Below
the surface an imperfect combination only takes place, which,
in the case of alumina, results in brick-colored reds, and with
iron mordants, in rusty snuff colors.   These form a bad basis
for the real madder colors and injure their shades.   Beyond a
small extent it is, therefore, injurious to the colors to have
them deep seated in the cloth; the whiteness of the cotton is
the best ground for them to show upon, and a color is better
in proportion as it is upon the surface, or even appearing to
stand in relief.   These remarks apply principally to madder
purples and pinks, but are equally true of all light dyed colors.
For dark shades it is necessary sometimes to penetrate the
cloth, or, at least, it seems necessary; but, I believe, with management, better and cheaper colors could be obtained by keeping on the face side only.

Only two extreme cases of thickness have been mentioned,
but there are many intermediate degrees from which the color
mixer can choose, to accommodate his colors to the necessities of
printing.   The second class of thickenings are those which
are best adapted for all light shades of color.   They appear to
be less subject to irregularity than paste and starch colors,
and give better furnished and fuller shades, at the same time
they are livelier and brighter, which may be attributed to their
being more superficial than the paste colors would be.   The
third class, or mixed thickenings, form a very useful class,
capable of valuable application.   By a judicious use of them
the valuable qualities of both paste and gum may be in a
great measure combined, and economy of thickening matters
secured, without waste of dyewood.   Gum manufacturers send
such gums or mixtures out into the trade, some of which are
good and some are bad.   A color mixer has it in his own power
to make nearly all useful mixtures of this class from the *bona
fide* gums, sold by respectable manufacturers, and the starch and
flour with which he is supplied.   In making mixtures of various
thickening substances, it should be taken as a principle that
those will work best together which have separately the greatest
similarity in properties.   It is not well, for example, to mix
starch with calcined farina; they are at the opposite ends of

the scale, and are too widely different in their properties to work smoothly together. The mixture will separate, the thick part leave the thin, and work rough and curdy in the machine. If a mixture of starch and gum is wanted, some gum should be chosen whose thickening power is about four pounds per gallon, and there ought to be more of it by weight than the starch to form a good mixture. If it is required to strengthen calcined farina with some substance thickening further than it, a gum should be chosen thickening at from four to five pounds per gallon. A mixture of calcined farina and flour will work well if it contain also a certain quantity of gum whose actual thickening point is about the same as the calculated mean of those two substances. It acts as a medium, holding the extremes together. Two gums which are perfect gums will mix and unite in any proportion without subsequent separation, and whatever their respective thickening powers may be. In mixtures of various thickenings it will be found that flour and starch give body and firmness—the natural gums give tenacity, and among these tragacanth is conspicuous—the true artificial gums give solidity or density. There is much good to be done by this method of mixtures; it often happens that two or three individual gums of an inferior quality will, give, by proper mixture, a good useable gum.

The adaptation of the thickening to the design is so much a practical matter that it is scarcely possible to go into details. The lighter and finer the engraving the smoother and solider the thickening should be; for fine outlines a good paste color is suitable, or else a smooth dense gum color; a soft puffy color would not answer, although this latter might pass for open blotch work. For fine covers, on light shades, gum colors are best adapted; but paste colors, made from finely sieved flour, can be often used with advantage.

Tin.—Tin is a very important metal to the dyer and printer from the affinity which it shows both for fibres and colors, and the brilliancy of the shade it gives. As a metal its expense prevents it being much used in the way of vessels or utensils; what is commonly called tin, being only iron coated with tin: tin vessels are usually distinguished as of block-tin. It is used, however, in some dye-houses, especially where scarlets from cochineal or lac dye are dyed. It is found that copper does very well except where the air comes in contact with the liquor, and I believe the pans are made of copper at the bottom to stand the fire, and of block tin on the upper part where the air acts and where the pieces would come in contact with the bare metal. For experimental purposes of dyeing, and especially for mordanting in tin solutions with heat, block-tin vessels are by far the best. Tin melts at low heat, and consequently vessels made of it must never be exposed to fierce fires.

Tin combines with oxygen in two proportions, forming the protoxide, which has one atom of tin to one atom of oxygen, and the bioxide or peroxide, which has two atoms of oxygen to one of tin; this last, from its sometimes acting the part of an acid, is called also stannic acid, the Latin name for tin being stannum. Tin in the metallic state is used in preparing one color for printing, that is, a kind of indigo blue called " fast blue." The tin for this purpose, and for all purposes of dissolving, is first granulated by pouring it from a height when melted, into cold water; in this case the granulated tin is boiled with caustic soda and the powdered indigo, the metal takes oxygen from the indigo under the influence of the alkali, and the coloring matter is thus brought into solution; the same, or an equally good color, can be prepared in other and better manners. The protoxide of tin, like many other protoxides, has an affinity for more oxygen to change itself into the peroxide, and it will take this oxygen, under favorable circumstances, from bodies put into contact with it; it is this property which enables it to dissolve indigo when mixed with caustic alkalies. The protoxide may be made by adding the proper quantity of caustic soda to solution of crystals of tin: too much caustic will redissolve the oxide; it falls down as a white pulp, which can be drained and washed on a filter. It is not necessary to separate the oxide of tin from the liquor for most practical purposes, and the necessary quantity of indigo, caustic, and tin solution are mixed and all heated up together. The peroxide of tin has no inclination to take any more oxygen than it already possesses, and cannot aid in the solution of indigo. Both of these oxides combine with acids, and they have different properties, though both answer nearly equally well as mordants.

Chloride of Tin, Muriate of Tin, Crystals of Tin.—The crystals of tin are a compound of chlorine, tin, and water; they are made by dissolving tin, by means of heat, in spirits of salts, boiling down and crystallizing. They are supplied by respectable chemical manufacturers in a state of almost chemical purity, but they are said to be sometimes adulterated with zinc. I doubt this with regard to the crystals. I never found any such adulteration, and I think it would spoil the appearance of the crystals so much as to make it apparent to any one that something was wrong. The crystals may be bad by age and by being too wet, and their apppearance shows this; they should be clean and glistening, slippery and greasy feeling to the finger, and have no white powder or white slime about them. When a couple of ounces are mixed with a gill of common water, the liquor should be clear, but when the same weight is mixed with a gallon of water, the liquor ought to become white with a white sediment forming; this test shows in the first case that they are acid enough, and in the second that they are not too acid.

Crystals of tin are largely employed in dyeing and printing; in dyeing, as a mordant; in printing, as helping to form colors and to communicate to them peculiar properties; as, for example, mixed with strong red liquor, it enables the red mordant to resist light covers of chocolate or purple mordants; it acts as a discharge on some colors, as iron buffs, manganese browns, etc., and has many other uses.

Liquid Muriate of Tin.—This compound is chemically the same as the crystals of tin, and is made in the same manner, except that it is not boiled down to the crystallizing point. It is decidedly more acid in its behavior than the crystals, and herein lies the only difference, if the liquid muriate is a genuine article. Weight for weight is not much more than one half of the strength in tin of the crystals. It is liable to be adulterated with several cheaper solutions, and I do not know any good practical test, except trying it in colors against a sample of known goodness. By chemical analysis it is easy to ascertain its exact value. It is used in dyeing for much the same purposes as the crystals, in color mixing also; this is the salt mostly used for preparing tin pulp or prussiate of tin for steam blues.

Sulphate of Tin is hardly at all used in dyeing as a pure salt, but there are cases in which a mixture of vitriol and crystals of tin is employed, and probably sulphate of tin is formed here; but the process never goes to the extent of driving off the muriatic acid, so it is not certain what the resulting product is, probably a sulphate of tin kept in solution by the muriatic acid. Sulphate of tin itself is not easily made, and is too expensive for general use. It can be made by stirring up granulated tin in water with sulphate of copper; the copper falls down, and the tin takes its place with the sulphuric acid. There are solutions of tin made by dissolving the metal in mixtures of vitriol, water, and common salt, or, instead of common salt, sal ammoniac, sometimes spirits of salts and vitriol; there is very likely formation of sulphate of tin in these cases. When strong sulphuric acid acts upon tin with the assistance of heat, various gases are evolved, and a compound of the peroxide is left, which requires a good excess of acid to keep it in solution; I have found no protoxide salt even when a good deal of the metal has been left unacted upon.

Bichloride of Tin; Double Muriate of Tin.—This differs from

common muriate of tin by having the metal in the higher
state of oxidation; it requires a special method of preparing,
and possesses quite different properties from the muriate.
When a few drops of this bichloride of tin are mixed with a
solution of bichrome, it ought not to change the red color to
green, which it will if it is not well made, or if it contains any
of the common muriate. The bichloride of tin is not largely
used either in dyeing or printing. It serves as a prepare for
woollens and delaines to make them take colors better in printing; it is for this purpose mostly mixed with some of the common muriate, and is an ingredient in some colors and in some
dyes. It can be made by dissolving grain tin in a mixture of
two parts spirits of salts, one part aquafortis, and one part
water, till no more dissolves.

Oxymuriate of Tin (Proto-per Chloride of Tin.)—This liquor is
made in general by dissolving granulated or block tin by degrees, in a mixture of nitric and muriatic acids; it is for the
most part a bichloride of tin, but frequently containing some
of the protochloride or common muriate; it is generally well
saturated, that is, with little excess of acid. It is also generally
of a milky appearance, from a quantity of undissolved oxide
of tin held in suspension. It can be made from the common
muriate of tin, or from the crystals, by heating them in a mug,
and adding strong aquafortis by degrees, so long as red nitrous
fumes are given off. It is extensively used in making spirit
colors by calico printers; it is employed in Turkey red and
madder pink dyeing, to reduce the shade to a bluer tone, and
in several cases of dyeing.

Dyers' Spirits.—These are solutions of tin in endless variety,
and with hundreds of modifications; every dyer or maker
thinks he has the best method of preparing the spirits, and
usually guards his secret as valuable. They are all of them a
mixture of protochloride and perchloride of tin, some with sulphuric acid; nitric acid almost always enters as a constituent,
but is doubtless all destroyed in the oxidation of the tin.
Common salt, sal ammoniac, saltpetre, nitrate of soda, and other
salts, are used in conjunction with the acids, and possibly
modify the product, so as to make it better adapted to the peculiar office it has to fulfil; more frequently these additions are
the effects of caprice, and the advantages they confer entirely
imaginary.

Stannate of Soda (Preparing Salts).—This salt is compounded
of the peroxide of tin, mentioned above, and caustic soda; its
value and its applications depend upon its giving up the
stannic acid, or peroxide, when an acid is mixed with it. The
method of preparing pieces with it, either for printing or dye-

ing, is to pad them in a solution of it, and then pass them into sours; the sulphuric acid takes the soda, forming sulphate of soda, while the stannic acid remains attached to the cloth. The value of the preparing salt depends upon the quantity of tin which it contains, which quantity can only be ascertained by means of analysis. A practical test is to prepare fents with it against other fents prepared with a salt of known quality. The strength of the hydrometer, which a certain quantity gives to a gallon of water, is little or no test. I had occasion to analyze a sample of preparing salt said to be of double strength ; it only contained as much tin as the regular quality, but it had about twenty-eight per cent. of common salt in it, and was quite dry, whereas the ordinary good stannate has from twenty-five to thirty per cent. of water in it, and no common salt ; the sample, which was of double strength, and was to be charged considerably higher, had been made by driving the water away from the ordinary quality, and putting salt instead ; one pound of it in a gallon of water stood several degrees higher than a pound of the common stannate—and this was the test the manufacturer desired to be applied—but chemical analysis demonstrated that it was not worth any more than the stannate with water, and was likely to be worth much less, as so great a quantity of useless substance might impede the fixation of the tin upon the cloth. Some very good preparing salts contain a portion of arsenic, in the state of arsenic acid combined with soda, and the makers consider it as important in yielding good results. It appears to combine with the tin, and fix with it upon the cloth. I have made experiments with preparing salts with and without arsenic in them, and, if the conditions were otherwise equal, I could never find that arsenic was an improvement. I consider that the good quality of these salts is attributable to the care exercised in their manufacture, and not in any manner to the arsenic present; at the same time it should be understood that, while in some cases a certain ingredient seems unnecessary and useless, it acts an essential part in other cases, where perhaps the general conditions are not so favorable, or, at any rate, not the same. Tungsten in the shape of tungstate of soda is sometimes mixed with stannate, and is thought by some printers to give improved results. Many experiments I made on that point gave only negative results, no improvement was visible.

I give here a number of receipts for preparing tin solutions as used in various cases. A great deal depends upon the fitness of the tin solution in producing the best shades of color, but it is not clear in what this fitness itself consists; it is only known that tin solutions are very much changed in their

bearings towards cloth by slight alterations in the manner of
their preparation. Chemistry teaches us that the oxide of tin
is capable of assuming several widely different physical aspects,
according to the manner in which it is produced; but we are
not yet informed as to the exact conditions which govern the
formation of the different isomeric oxides, and can only therefore exercise a general precaution against accidents.

Red Spirits.—Three parts muriatic acid, one part nitric acid,
one part water; granulated tin, to the amount of 2 oz. per
pound of the mixed acid, added in small portions, so that the
liquid does not get hot.

Yellow Spirits.—Three parts muriatic acid, one part sulphuric acid, one part water; add granulated tin as much as it
will dissolve, not allowing the heat to rise above 60°.

Barwood Red Spirits.—Five parts muriatic acid, one part
nitric acid; add granulated tin to the extent of 1 oz. metal to
1 lb. mixed acids.

Plum Spirits.—Six parts muriatic acid, one part nitric acid,
and one part water; add granulated tin to the extent of $1\frac{1}{2}$ oz.
for each pound of the mixed acids.

The above quantities are by measure, and are for the ordinary strength of acids sold under the above names.

The following oxymuriate contains, added sal ammoniac:
20 lbs. muriatic acid, 20 lbs. nitric acid, in which has been previously dissolved 5 lbs. sal ammoniac; dissolve in 10 lbs tin.
This preparation is suitable for printers' purposes.

Oxymuriate for Cutting Pinks.—16 lbs. crystals of tin, melted
in a mug, placed in hot water, 20 lbs. nitric acid, added by degrees.

Another.—60 lbs. crystals of tin, one quart water, heat in a
water bath until melted, then add by portions 92 lbs. nitric
acid at 60°.

Woollen Dyers' Spirits.—Two gallons water, 15 lbs. nitric
acid at 62°, 12 oz. common salt, $1\frac{1}{4}$ lb. granulated tin. There
should be no effervescence, and the liquor should be perfectly
clear and of a pale yellow color. Not safe after a week old,
and should be kept cool.

For Spirit Colors.—11 lbs. muriatic acid at 34°, 5 lbs. nitric
acid at 62°, 2 lbs. granulated tin, added by degrees.

Sulpho-muriate of Tin Spirits.—2 lbs. sulphuric acid, 3 lbs.
muriatic acid. Pour the muriatic acid upon an excess of granulated tin, and when it has ceased to act, add the sulphuric acid
and leave them for a day or two upon the tin without heat. If
heated, the sulpho-muriate may be formed in less time, but
does not seem to be so good or regular in its results.

Tin Salts as Mordants.—The affinity of the oxides of tin for coloring matters, and for textile fibres, is not inferior to that of any other oxide, and in some respects seems superior to all others. There are, as already stated, two distinct oxides of tin, the protoxide and the peroxide, the latter containing twice as much oxygen as the former; and although the protosalts are generally applied as mordants—as the crystals of tin and common liquid muriate of tin—it seems probable that it is as the higher oxide that it acts eventually as the bond between the coloring matter and fibre.

Each of the oxides of tin has a general affinity for all the varieties of fibre, and combines equally well with cotton, wool, and silk; but the combinations are not of the same permanency in each case. They are more permanent on wool than on silk and cotton, and more powerful on the former than the latter. Tin mordants, upon cotton, give a class of colors which are called "spirit colors," from the old name for solutions of tin; they are easily made, look very well, but are not fast. Tin, upon cotton, should be employed rather as an useful auxiliary than as a sole mordant; it serves to brighten colors, but it ought not to be depended upon for giving permanent colors. In dyeing, various salts of tin are largely used for producing fancy shades upon cotton. The method of application usually consists in mixing the solution of tin with the coloring matter, and running the piece through the mixture until it has taken up as much as is required. Colors so produced do not possess much stability. A better method would be to prepare the cloth first with tin, by means of the stannate of soda, and then run it through the mixture of tin solution and dyewood. The basis of tin being more intimately connected with the fibre in this case, the adhesion of the colored compound is the more perfect.

Salts of tin cannot be advantageously used as mordants in the way that red liquor and iron liquor are used in calico printing. They take colors, but not in a satisfactory manner, and the shades which are produced are very loose indeed, so that they will not even resist the clearing operations necessary to obtain good whites. Strong red mordant is often mixed with crystals of tin, sometimes with a view of enabling it to resist colors printed over it, sometimes with the intention of brightening the color. Such a mixture is subject to irregularities in the dye, depending apparently upon the quality of the red liquor used. There are cases in which the alumina seems to be nearly all displaced by the tin. The red looks very well out of the dye, but is much injured in the soaping and clearing.

Salts of tin are much used in woollen dyeing and printing; and, owing probably to the difference in structural arrangement of the fibres, produce colors which for permanence leave little to desire. The stannate of soda cannot be beneficially applied to woollen goods as a prepare, on account of its alkalinity, which is detrimental to the fibre. It is usual to employ the acid solutions of tin, as the oxymuriate, sulpho-muriate, etc., from which the woollen fibre easily abstracts the required amount of oxide. The same remarks apply to silk, which, however, is very rarely submitted to such a preparatory mordanting.

The colors produced by tin oxides differ from those of alumina and iron, bearing more analogy to those from alumina than to the iron colors. Solutions of tin decompose with greater facility than those of either iron or alumina, and the oxide will become attached to cloth under circumstances in which not a particle of the other oxides could be deposited. But cotton cloth does not seem capable of holding it in a large quantity; a small quantity it retains with a pertinacity which has no analogy in the cases of the other mordants. Similarly, tin has no great capacity for colors, or it has no strong retaining hold of them. A tin mordant can be dyed up and the color almost soaped out, and the mordant left able to dye again. Tin is most useful, and forms the best colors, when in presence of a large excess of coloring principle.

Tobacco Color.—Any shade of brown resembling the color of tobacco may be so called. In calico printing the name has been used for a brown produced by a double dyeing in bark and madder. The mordant is a mixture of iron and red liquor, printed, dunged, and dyed in bark, and then dyed over again in a small quantity of madder. The bark and madder may be mixed at once, instead of using them separately, but then the results are not so certain.

Turpentine, *Spirits of Turpentine.*—Turpentine is somewhat largely used in calico printing, at least it is employed in many colors. Its action is supposed to be entirely of a physical nature, giving smoothness to the paste, preventing frothing, etc. In some colors where spermaceti or oil of any kind is an ingredient, the use of turpentine is to assist in diffusing the fatty matter through the thickening. In some few cases, as in albumen colors, and in cochineal liquor, turpentine acts as an antiseptic, and preserves the animal matter from putrefaction. The addition of turpentine to steam colors is thought also to prevent the "coppering" alluded to on page 422. The quantity of turpentine added to steam colors seldom exceeds $\frac{1}{3\cdot2}$ part of their bulk.

Turmeric.—This yellow coloring matter is the root of a plant, *curcuma longa*, growing in the East Indies. Its color.

ing matter does not dissolve readily in water, but is very soluble in alkaline solutions. It is chiefly used in silk printing and dyeing; it is also employed in woollen dyeing for dark and full shades of color. It is valued for a rich yellowish reflection it communicates to such colors. Upon cotton it dyes without any mordant, but produces one of the most fugitive and changeable of colors, and is only employed in the extreme fancy styles. Its pure coloring matter is called curcumine.

Tungsten.—The ore of this metal exists in tolerably large quantities in Cornwall in conjunction with tin ore. All through its compounds it has a resemblance to tin. Some years ago tungstate of soda was sent into commerce to be used as a substitute for stannate of soda, but it would not answer, and all attempts to apply it as a mordant for colors failed. Some of the compounds of tungsten have good colors, but they vanish in trying to fix them. If some tungstate of soda be put into a vessel, an excess of muriatic acid added, and then some slips of thin zinc, a fine blue powder will be formed in a short time, the composition of which is not very well known. If this be collected on a filter and washed it soon loses color, and in a few hours will be nearly white. Many attempts which I made to put this blue compound into something like an inactive state, or to produce it upon the cloth itself, were without any practical result. It appears from all my trials that tungsten would be of no use in dyeing or printing, however cheap or plentiful it might be.

U.

Ultramarine Blue.—An artificial preparation closely resembling both in hue and chemical composition; the natural and high priced ultramarine blue is now extensively manufactured. It is a powder quite insoluble in water or any other known menstruum, withstands exposure to air and light without any injury to its brightness; is not altered by alkalies, but its color is immediately destroyed by acids with evolution of a sulphuretted gas. It is only a fast color when worked with albumen or lactarine; for the calico printers' purposes it must be in a very fine powder, and when mixed with albumen or lactarine solution, it must be well stirred up with it so as to thoroughly incorporate every particle of the powder with the thickening. Not being a solution, it will, of course, gradually deposit unless the mixture is very thick; it is necessary therefore to keep it occasionally stirred when working either by machine or block.

29

Ultramarine blue has also been employed in finishing, where it gives an agreeable purplish blue cast to the white pieces.

Urine.—Urine is of very ancient use in connection with dyeing, and is yet employed, but not to the same extent as formerly; because for many of its uses substitutes are found in chemical drugs, which, if not cheaper, are more regular, more easily kept, and pleasanter to work with. Fresh urine is of no use for the dyer's purposes; it is only when it has been kept for some time, and after fermentation or putrefaction has commenced, that it begins to possess the properties on account of which it is valued. In this state it is called lant. Urine contains a substance known as urea; it is a crystalline matter, neither acid nor alkaline itself, but of a basic nature, combining with acids. It is of a complex composition, and its elements are so arranged that it takes very little to change their order. This is done by the fermentation which naturally commences in urine, and all the urea is changed into carbonate of ammonia, which, remaining in the liquor, communicates to it its soapy or alkaline properties. The strong smell of old urine is due to this carbonate of ammonia, which is in composition the same as ordinary smelling salts; and if the smell of old lant is not quite so agreeable as these salts, it is because there are animal matters of another nature present in it, which mix their odors with that of the ammonia. Lant is therefore of an alkaline nature, and its action upon substances can be partly predicated from that knowledge; it will tend to neutralize and kill acids, and generally to act as a weak solution of crystals of soda would. It is employed in bleaching wood, in several cases of dyeing to modify and change the shade, and to moisten dyewoods with before using. It is used to develop some coloring matter, as those of archil, litmus, and cudbear, and is still employed in the composition of the indigo vat.

Uric Acid.—This acid may be mentioned in connection with urine, because it is very frequently present in it, but always in small quantities, except in cases of disease. It is not worth while extracting it from urine, but it can be obtained in good quantity from the excrement of birds and serpents. Guano, in a genuine state, contains a good deal of uric acid, and the white excrement of the boa constrictor is a nearly pure compound of uric acid and ammonia. Uric acid has created a high interest within these few years, on account of a splendid purple color which can be obtained from it, and which has been applied to some extent upon silk, wool, and cotton. The coloring matter is called MUREXIDE, which see.

Uranium.—This is the name of a metal which, up to the present time, has been found but in small quantities, and is, consequently, very expensive. It has some points of resemblance to iron, and, like it, is capable of acting as a mordant for coloring matters. It seems probable that it would receive some applications in dyeing if it were more plentiful.

V.

Valonia, *Valonia Nuts.*—These nuts are the acorn-cups of a species of oak, *quercus ægilops,* which flourishes in the Levant. Under the names of *Camata* and *Camatina,* the dry and immature acorns of the same tree are imported, and, with the cups, extensively used in tanning. The attempts which have been made to employ the valonia nuts in dyeing do not appear to have been very successful. They contain tannin and astringent matters, and can be used as substitutes for galls and sumac; but, so far as cotton dyeing is concerned, they do not answer very well. They are supposed to be employed with more success in some branches of silk dyeing.

Vanadium.—This metal is said to have some resemblances in its combinations to the compounds of chromium, and hopes are entertained that if ever a plentiful supply of its ore can be obtained, it will be of some use in printing and dyeing. The late Dr. Ure stated that an ink could be prepared from it of very superior color and durability. It is exceedingly rare under any form, and only kept as a chemical or mineralogical specimen.

Vermilion.—This brilliant pigment is a compound of sulphur and mercury. It is too dense, and too deficient in covering power, to be employed in calico printing.

Vitriol.—This is the old name given to sulphate of iron or green copperas, and seems also to have been generally applied to the metallic sulphates, as vitriol of copper, of zinc, etc.

White Vitriol is sulphate of zinc.

Blue Vitriol is sulphate of copper.

Green Vitriol is sulphate of iron.

Oil of Vitriol, or *Vitriol,* is the still common name of sulphuric acid.

W.

Walnut Peels.—The rinds or husks of fresh walnuts are well known to contain a colorable matter, for though white when freshly opened the air soon causes them to turn brown or black; and if the juice fall upon the skin it speedily dyes

it a dark brown color not easily removed. It does not appear
that the British dyers make any general use of walnut peels,
but on the continent they are much employed for saddened
shades upon wool. Berthollet says the peels are gathered
when the nut is entirely ripe, if taken from unripe nuts they
are still applicable, but do not keep so long; they are stored
in casks which are filled with water, and not used until one or
two years old. Woollen requires no mordant for dyeing in
the decoction, and is simply wetted out and worked in until
the desired shade is obtained. The shades obtained from walnut peels are esteemed on account of their softness, and the
absence of a harsh feel, which colors saddened with green copperas always possess.

The root of the walnut tree gives the same colors, but
being less rich in coloring principle a greater quantity has to
be employed.

Water.—Water is the most important substance which is
used by the dyer and printer, it enters into all his processes,
and its quality so much influences the results which can be
obtained, that every precaution should be used to procure a
good supply in the first instance, and to provide against the
entrance of any contaminating matters into it. If the situation
of the print or dye works compels the use of an inferior water,
great pains should be taken to ascertain its composition, the
nature of its variations, and the probability of being able to
improve it by chemical treatment. A volume might be written upon this subject, but its applicability would not be general. There are some print and dye works so fortunately
situated as never to have any difficulty upon this matter,
requiring no lodges, no reservoirs, no filters, and hardly any
pipes beyond wooden spouts to supply all the needs of the
works. There are others who are never for a day without
anxiety about the state of their water, and to whom the filtering and purifying of a supply, over the quality of which they
have little or no control, is a constant source of trouble and
expense.

Perfectly pure water does not appear to exist in nature; it
is artificially prepared by distilling ordinary water, with
several precautions, in vessels of silver or platinum. Steam
is pure water in the gaseous state, and when it is condensed,
the water is obtained pure. Ordinary steam, or condensed
water, is subject to be contaminated by several impurities: iron
from the pipes is sometimes found; ammonia, from vegetable
or animal matter present in the water; and very generally a
certain volatile oily matter, probably originating from the tallow or other grease used in the boiler, as well as impurities

derived from the packings of the steam pipes. But when the pipes are well seasoned and the original water not very impure, steam water is nearly pure water, and for all practical purposes may be taken as such. All the natural sources of water are derived from rain, which is produced by the condensation of vapor originally rising from the waters of the ocean. This being a great natural distillation, it will be anticipated that rain water is the purest of all kinds of water. When carefully collected at a distance from human habitations, rain water only differs from the purest distilled water by containing minute quantities of organic matter, and certain gaseous bodies which it has imbibed from the atmosphere. But the moment it touches the earth, its solvent powers are so considerable, that it is instantly contaminated with earthy matters to a greater or less degree, depending upon the nature of the ground. The impurities of water consist in the mineral or vegetable matters which it has extracted from the earth over or through which it has passed from its first fall. If the ground be composed of hard insoluble rocks, the water passes over or through them, taking up very little of their constituents in its passage, even though the contact may have been a prolonged one; such a water will be a soft and pure water, whether it flow as a river or rise as a spring. If, on the contrary, the ground be composed of substances dissolvable by water, the water will soon become saturated with the soluble principles; if a current of pure water meet a bed of rock salt it becomes brine; if it pass through or over a strata of limestone or gypsum, it soon becomes charged with those substances constituting a hard, limy, or calcareous water. If the land from which the water drains be peaty or boggy, containing much vegetable matter, a small portion of this will be dissolved. Though the quantity of these adventitious matters is small when compared with the water itself, they influence its quality to a remarkable extent as a medium of communicating colors. Perfectly pure water would be the best for all manufacturing purposes, if it were procurable. The preference for any particular source of· water is always traceable to the absence of injurious components, or to the fortuitous occurrence of some substance which acts as a corrective of a natural impurity, and not to the existence of the impurity itself. There are, however, some colors which can be dyed very well in water strongly impregnated with mineral matters, and it is thought with better results than in pure water; but in all known cases it is possible to add such chemicals to pure water so as to produce equally good results. But it is not possible so to remove the impurities from water, as to make it in every

case equal to a naturally good supply. In making choice of a
source of water, not only the quantity but the peculiar nature
of the impurities must be taken into account. Some impurities are readily removed by filtration and exposure to the air,
others are not affected by such a treatment; some can be
readily purified by lime, others are not in the least improved
by it; one class of impurities is injurious to one style of production, but not to another, and so on. As a general rule it
may be laid down that river water, that is, surface drainage
water, contains the least amount of mineral matters and the
largest amount of vegetable matter, and that spring or well
water contains the greatest quantity of mineral substance, with
a minimum of organic matter.

Up to the present time the following substances have been
detected in natural waters:—

Acids.—Carbonic, sulphuric, sulphurous, nitric, phosphoric,
   boracic, silicic, and hydro-sulphuric.

Bases.—Soda, potash, lithia, ammonia, lime, magnesia, strontia,
   baryta, alumina protoxides of iron and manganese, oxides of
   zinc and copper, tin, lead, silver, antimony, arsenic, nickel,
   and cobalt.

Also, *Not Being Bases*, are found chlorine, bromine, iodine,
fluorine, sulphur, and hydrogen.

Any given sample of water would only contain a few of the
above substances, but it is possible for any of them to be there.
The mineral substances mostly found in river, spring, and well
water, are as follows:—

Lime, combined with carbonic acid, and with sulphuric acid,
as bicarbonate and sulphate of lime; *magnesia* combined frequently with muriatic acid, sometimes with carbonic acid;
potash and *soda* usually in very small quantities; *iron* in the state
of a carbonate held in solution by carbonic acid, or in some
other form; *silicic acid*, either in the free state or combined
with the potash, generally the former. Those which chiefly
concern the dyer are the lime, the iron, and the magnesia, the
remainder are of little consequence.·

Lime in Water.—A practical man knows when his water
contains lime by several characters, the most striking of which
is the way in which it acts with soap: a calcareous or limy
water destroys the soap, throws it up as a curd, and does not
give a lather until as much soap has been put in as takes up
all the lime. All this soap is wasted, and may even injure
cloth by the earthy soap produced, not being washed off easily.
But the soap test is not precisely a chemical test for lime,
because there are many other substances which would act in the

same manner; but these substances are not likely to be present in water, unless it be magnesia, and this not often, so that whenever a water does curd soap, it may be looked upon as a sure indication of the presence of lime. The quantity of soap which a given bulk of water can destroy, is also a good test of the quantity of lime in the water; it may be used roughly to compare two different waters, or the same water at different times; an exact method of doing this is given further on. The reason why soap is destroyed by a hard water is, that being a compound of fatty matter with soda, the whole dissolves in pure water, but the fatty matter is said to have a stronger desire to go to the lime than to remain with the soda; and when in contact with the lime it combines with it, leaving the soda, and because the compound of lime and fatty matter cannot dissolve in water, it rises as a greasy curd until all the lime is combined with the fatty matter. A hard water can be made soft for some purposes by the addition of soda ash, or soda crystals: if the water be brought to a boil after addition of soda, most of the lime will rise up as a scum to the surface, which must of course be taken off before any goods are entered. The proper chemical test for lime in water is the salt called oxalate of ammonia; when a clear solution of this is poured into a water containing even a very small quantity of lime, it indicates it by producing a milkiness, which on standing, settles down as white sediment, the quantity of this sediment or precipitate will be proportionate to the amount of lime in the water. But neither this nor the soap test will show what kind of salt of lime is in the water; they show the same characters whether it be gypsum (sulphate of lime), chalk, or muriate of lime. Further tests are required to ascertain which is really present. If about a pint of water, containing lime salts, be boiled down in a glass flask until it is reduced to half a noggin, it will be seen that the water is turbid, and that the sides of the flask are covered with a white pellicle. This indicates that either sulphate of lime, or carbonate of lime, or both, are present; to the fluid in the flask let a few drops of spirits of salts be added; if there is an effervescence, and the liquid becomes quite clear, it shows that all the lime is present as carbonate of lime; if there is no effervescence and no clearing of the liquid, it may be judged that it is sulphate of lime; and if, as usually happens, there is an effervescance and only a partial clearing, it proves that there is a mixture of both salts of lime; to determine how much of each requires analytical processes. If there is no deposit or milkiness in the flask after the boiling down, either there is no lime at all, or else it is in the very unusual condition of muriate, or nitrate of lime.

With regard to the different actions of these two salts of lime, that is, the carbonate of lime and the sulphate of lime, in dyeing, it may be said that they are both injurious to fine colors, but neither of them hurtful to saddened or dark colors, unless the water be impregnated to a very great extent. Carbonate of lime in water is thought to be advantageous in madder dyeing, especially with certain kinds of madder, and very frequently ground chalk is added to the water before dyeing with it, to make up any deficiency. Sulphate of lime in considerable quantity is injurious to madder dyeing, and indeed to all kinds of dyeing with woods, causing a waste of coloring matter and giving inferior results; it is hardly possible to dye good bright light shades in such a water.

Magnesia in Water.—If the magnesia be in the water in the state of muriate or chloride, and not inclined to become carbonate by boiling (which it does if much carbonate of lime be in the water), it will not be hurtful, the same if it exist as sulphate. But if it exist in the water as carbonate, or can be transformed to such, it may prove very detrimental indeed, completely preventing the dyeing of certain colors, and spoiling others. It appears from my experiments that magnesia is much more injurious in water for madder dyeing than lime, and that, in fact, in the state of bicarbonate or carbonate, either actual or possible, it is more to be feared than any other common ingredient of natural water.

Iron in Water.—Most waters contain a small quantity of iron; it usually exists as a bicarbonate of iron; some waters contain so much that the stream is quite of a rusty color, and the bottom and sides perfectly yellow with the iron rust which has deposited from the water. It is fortunate that this metal has a great tendency to fall out of water spontaneously, that is, simply by contact with the air, because many spring waters contain iron and are quite unfit for dyeing and bleaching purposes; but by a process of filtration and exposure to the air the iron may be completely separated. The best test for iron in water is tincture of logwood or logwood liquor; when a single drop of strong logwood liquor is put into a wineglass full of water free from iron, it gives either a sherry color or a claret, depending upon the amount of lime present; but if there be any iron the color changes to blue, blue black, and finally to inky black, depending upon the quantity of that metal contained in the water. This test must be judged of within a short interval of time, for if the glasses be left exposed to the air, the logwood becomes altered, turning darker, and might be supposed to indicate iron when in reality there was none. In testing a spring water for iron in this way, it is necessary that

the water should be freshly drawn; if it be some days or even
hours old, especially if it be carried far in a bottle, all the iron
is thrown out as insoluble oxide, and the change of color does
not take place. The influence of a water holding iron in solution, upon dyeing, is very marked; it turns pinks into drabs,
and reds into dull browns and chocolates; if much iron be
present it prevents dyeing altogether, for the iron in the water
combines with the coloring matters and the cloth receives none,
or only a feeble proportion. An exaggerated specimen of a
ferruginous water may be obtained for experiments from an
iron steampipe where condensed water has stood for a day or
two; it will be quite clear and bright looking, but if left in an
open basin for a few hours, the iron rust will be seen to separate,
or if a bleached fent be dipped in it, and hung up, it will soon
become of a light buff color; in such a water madder refuses
to dye, all colors are injured except blacks and saddened shades,
and under no circumstances will it leave the whites of printed
cloth clear, or possible to be cleared, without almost destroying
the dyed color.

Other Substances in Water.—The potash and soda salts which
often exist in water are usually without any marked effect in
dyeing or bleaching, being present in very small quantities and
generally of a neutral nature. The silicic acid, which is a frequent constituent of water, but in small quantities, is likewise
without any marked action in dyeing; from the author's experiments, it appears that pure hydrated silicic acid is injurious
in madder dyeing, but not strikingly so. The organic matter
which is contained in some waters, and which has received the
names of *crenic* and *apocrenic acid*, is not without influence in
dyeing; but it is more in bleaching, and especially in clearing
printed goods that the presence of organic matter is troublesome. It prevents the brilliant white finish on bleached
goods, and often in the *chemicking* causes them to take a yellow
cast very difficult to remove. In clearing madder prints in
the beck with solution of bleaching powder, and water charged
with organic matter, the reaction between the chemic and the
vegetable matter when the temperature is raised causes the
formation of a yellow or brownish matter which falls upon the
cloth and spoils the whites, and is not easily removed; at the
same time the color is injured. A good test for this condition
of vegetable matter in water consists in taking about a pint of
the water and mixing with it about half an ounce of clear
bleaching powder solution standing at 2° Tw., and warming in
a glass flask to about 160° F.; if it goes yellowish, and in a
short time deposits a buff colored deposit, it may be considered
certain that it is not fit to clear pieces by the old plan; for

clearing in the padding machine and steam box, much less water being used, it does not matter so much. Nitrate of silver, chloride of gold, and permanganate of potash are also used as tests for the presence of vegetable matter in water.

Purification of Water.—Few works are so fortunately situated as to be able to use water without some purifying process, and none should be without the means of purifying in case of necessity, for the best streams are at times unfit for working with, and unless filters are at hand, the works must stop until the water becomes clear again. The purifying agents to be employed are principally those of nature—exposure to air and light, and a straining out of suspended matters by filtration. The methods employed are too well known to need description in detail. The water from the source is led or pumped into a reservoir of size proportionate to the wants of the works, the larger the better, and preferably long and narrow, so that as much distance as possible may exist between the water coming in and that flowing out to the filters. This reservoir is meant to keep up a stock, and to allow mud, etc., to settle out; sometimes two are used alternating, to allow the water some hours of quietness. From this depositing reservoir the water goes on to the filters. In many works, where the stream of water is not clear enough to be used without settling, and yet settles very clear, filters are not thought necessary. This will answer for many kinds of water, but not with all, because even perfectly clear water may contain so much iron or organic matter as to give very inferior results in the dyehouse. The object of filtration is something more than to remove visible foreign matters in the water, and if it should turn out that the water, even when quite clear, is not giving good work, filtration should be tried. The filter is essentially a bed of coarse sand, supported upon pebbles and boulders. The filtration being downwards, the water passes through the sand and on to the works by proper arrangements, as is well understood. The surface of the filters should be as large as the space at command will allow, both because they will then filter more water and filter it better. The sand acts partly as a strainer to keep back the mechanically suspended impurities, but its most important action is of another nature—the exposition of the water more completely to the action of the air. Each particle of sand becomes covered with a film of water so thin that the air can act very completely upon it, penetrating it as it were through and through. The vegetable, and some of the mineral impurities, are changed by the action of the air; they become insoluble in the water, and, instead of passing through the filter, they adhere to the particles of sand, in the form of a slimy, glairy substance. If the water be bad, this slime collects in such quantity as to stop the action of the filter, either not letting the water pass through at all, or letting it pass through without purifying it. The remedy consists in scraping the top sand off the filter deep enough to remove that portion saturated with impurities, when the filtration will go on again. The depth of sand required to filter well depends a good deal upon its quality or fineness, and the kind of water to be filtered. A bed twelve inches thick should be sufficient under all ordinary circumstances, and it should not be reduced to less than four inches by scraping. When the sand, after being used, is well washed with violent agitation, the impurities of the water are in a great measure detached, and it can be employed over again, but it is best to leave the washed sand several weeks exposed to the air before spreading it on the filters. When water is highly charged with vegetable impurities, the filter does not seem to remove them, but, by paying attention to a few points, the water may generally be made useable. First, as above stated, it is the air which purifies the water, the sand being the instrument or apparatus for exposing the water to the air: the worse the water, the greater pains must be taken to let the air have access to it; the filter bed must not be overcharged with water, that is, it must not have several inches of water, floating over the sand, but, on the contrary, the surface of the sand should be exposed to the air, the water only flowing on to it as fast as it passes through; and, secondly, the water should not be allowed to collect under the sand, but be drawn off nearly as fast as it filters; by this means the foundation of the filter is kept full of air, and, after passing through the sand, the water gets a subsequent purification in trickling over the pebbles and boulders below. If the water is not good after this treatment, it must be bad, indeed, and the lime treatment should be tried.

Purification of Water by Lime.—Lime has been used for a long time to purify water; and, though it was patented a few years ago, by Mr. Clarke, it was practised long previously, sometimes by throwing lumps of lime into the pits or lodges, sometimes by slacking the lime in a tub, and throwing it in, as milk of lime, with a scope, but it is probable that Mr. Clarke may have been the first to apply it to water for domestic purposes. The action of lime upon water is twofold; in the first place, water which contains the bicarbonate of lime is deprived of this substance by the addition of lime, for, though it may seem paradoxical that lime should throw out lime, it is nevertheless quite true, and easily proved, that such a water contains less lime after the addition of a proper quantity of lime than

previously, that is, after settling and filtration; the lime added
combines with the lime in the water, and both together fall
down as a sediment. If any bicarbonate of magnesia be in the
water, it will also be precipitated. The second action of lime
is to throw out vegetable matters, and whether it accomplishes
this by combining with them itself and carrying them down,
or whether it is that the bicarbonate of lime in the water
keeps them in solution, and upon its being destroyed, they
fall out, or whether the action is made up of both these is not
clearly known, but it is a fact that lime tends to diminish the
quantity of organic or vegetable matter in water. There need
be little fear of applying an excess of lime, or, at any rate, of
that excess doing any harm. It is not possible to say what
the quantity is that should be used, circumstances varying so
much; but the author has known seven hundred pounds
weight of quick lime applied, in ten hours, to about three
hundred thousand gallons of water, and with good effect, but
as a daily addition one quarter of this quantity should suffice.
The lime should be added to the water in such a state as to
ensure its utmost action, and in such a manner that it may be
thoroughly mixed, therefore it should not be thrown in in
lumps, but carefully slacked and mixed into a kind of cream
with water. If practicable, it should be added to the water as
it runs in the channel from the source to the first reservoir,
letting the milk of lime flow from a tub into the water in a
regular small stream, then, both running together, they will be
well mixed, and the lime exercise its full action. The lime
should be allowed time to perform its duty, and space enough
to settle in before arriving at the filter, or else it will be
troublesome, by choking the filter. As before remarked, it is
hardly possible for an injurious excess of lime to be added, or
to pass through the filter, the air removing an excess very
quickly; but if an excess is suspected it can be tested by red
litmus paper, which is turned blue by lime-water, or better, by
adding a drop of solution of nitrate of silver to a wine-glass
full of it, when, if it gives a brown precipitate, an excess of
lime may be considered as present. Lime should not be used
excepting the ordinary means of purification have failed. The
great point is to imitate nature, as far as possible, in her method
of purifying water, and that is exposing it to the air. The deep
and silently flowing rivers are never so brilliant and pure as the
shallow stream which runs amongst large boulders, over gravel,
and sand, the waters of which are continually broken into sheets,
and thrown up into contact with the air. The power of the
atmosphere, and a physical agent like sand, to remove substances from water is far greater than is generally supposed,

rendering bodies insoluble and inert which are little suspected of being acted upon by it; it is, on a large scale, what animal black is in the laboratory, or even that more powerful agent, platinum black, the properties of which all students of chemistry are acquainted with.

Other Methods of Purifying Water.—Several methods of purifying water have been patented within these few years, but none of them have been applied on a sufficiently extensive scale to enable us to see if they are real improvements. The only real improvement that seems possible to be introduced, will be some apparatus that shall purify and filter large quantities of water in a small space. At present, the filters and water lodges take up a good deal of ground, which could be otherwise employed if any such apparatus was to be introduced. There is not much inducement to spend money or time in such a matter as far as regards dyeing, bleaching, etc.; because the works are usually in places where land is cheap, and the present method of filtering requires very little attention and expense to keep it in action; but a compact and efficacious filtering apparatus would doubtless be a valuable property, and sooner or later be adopted by all who use large quantities of water.

Testing and Analysis of Water.—An accurate chemical analysis of water requires great care, and can only be undertaken by a skilful chemist who has access to all the appliances of a laboratory; but there are some chemical tests which may be applied and give useful information without pretensions to absolute accuracy. The quantity of solid matter in water can be ascertained by evaporating a thousand or ten thousand grains to dryness, and weighing the residue; five grains of solid matter per gallon of water is thought small, ten grains medium, and twenty grains a large quantity. Pure water contains no solid matter, and river or spring water is better the less it contains, consequently one containing five grains would be preferred to one containing twenty grains per gallon, but this is only strictly true when the contained matters are similar in chemical composition. If the five grains in the one case were mixed carbonates of lime and magnesia, and the twenty grains in the other common salt, then the five-grain water would be a bad one, and twenty-grain water a good one. Instances are common enough of water containing twelve to fifteen grains of solid matter per gallon being successfully used in printing and dyeing, and of other waters not containing five grains being very inferior. The quantity of solid matter alone is not therefore a good ground for comparing two samples of water. But it may be taken as a principle that that solid matter in water which, after having been dried, is again dissoluble by pure water is not a kind of matter hurtful in dyeing. Thus, sulphate of soda, carbonate of soda, common salt, muriate of lime, sulphate of magnesia, and salts of potash, when dissolved in water, may be evaporated to dryness, but will dissolve again if mixed with pure water. On the other hand, carbonate of lime, carbonate of magnesia, sulphate of lime, oxide of iron, and other bodies may be perfectly dissolved in the water before evaporation to dryness, but by the evaporation have become insoluble, and are no longer capable of forming a clear solution in pure water.

The former materials are not hurtful in dyeing, while the latter are generally very hurtful. An additional test, therefore, would be to ascertain how much of the solid matter was again soluble in cold water, and compare the amounts of the insoluble residues; but this would not be conclusive, because we may have any of the three substances—iron, magnesia, and lime, left insoluble; if the residue was all iron or all magnesia the water could not be good, while, if it were all lime, it might be suitable for many styles of dyeing. The task of distinguishing the qualities of these matters must be left to the analytical chemist, being entirely a laboratory matter. To obtain some information as to the comparative amounts of substances in two or more samples of water, glasses may be filled, and the following tests applied, comparing the effects produced:—

Oxalate of Ammonia.—A white precipitate indicates lime.

Nitrate of Silver.—A white precipitate, not dissolved by pure nitric acid, indicates chlorides.

Nitrate of Baryta.—A white precipitate, not dissolved by pure nitric acid, indicates sulphates.

Lime Water.—A white precipitate indicates carbonates.

Phosphate of Soda, with addition of Ammonia.—Produces a white crystalline precipitate if magnesia be present. Before testing for magnesia the lime should be all removed.

Yellow Prussiate of Potash.—A blue precipitate indicates iron.

Logwood Solution.—A dark purplish black indicates iron; a deep claret indicates carbonated alkalies, or earths in solution.

The greater or less abundance of the precipitates produced will be in ratio with the quantity of substance present.

Test for Hardness of Water.—The soap test for ascertaining the hardness of water is a tincture of the best curd soap, made by dissolving one part of it in 75 parts of warm distilled water, and then adding an equal volume of rectified alcohol. This strength of soap tincture is not the only one which can be used, but it is convenient ; it does not gelatinize at ordinary temperatures, which it would do if made stronger or without alcohol ; it keeps well if there be no acid in the alcohol. On account of the variable quality of the soap the precise quantity of lime required to curd it must be ascertained by experiment. Take five grains of marble in a capacious porcelain dish, and dissolve, it in pure hydrochloric acid, heat to expel the excess of acid, and then add forty ounces of distilled water : this produces an artificial calcareous water representing in its action upon the soap twenty grains of carbonate of lime per gallon of water ; it should be again mixed with one or two volumes of distilled water, because the action of the test is not clear in waters containing an excessive amount of lime. On this account it is often advisable to mix a natural water to be tested with an equal volume of pure water. The water to be tested should be put in a phial, which it should not fill more than a third, and the soap tincture added slowly from a graduated measure, with occasional stoppages for the purpose of violently shaking the mixture ; as soon as the soap bubbles remain permanent on the surface of the water the operation is finished. Of a hard water five hundred grains is sufficient to take at once, mixed with an equal volume of distilled water; of a less hard sample, a thousand grains may be taken without admixture.

A soap tincture made from a good specimen of white curd soap, gave the following results in the author's hands, which may serve for comparison with those obtained from any other good soap. The quantity of water used in each case was one thousand grains, and the numbers indicate the measures of soap tincture required to produce a permanent froth or lather, each measure equalling ten grains of water :—

Distilled water.	2 to 3
River water, Cheshire	8
Corporation supply to Manchester	9
River near Manchester	16
Water from the Thames	30
A spring water in Manchester	56
Same water after being treated with lime and filtered	32
Mixed spring and drainage water, neighborhood of Manchester	34
The same after treatment with lime	17
A spring water from a dye works near Manchester	85
Water containing chloride of calcium, equal to 16 grains carb. lime per gal.	26

Water which will not froth with less than thirty measures of a
soap tincture is not well suited for general dyeing purposes.
I would not recommend a dependence upon this test altogether,
and especially independent and separate observations are not
of much value ; it is only useful as a comparative test, and, as
before stated, indicates not only lime salts, but all other salts
of oxides which form insoluble soaps with fatty matters.

It has been stated that such and such substances are injurious
in a water for dyeing. It would be very satisfactory to be able
to define the precise action of these substances in the dyeing,
but a great deal must be left to conjecture on account of the
absence of exact information. All testimony seems to concur
in proving that any water, which upon heating or boiling produces a precipitate, is a bad water for dyeing; further than that,
water which is conspicuous for producing incrustations in
steam boilers is a bad water, both because it must contain a
large proportion of mineral matter to produce this effect, and
it must be of a nature to fall readily out of solution. Now,
all precipitates formed in a colored liquor combine with a
greater or less quantity of the coloring matter and render it
insoluble, forming a species of lake; and this, which is a characteristic and conspicuous property of some oxides, as iron,
alumina, and tin, when so precipitated, is true in a less degree
not only of all other oxides but also of insoluble saline compounds. Upon these grounds, I am led to believe that the
injurious action of these substances in water is owing to their
abstracting the coloring matter from the solution, which they
are enabled to do by being thrown into an insoluble and basic
state, and not to their destroying the coloring matter. Thus
magnesia and lime in the state of bicarbonate lose carbonic acid
upon heating and become precipitates, combining with the
coloring matter and impoverishing the bath; in the same way
carbonate of iron acts, similarly also sulphate of lime, which
seems capable of forming insoluble compounds with coloring
matters, with or without decomposition. So I explain why
salts such as sulphate of magnesia and muriate of lime do not,
under certain circumstances, appear to be at all injurious : they
are not capable of being rendered insoluble. Beyond this view,
however, it is evident that coloring matters are not so soluble
in certain saline solutions as in pure water, or in other saline
solutions; and then, consequently, impure waters are often
defective by not dissolving the coloring matter from the root
or wood.

Some waters may be corrected by chemical means, but there
is always a risk in attempting it, because the remedial agents,
if used in excess, would prove more injurious than the original
defect. Thus, an extremely calcareous water, which contains

only bicarbonate of lime, may be mixed with sulphuric acid, or oxalic acid, to neutralize the lime, and this is frequently done. With some qualities of madder a limy water dyes very well, because the madder is of a very acid nature and neutralizes the lime; but with another kind of madder or with garancine, the same water would yield very bad results.

Solvent Powers of Water.—Water is a physical rather than a chemical agent in bleaching and dyeing; it is the vehicle which carries the chemical substance to the cloth to be operated upon, or which removes the matters necessary to be removed from it. When a substance is mixed with water, it may either be dissolved by it, and disappear, as salt does; or it may remain in suspension, as chalk does. Nothing is considered to be actually dissolved in water if it can settle out again, or if it will not pass with the water through a filter made of paper or calico: thus, to talk of dissolving ground chalk in water is incorrect, for if allowed to stand it would settle out; or, if the mixture were filtered, the water would pass clear while the chalk would remain upon the calico; but blue vitriol (sulphate of copper), for example, does really dissolve in water, and the liquor all filters through together: to deprive the water of the blue vitriol would require chemical means different in kind from filtration. Water, therefore, dissolves some substances and not others. Water does not dissolve the same quantity of all soluble substances; of some it can dissolve its own weight, and more; of others, a smaller portion; and of some, extremely little. As a rule, hot water dissolves more than cold, and more quickly than cold; but, upon cooling, the excess mostly falls out as crystals. This point deserves notice, for a liquor, which is of right strength when a little warm, may be too weak when it becomes cold: left in a carboy, for example, in a cold place, because the salt crystallizes out; this is the case only with those salts that are but sparingly soluble, as chlorate of potash, cream of tartar, sulphate of potash, etc. This crystallizing is sometimes troublesome in steam colors which, right enough when freshly made, become filled with small crystals on cooling, and work rough in the machine: it is felt in the case of an ageing liquor, which contains chlorate of potash, as an active agent, which, crystallizing out, leaves the liquor weak and not able to do its work. As an usual thing, the drugroom upon a printing or dyeing works should be cool, but there are some liquors better in a moderately warm place; brown vitriol, for example, in winter time is apt to go solid in the carboys, if kept in an exposed place. In the following table will be found most of the substances used in bleaching, dyeing, and calico printing, with useful information as to how they behave themselves with

30

water: the results are from the author's experiments, and exact
enough for pratical purposes. The second column gives, in
comparative expressions, the degree of solubility of the substances; the third, gives the degree of Twaddle which 16 oz.
of the substance dissolved in a gallon of water stands at; and,
in the fourth column, are remarks proper to the particular
substance. By this table one can calculate, in a rough practical
kind of way, how much of a salt there is in a gallon of water
by knowing its strength; and, on the other hand, can tell how
much of a drug to use to make a liquor at a certain given
strength :—

*Tables showing the Action of Water upon Various Bodies, and
Strengths of Solutions of some of them.*

Substance.	Action of water.	Strength of a sol. at 16 oz. per gal.		General Remarks.
		Tw.	sp gr.	
Acid, arsenious............	little sol.	...	......	Some varieties dissolve easier than
" citric.................	very sol.	6½	1034	[others.
" oxalic.................	soluble	6	1031	Saturated solution.
" tartaric	very sol.	10½	1052	
Alum (potash)...............	soluble	10	1050	
Alum (ammonia)..........	soluble	10	1050	Nearly saturated in the cold.
Alumina, sulphate........	very sol.	10	1050	
Albumen..................	very sol.	...	......	Coagulated by hot water.
Ammonia, muriate........	very sol.	5½	1027	
" carb.	very sol.	8½	1042	
" sulph...........	very sol.	10½	1052	
" oxalate.........	soluble	7	1035	
" nitrate.........	very sol.	8	1040	
" tartrate	soluble	9½	1047	
Barium, chloride..........	very sol.	13½	1067	
Baryta, nitrate...........	soluble	12	1060	
" sulphate..........	insoluble	...	......	Mineral white, heavy spar. [ment.
Bleaching powder.........	soluble	11	1055	Does not dissolve clear, always sedi-
Copper, acetate...........	soluble	...	......	Some kinds do not dissolve without
" chloride...........	very sol.	12	1060	[acid.
" nitrate	very sol.	10	1050	Mostly sold as a liquid.
" sulphate	soluble	11	1055	
Dyewoods and Stuffs :—				
Archil	soluble	...	......	Sold as a liquid.
Alkanet...................	insoluble	...	......	Dissolves in turpentine, spirits, oils.
Annatto...................	insoluble	...	......	Dissolves in alkali, soda.
Berries...................	soluble	...	......	Coloring matter very soluble.
Catechu...................	soluble	...	......	Dissolves completely in water.
Cudbear	soluble	...	......	
Camwood	soluble	...	......	
Cochineal	soluble	...	......	
Fustic	soluble	...	......	Of these dyeing matters not one dis-
Litmus...................	soluble	...	......	solves completely in water. Water
Logwood	soluble	...	......	only extracts certain soluble prin-
Quercitron bark...........	soluble	...	......	ciples, including the actual color-
Peachwood	soluble	...	......	ing matter, and leaves undissolved
Gall nuts.................	soluble	...	......	the great bulk, which is fibrous
Madder...................	soluble	...	......	and woody matter.
Sapan wood...............	soluble	...	......	
Sumac....................	soluble	...	......	

Substance.	Action of water.	Strength of a sol. at 16 oz. per gal.		General Remarks.
		Tw.	sp. gr.	
Farina.	insoluble	...	......	Soluble in boiling water.
Flour..........................	partly sol.	...	......	Soluble imperfectly in boiling water.
Glue..........................	soluble	...	......	Some kinds thick, and some thin, at
Gums (foreign)	soluble	...	......	[1 lb. per gallon.
" (British).............	sol.& insol.	...	......	Depends upon method of manufacture.
Iron, acetate	soluble	...	......	Sold as liquid at from 24° to 28° Tw.
" muriate	very sol.	...	......	Sold as a liquid, at 80° Tw.
" nitrate.................	very sol.	...	......	Sold as a liquid, at 90° Tw.
" sulphate.............	very sol.	10	1050	
Lead, acetate................	very sol.	12	1060	Gives a milky solution with common
" nitrate.................	very sol.	17	1084	[water.
" sulphate............	insoluble	...	......	Dissolves in some saline solutions.
" red....................	insoluble	...	......	Partly dissolved by nitric acid.
" litharge..............	insoluble	...	......	Partly dissolved by nitric acid.
Lime, quick	little sol.	...	......	Lime water contains very little lime.
" muriate	very sol.	10½	1052	
" carbonate............	insoluble	...	......	Ground chalk; dissolves in acids.
" sulphate.............	very lit. sol.	...	......	One part dissolved by 400 parts water.
" acetate	soluble	...	......	Mostly sold in solution.
Magnesia sulphate........	very sol.	10	1050	
Manganese, acetate.......	very sol.	...	......	Sold in the liquid state.
" muriate......	soluble	13	1064	Sold in the liquid state.
" sulphate	soluble	...	......	
" blk. oxide..	insoluble	...	......	Not dissolved by cold acids.
Mercury, metallic.........	insoluble	...	......	Dissolved by acids.
Mercury, bichloride......	little sol.	17	1085	Quite saturated at 1 lb. per gallon.
Murexide	soluble	...	......	
Potash, hydrate............	very sol.	15½	1077	
" carbonate............	very sol.	13½	1067	
" bichromate.........	soluble	13½	1068	Solution nearly saturated.
" yel. chromate.....	very sol.	15	1075	
" bitartrate	little sol.	...	......	Cannot be made stronger than about
" chlorate.............	little sol.	6	1031	[two degrees.
" nitrate..............	very sol.	11½	1057	Very soluble in hot water.
" red prussiate......	very sol.	10	1051	
" yel. prussiate.....	soluble	11	1055	
" sulphate............	little sol.	...	......	Water does not dissolve one-tenth its
" bisulphate.........	very sol.	13	1065	[weight.
" oxalate	little sol.	...	......	Saturated, marks 2½° Twaddle.
" iodide................	very sol.	14½	1072	
" arseniate...........	very sol.	12	1060	
Soda, ash.....................	very sol.	...	......	Varies in its composition.
" crystals	very sol.	7	1035	
" bicarbonate	soluble	13	1065	
" borate................	soluble	9	1045	
" muriate...............	very sol.	13	1065	
" nitrate	soluble	13	1064	
" sulphate	soluble	10	1040	
" stannate	soluble	12	1061	Varies according to its composition.
" tungstate............	soluble	15½	1077	
" hyposulphite	very sol.	10	1050	
" sulphite.............	very sol.	9	1046	
" phosphate..........	soluble	6½	1032	
Tin crystals	very sol.	12	1060	Decomposed by much water.
Uranium, nitrate..........	very sol.	13	1065	
Zinc, acetate	very sol.	...	......	Sold generally as a liquid.
" chloride	very sol.	15½	1077	
" sulph.	very sol.	11	1055	
" oxide.................	insoluble	...	......	Dissolved by acids.

Weld, *Wold.*—This was the chief yellow dyeing substance employed in Europe before the introduction of quercitron bark; it is still cultivated on the continent and used in dyeing, but it is nearly unknown in England. It is a reedy plant, and sold in the sheaf like straw: the whole of the plant except the roots were employed in dyeing, but the greater part of the color resides in the seeds and upper extremity. In dyeing with it, its coloring matter is extracted by boiling in water, and the decoction only added to the goods. With alumina it dyes up a very fine clear yellow color, tolerably permanent in soap, but not well resisting air and light. It has not more than one-fourth the power of quercitron bark as a color, and on this account, as well as the difficulty and cost of carriage, it has been driven from the English market. Its pure coloring matter is called *luteoline*, from the botanical name of the plant *reseda luteola.*

Woad.—This dyeing matter, which was employed from the most ancient times, is now nearly unknown in this country. It is yet cultivated in some parts of Europe, where it goes under the name of pastel. The coloring matter it contains is chemically and practically the same as indigo: it is still used in setting the indigo vats for dyeing woollen, but always in conjunction with indigo. It appears that the woad plant, as sold to the indigo dyer, readily enters into fermentation, and in that state is useful in deoxidizing or reducing the indigo to the soluble condition; but it contains very little coloring matter itself, so that it was hardly possible to dye a deep blue with it. The blue colors, however, which it did yield to cloth were very durable and permanent. Its principal use was in giving a fast blue basis upon broad cloth which was to be afterwards dyed black. From its name came the term *woaded colors*, still in common use for colors which are supposed to be dyed upon a basis of woad blue.

Wongshy.—A new coloring matter under this name has been reported upon the chemical journals. It dyes up shades of yellow and orange upon woollen and silk, which do not appear to be possessed of much stability. In some of its reactions it bears a resemblance to anotta, but in its general properties it is very distinct. It does not appear to have been put into practical use.

Woods.—The term wood is used among the dyers to indicate the dye stuffs, like logwood, peachwood, etc., which are hard and solid; but many dyers use the term loosely, and include all the dye stuffs, as cochineal and indigo, under this term.

Wool.—The fibre of wool is different in many respects from that of cotton or silk. Its quality varies greatly; its length is between three and eight inches, and the diameter of single fibres is from the thousandth to the fifteen hundredth part of an inch. Under the microscope it appears as a tube, circular and hollow, and at intervals, of which there are three hundred in an inch, are seen rings or projections in regular order, which, in arrangement, have been compared to the scales of a fish, the skin of a serpent, or as if a number of hollow cones were placed one inside of the other. It is as if the growth of wool were not in one even, constant progression, but more rapid at one time than another, the concentric rings representing a state of rest or inactivity, which follows on the active period. Several peculiar properties of wool are attributed to the character of the fibre—the felting or adhering together of fibres of wool by simple working together, the harshness which is felt by the finger or lips when a fibre of wool is drawn in one direction but not in another; a property stronger in hairs than in wool, but the same in character and origin. In working woollen cloths, they are, as is well known, liable to run up, contract in certain dimensions, becoming thicker at the same time. This is what takes place purposely in fulling, and accidentally in too much or too roughly handling woollen goods in washing and dyeing; such runnings up are familiar in domestic economy. They are attributable to this construction of the fibre of wool; each fibre may be looked upon as barbed like an arrow or a fish hook, easily going one direction, but not able to return on account of these projections holding it, and generally all kind of motion among the fibres, as rubbing, beating, or stamping, causes them to advance in the direction of the small end of the cone, and remain there unless pulled back by force. In woollen goods and muslin delaines much injury may be done to the general appearance of the cloth and goodness of the colors if the pieces are too roughly used, or allowed to remain loose when in a wet state.

It is on the same account that wool is soaked in oil in order to spin it; the oil appears to fill up the concentric ridges to a certain extent and facilitate the working of the fibre. Raw wool contains a large quantity of fatty matter, which is natural to it; but this is not the same as that which has to be removed from spun or woven goods before they can be dyed. The affinity which wool exhibits for coloring matters, and other substances, is treated of in the articles on FIBROUS SUBSTANCES, page 213.

The superior affinity which woollen cloth enjoys for many colors may cause it to be looked upon as a natural mordant, but that would be a loose way of considering its properties. It does not apparently contain anything like a mordant, any more than cotton does. If it can be looked upon as containing a mordant it must be considered as wholly a mordant, and many have fallen into this error, and have imagined that by dissolving the wool in alkalies, and impregnating cotton with it, they could indue cotton with the stronger affinities of wool. The results have shown the fallacy of this line of reasoning. Cotton may·be as reasonably considered a mordant because it takes the indigo blue from the lime and copperas vat, as wool because it can take blue from sulphate of indigo. The explanation of these differences must be looked for, not only in the various chemical constituents of the fibrous matter, but also in the physical structure of the fibre itself. There recently appeared an account of the possible application of the newly discovered solvent for wool and silk, viz., the ammoniuretted solution of oxide of copper and nickel. The statement was to the effect that the solution of silk or wool could be applied to cotton fabrics, to give them the appearance and properties of silk and wool. This is quite false; the appearance and properties of wool do not depend upon the amount of carbon, hydrogen, oxygen, and nitrogen which it contains, so much as upon its physical structure, and it would be as true to say that a heap of sawdust was a piece of timber, as to say that dissolved wool was the same as fibrous wool; the chemical elements are there, but the structure is for ever gone.

That the whole of the affinity of woollens and silk, and some other animal matters, for colors is not due to their physical organization, is proved by their possessing powers of withdrawing coloring matters when all trace of structure has been destroyed by acids, alkalies, and solvents. I mean all structure which has been owing to growth and gradual development. But in this state of disorganization they approach in properties to many other of the neutral and insoluble bodies. It is possible that if dissolved wool or silk had any affinity for the fibrous matter their powers of attracting color might be utilized, but they do not adhere in the slightest degree when deposited from alkaline solution upon cotton; as soon as the fluids have dried, the precipitated animal matter can be shaken or brushed off.

Y.

Yellow Colors.—Yellow is not an important color in dyeing or printing on account of the little demand existing for it: it is obtained by the following processes:—

Yellow Colors on Cotton by Printing.—The coloring matter

chiefly used is from Persian berries, and the mordant, alum or salt of tin.

Steam Yellow for Calico.

2 gallons berry liquor at 6°,
3 lbs. of starch; boil, and add
4 oz. crystals of tin,
1 oz. oxalic acid.

This is to be printed upon cloth: an excess of tin makes the shade more orange.

Steam Yellow from Bark.

1 gallon bark liquor at 7°,
1 lb. alum,
1 quart hot water, }
3 lbs. gum.

Steam Yellow from Berries and Alum.

1 gallon berry liquor at 4°,
1 lb. alum; dissolve, and add
5 lbs. gum.

Another Steam Yellow.

1 gallon berry liquor at 11°,
1 quart red liquor at 18°,
3 lbs. gum.

For CHROME yellows, see page 142.

See also a steam yellow from FLAVINE, p. 226.

Yellow Colors on Cotton by Dyeing.—The chrome yellows are those principally in demand (see page 142). Bright, but unstable yellows, are also obtained from quercitron bark.

Spirit Yellow on Cotton.—Saturate the goods in sumac liquor by steeping; then mordant in oxymuriate of tin (yellow spirits) at 2° for thirty minutes; dye up in a clear decoction of bark until the proper depth of shade has been obtained; then add a quantity of the yellow spirits to raise the color.

More solid, but less brilliant yellows are obtained by mordanting in alumina, and dyeing in bark.

Yellows upon Wool by Printing.—The chief yellow colors upon wool are from Persian berries and tin salts; bark gives orange yellows, which are sometimes used, and more rarely fustic and turmeric.

Yellow for Wool.

1 gallon berry liquor at 10°,
5 lbs. gum,
14 oz. crystals of tin.

Yellow for Wool—Orange Hue.

1 gallon berry liquor at 14°,
1½ lb. starch ; boil, and add
12 oz. alum,
8 oz. crystals of tin,
3 oz. oxalic acid.

Spirit Yellow on Wool.

1 gallon bark liquor at 30°,
3 lbs. gum,
12 oz. alum,
12 oz. bichloride of tin at 120°.

This color is very strong and suitable for small objects, or, as an ingredient in those compound shades where a yellow part is required.

Turkish Yellow for Wool.

2 quarts bark liquor at 18°,
8 oz. archil liquor at 10°,
1½ lb. gum,
3 oz alum,
1 oz. tartaric acid,
1 oz. oxalic acid,
3 oz. bichloride of tin at 110°.

The yellow colors upon muslin de laine are precisely the same as those for all wool.

Yellow Colors upon Wool by Dyeing.—The chief yellow coloring matter employed in wool dyeing is fustic—for the orange or maize shades a tin mordant is employed, but for the lemon shades the aluminous mordant is prepared. Weld is yet used for dyeing yellows, which have considerable permanence and durability ; quercitron bark is but little employed for yellows upon wool. Picric acid gives a fine lemon yellow, but is hardly used in general dyeing.

For 10 lbs. wool, mordant in 3 ozs. bichromate and 2 ozs. alum, and dye in 5 lbs. fustic ; or,

Mordant in 8 ozs. tartar, 8 ozs. alum, and dye in a mixture of bark and fustic, raising with oxymuriate of tin.

To dye in weld, 20 lbs. of woollen cloth are mordanted in

4 lbs. alum and 1½ lb. tartar, and dyed in the decoction made from 15 to 20 lbs. of weld.

In merino dyeing young fustic is extensively used for shades of golden yellow. The wool is not subjected to a previous mordanting, but entered at once into the dyeing bath, which is made up with tartar, oxymuriate of tin, and the decoction of young fustic. 12 lbs. of wool require about 15 lbs. of young fustic to dye a full and deep shade.

Yellow Colors upon Silk by Printing.—Persian berries yield the coloring matter which is generally used in silk printing; bark liquor may also be employed, and decoction of turmeric.

Yellow for Silk.

1 gallon berry liquor at 11°,
8 oz. alum,
8 oz. crystals of tin,
3 lbs. gum.

Another Yellow for Silk.

1½ lb. turmeric,
1½ lb. Persian berries;

Boil these in water and reduce to two quarts, and add

2 oz. crystals of tin,
4 oz. alum,
1 lb. gum.

Another Yellow for Silk.

3 pints turmeric liquor,
1 pint berry liquor at 5°,
4 oz. alum,
8 oz. oxymuriate of tin,
1½ lb. gum.

Yellows on Silk by Dyeing.—For pure and bright yellows of a golden shade, weld seems the most suitable coloring matter. The silk is mordanted by working in a solution of alum for about an hour, and then worked in decoction of weld, and raised by adding solution of alum. By substituting bark or fustic, or mixtures of the two, and by raising in tin spirits instead of alum, modified shades can be readily obtained.

Picric acid gives very bright lemon yellow colors upon silk without mordant.

Z.

Zinc.—The metal zinc is but little employed in dyeing or printing operations. It is not, like iron, actively injurious to colors or mordants, but it is rapidly corroded under the influence of acids or alkalies, vessels made of it wearing out in a short time. Zinc combines with oxygen to form a white oxide, which is of a brilliant lustre; it has been used as a pigment color in calico printing, being fixed by albumen. The oxide of zinc, made by burning, is the most suitable for this purpose; that which is produced by precipitation being defective in softness and lustre, probably owing to a different molecular arrangement. Oxide of zinc is soluble in ammonia, and nearly all the acids, yielding colorless salts, unless the acid be colored. The only zinc salts used in dyeing or printing are the sulphate, chloride, and acetate.

Sulphate of zinc, or white vitriol, can be prepared by dissolving zinc scraps in weak oil of vitriol. As zinc mostly contains a small quantity of iron, it should be removed from the solution. This is done by adding a quantity of a mixture of chemic (chloride of lime) and water to the liquor when it is saturated with the metal; this mixture oxidizes the iron, and throws it down at the same time as an ochry powder. Pure sulphate of zinc gives only a white precipitate with yellow prussiate, and when mixed with strong ammonia gives a precipitate at first, which dissolves when sufficient ammonia is added. It is especially when sulphate of zinc is to be used for adding to red liquor mordants, or for mixing with the dung in cleansing or fixing alkaline pinks, that it should be free from iron. Sulphate of zinc serves as a resist in several styles, and is a constituent in what is termed "mild paste." A new use of sulphate of zinc has been proposed by Balard and Sacc, by which, if it turns out successful, this salt may be employed instead of tartaric acid for discharge upon dyed grounds. They have found that if sulphate of zinc be mixed in certain proportions with solution of bleaching powder, it increases its power in about the same way as if acid was added; and they have found that if dyed cloth (Turkey red for example) be printed with sulphate of zinc, and passed into bleaching powder, it discharges the color wherever the zinc salt was printed. The applications of this discovery have yet to be made upon the large scale, and it remains to be seen whether it will prove economical or practicable.

Chloride of Zinc (Muriate of Zinc).—This salt is easily obtained by dissolving metallic zinc in spirits of salts. It is not much used either in dyeing or printing. It is employed to fix the alumina of the alkaline pink mordant, and is added to some colors to keep them moist or soft, the muriate of zinc having a great tendency to attract moisture from the air.

Nitrate of Zinc has been employed for the same purpose as the muriate of zinc, and especially in the case of red liquor pinks. It is made by dissolving the metal in weak aquafortis.

Acetate of Zinc.—This salt is very little used. It may be made by dissolving the oxide of zinc in acetic acid, or from the sulphate of zinc, by means of acetate of lead. It gives a beautiful orange yellow on silk and cotton with murexide.

Zinc yields no colors except the white from the oxide; it does not form colored compounds, and it has hardly any affinity for either vegetable or animal fibre.

APPENDIX.

DYEING AND CALICO PRINTING AS SHOWN IN THE UNIVERSAL EXPOSITION, PARIS, 1867.

Extracts from the Reports of the International Jury,* and from other Sources.

IT is well known that silk, by the process of dyeing, can have its weight increased 10 to 40 per cent., and yet·give products of a good quality. The competition and the dearness of silk have been so great of late years, that, often, the weight of silk is increased 150 to 200 per cent. by dyeing, especially for blacks.

Such silk is rough to the touch, without lustre, easily cut, and will not last. Heated to about 230° Fah., it will fall to pieces.

By this process of over adulteration, silk increases much in volume, and the fibres, viewed under the microscope, are swollen. The swelling is also sensibly in proportion with the increase of weight.

With mordants of tannin, tin, and oily substances, nearly all the new coal-tar colors have been fixed on vegetable fibres.

Mr. Reimann, of Berlin, dyes cotton yarn with aniline colors, and without mordant, by effecting the operation in closed vessels, heated up to about 300° F. The shades, on leaving the apparatus, are said to be fast, but not bright. They are raised by another dyeing operation conducted in the open air.

Such a process requires costly apparatus, does not allow an easy dyeing to a given shade, and, granting that the dyed ground is fast, it does not appear that the raising given afterwards will be faster than by the ordinary process. Neverthe-

---

* Rapports du Jury International, publiés sous la direction de M. Michel Chevalier, Membre de la Commission Impériale. 13 vols. 8vo. Paris, 1868.

less, the application of dyeing under pressure in closed vessels, is a curious one, and might be used to advantage in other cases.

Since Messrs. Tessié du Motay and Maréchal have succeeded in producing cheaply alkaline permanganates, these salts begin to be used for bleaching goods. By the decomposition of the permanganate, its oxygen destroys or modifies the substances foreign to the cloth, which are washed out. At the same time, oxide of manganese is precipitated upon the cloth, and is removed by washing in a dilute sulphurous acid solution. The solution of permanganate of soda is also to be employed in a dilute state.

Feathers may be bleached by the process of Messrs. Viol and Duflot, as follows: Steep the feathers for from three to four hours in a tepid and diluted bath of bichromate of potassa with nitric acid, then pass through another bath holding a very weak solution of sulphurous acid, and rinse.

Dyeing aniline black on wool has not been entirely successful, notwithstanding the chlorine process of Mr. Lightfoot. Some recent experiments, however, permit us to hope that aniline black will be employed for wool as well as for cotton.

Casein (curd of milk), as a mordant, is better dissolved in crystallizable acetic acid, or in a milk of lime. In the latter case, the colors are said to be faster than when using casein dissolved in ammonia water, or even albumen. But printing should be effected rapidly, because the paste loses its fluidity very rapidly, especially with ultramarine.

By means of a metallic engraving in relief, which distributes drops of colored and melted resin on silk goods, Mr. Petitdidier imitates embroideries.

Light tissues, like tulle or bobbinets, are also covered with drops of gelatin, or gum, which fall from rows of pins, variously arranged, according to the processes of Messrs. C. Depouilly, Meyer, and Agnelet brothers.

By printing, in a peculiar way, silk warps previous to weaving, various combinations of figures and designs may be effected on the loom, without the expense of the cartoons of the Jacquard loom.

The various aniline blacks, prepared whether by the bichromate of potassa, or by the chlorate, are soluble in a mixture of alcohol and sulphuric acid. This solution, thrown into a large quantity of water, dyes animal fibres a fast gray.

For dyeing black on cotton, Messrs. Paraf and Javal, pass the cloth through a bath containing a mixture of sulphate of aniline and bichromate of potassa. The color appears on the fabric immediately after it leaves the bath, the temperature of which must be kept a little below the freezing point, not above.

Another method consists in mordanting the cotton cloth with chromate of lead, and then passing it through an acidulated bath of oxalate of aniline. In this case, the reaction taking place only on the cloth, the temperature has not to be so strictly low as in the former method.

Mr. Dumas frees the indigo from its red and brown coloring substances by aniline. Indigo thus purified gives very good results when used in printing on cotton.

One of our cotemporaries speaks of chloroform as being a solvent of indigo. Not having tried the process, we can but believe that the chloroform may be a solvent of the impurities of indigo, rather than of indigo itself.

From the same source we find for dyeing animal fibres a silver gray color: Boil 10 pounds of wool in a bath containing 4 ozs. of sulphuric acid, and 4 ozs. of glauber salts (sulphate of soda). Then dye to the shade by means of iodine violet and some carmine of indigo.

There are many recipes for the preparation of the printing paste for aniline black; they can be summed up into a composition of tartrate of aniline, sulphide of copper, chlorate of potassa, and sal-ammoniac, the whole thickened with a mixture of starch and torrefied starch, with enough water to make the volume of the aniline about one-tenth of the whole.

Aniline black succeeds very well when printed with or under chrome orange. In this case the lead mordant is basic.

Mr. Horace Koechlin has succeeded in printing aniline greens on silk and wool by adding alkaline sulphites to the color.

For cotton goods, besides the sulphite, some tannin is necessary.

The following are the values in coloring power of several madder extracts:—

That of Professor Rochleder, of Prague, is dry and equal to 140 times its weight of madder; that of Messrs. Pernod and Picard, of Avignon, is in paste and equal to 16 to 20 times its

weight of madder; that of Mr. Schutzenberger, manufactured
by Mr. C. Meissonnier, is also in paste and equal to 30 times
its weight of madder.

These extracts are free from resin, and therefore, can be
thoroughly mixed with water, but they require a nice adjustment in the proportion of mordants. The steaming process
lasts two or three times as long as with ordinary steam colors.
The shades are also to be raised by drawing the printed goods
through soap baths. No mixture of acids, oxidizing agents, or
ageing is necessary. The principal mordants still used are
those of alumina and iron.

On the other hand, some persons assert that it is possible to
print with these extracts, on tissues which have not been mordanted.

INDEX.

31

www.ingramcontent.com/pod-product-compliance
Lightning Source LLC
Chambersburg PA
CBHW032009110726
47901CB00004B/1021